# Introduction to
# Experimental Statistics

# McGraw-Hill Series in Probability and Statistics

## David Blackwell and Herbert Solomon, Consulting Editors

# Introduction to
# Experimental Statistics

**C. C. Li**

*Professor of Biometry*
*Graduate School of Public Health*
*University of Pittsburgh*

**McGraw-Hill Book Company**

New York   San Francisco   Toronto   London

**INTRODUCTION TO EXPERIMENTAL STATISTICS**

*Library of Congress Catalog Card Number* 63-23045

37706

*To*
**Professor Harry Houser Love**
*who taught me*
*Biometry and Genetics*

# Preface

An author writing a new book in a field where several good texts already exist inevitably wishes to explain why his book is necessary and how it differs from the texts that already exist. My explanation is as follows.

A student of mathematical statistics has a large array of books on probability or statistical theory from which to choose; he can find collections of mathematical theorems on the analysis of variance readily available. But the student whose field is not mathematics—the biological or medical research worker, for example—is in genuine need of a short, nonmathematical course on the design and analysis of experiments, written in a rather informal style. This book is offered, then, as an answer to that need.

I have tried to make the book useful to the practising experimental worker as well as to the student. A researcher who cannot spare the time to take a formal course in experimental statistics can profit from studying this volume without benefit of a teacher. Yet the book will also be found suitable for a short, formal course at a college or university.

The artificial nature of my numerical examples may be criticized, and my answer would be this. How does one first learn to solve quadratic equations? By working with terms such as

$$242.5189x^2 - 683.1620x + 19428.5149 = 0$$

or with terms like $x^2 - 5x - 6 = 0$. When the numerical examples are simple, the methodology and procedure for interrelating various quantities can be learned quickly, without the burden of complex arithmetic. With so-called realistic examples, one calculates more and enjoys it less.

The first fourteen chapters of the book (Part 1) are essentially the same as my earlier "Numbers from Experiments." These chapters deal with the arithmetic, algebra, and mechanics of the analysis of variance. Enough theory is given to show the reader that there is nothing mysterious in the operation. Part 2 deals with basic experimental designs. The last few chapters (Part 3) take up some special types of data: problems of unequal variances, proportional data, and multiple measurements, which come up very frequently in biological experiments.

Many of the chapters are practically self-contained, so that the teacher and the student may use them in any order they prefer, rather than following the sequence of the book. Since the original fourteen chapters from "Numbers from Experiments" are preserved as independent introductory material to the analysis of variance, their connection with later chapters (Part 2) may be noted: Chapters 5 to 8 go with Chapter 15; Chapter 10 goes with Chapters 16 to 19; Chapters 11 and 12 go with Chapters 21 to 33; and Chapters 13 and 14 with Chapter 20. Indeed, for a very short course, the teacher may begin with Chapter 15 and treat the first fourteen chapters as supplementary notes.

The exercises at the end of each chapter are intended to review the text material and to show how to carry out the calculations correctly. For numerical problems either the partial or the complete answer is given, as reward, like justice, should be prompt.

No appendix of statistical tables has been furnished; the reader is expected to use a separate volume, such as Fisher and Yates, "Statistical Tables for Biological, Agricultural, and Medical Research." A small part of the available $F$ table is reproduced in Chapter 8, but only for teaching purposes. This is true of several other small tables.

Criticisms of the book or suggestions for corrections, additions, or new presentations will be greatly appreciated.

*C. C. Li*

# Contents

## part 2   Experimental designs

## part 3   Some related topics

part **1**

# Basic mechanics
# and theory

# Mathematical expressions

Before saying anything about statistics or experiments, I find that a few words about the nature of mathematical expressions are very helpful to nonmathematical students in reading and understanding the rest of the book or, for that matter, any other book containing mathematical expressions. When a beginner sees writings in symbols, he usually makes no distinction between the different kinds and calls them all "formulas." In the following we shall first distinguish the three major types of mathematical expressions (equations, identities, and definitions) and then mention some others.

## 1. Equations

An *equation* is a statement of a condition. The statement involves some unknown quantities. We seek some appropriate values for the unknowns so that the stated condition is satisfied. Suppose that for some reason or other, we are looking for a number $u$ which will satisfy the condition

$$84 - 12u = 0$$

This expression is then an equation.   We see that $u = 7$ satisfies the condition, and hence we say that $u = 7$ is a *solution* of the equation.   The equation is then said to be solved.   In this particular case, $u = 7$ is the only solution.

An equation may be simple and easy to solve or complicated and difficult to solve, but its nature remains the same: it is a statement of a condition to be satisfied.   The equation $84 - 12u = 0$ is called a *linear equation* in $u$, because it involves only the first power of $u$.   As a second example, suppose that we want to find a number $y$ such that

$$y - \frac{60 + y}{4} - \frac{187 + y}{9} + \frac{700 + y}{36} = 0$$

This is a linear equation in $y$.   The simplest way to solve for the value of $y$ is to multiply the entire expression by 36, so that it becomes

$$36y - 9(60 + y) - 4(187 + y) + (700 + y) = 0$$

Simplifying, we obtain

$$24y - 588 = 0$$

and the solution is

$$y = \frac{588}{24} = 24.50$$

Although the original form of the equation is slightly more complicated, it is in principle just as simple as $12u - 84 = 0$.

Sometimes we have to solve two simultaneous linear equations involving two unknowns:

$$12y_1 - \ 3y_2 = \ \ 66$$
$$-3y_1 + 12y_2 = 231$$

The reader may verify that the solutions are $y_1 = 11$ and $y_2 = 22$.

Finally, we may have $n$ simultaneous linear equations involving $n$ unknowns to be solved.   An extraordinarily simple set of five equations in five unknowns is given in the accompanying table.   In this book we shall deal most of the time with sets of linear equations as exemplified above.

| Set of equations | Solutions |
|---|---|
| $144 - 4u \qquad = 0$ | $u = 36$ |
| $100 - 2u - 2r_1 = 0$ | $r_1 = +14$ |
| $44 - 2u - 2r_2 = 0$ | $r_2 = -14$ |
| $88 - 2u - 2t_1 = 0$ | $t_1 = +8$ |
| $56 - 2u - 2t_2 = 0$ | $t_2 = -8$ |

## 2. Identities

An *identity* represents two different ways of writing the same thing—two different forms of expressing the same quantity. For example,

$$(a + b)^2 = a^2 + b^2 + 2ab$$

is an identity, by which we mean that the two sides of the expression are *always* equal, whatever the values of $a$ and $b$. If we let $a = 3$ and $b = 4$, we then have

$$(3 + 4)^2 = 3^2 + 4^2 + 2(3 \times 4) = 49$$

which is correct. If we let $a = 6$ and $b = 5$, we then have

$$(6 + 5)^2 = 6^2 + 5^2 + 2(6 \times 5) = 121$$

which is also correct. The reader should convince himself that the above expression is always correct whatever the values of $a$ and $b$. You cannot "solve" an identity and obtain "solutions." Since the two sides represent the same quantity, you can use either form to denote the quantity, depending upon which is more convenient for your purpose. Much of algebraic "manipulation" is based on identities. To emphasize the nature of an identity, the identity sign $=$ is used in formal cases instead of the ordinary equality sign. Although an identity cannot be solved (in the sense that an equation is solved), it should be established or proved so that we know it is actually true.

Another simple algebraic identity is

$$(a - b)^2 = a^2 + b^2 - 2ab$$

There are many reasons for establishing and using identities. One is to simplify mathematical expressions. Another is to see certain relationships between quantities. For example, since $(a - b)^2$ must be a positive quantity, we conclude that

$$a^2 + b^2 \geqslant 2ab$$

no matter what two numbers we take as $a$ and $b$. The expression says that the sum of squares of any two numbers is *always* larger than or equal to twice their product.   The reader is advised to try out a few instances numerically and see that this is always the case.

A simple identity may often be generalized.   Thus,

$$(a + b + c)^2 = a^2 + b^2 + c^2 + 2ab + 2ac + 2bc$$

is a more general identity than $(a + b)^2 = a^2 + b^2 + 2ab$.   In fact, it may be generalized to any number of items: $(a + b + c + d + \cdots)^2$.

When a number of simple and basic identities have been established, it is possible to build or deduce from them more complicated ones, just as in geometry very complicated theorems (statements of relationships) are built or deduced from some very simple and basic relationships.   For example, from the two simple identities

$$(a + b)^2 = a^2 + b^2 + 2ab$$
$$(a - b)^2 = a^2 + b^2 - 2ab$$

we obtain a new identity by adding the two together,

$$(a + b)^2 + (a - b)^2 = 2(a^2 + b^2)$$

or

$$a^2 + b^2 = \left(\frac{a + b}{\sqrt{2}}\right)^2 + \left(\frac{a - b}{\sqrt{2}}\right)^2$$

which expresses a very interesting relationship.   For example, let $a = 14$ and $b = 6$,  so that $a^2 + b^2 = 196 + 36 = 232$.   On the other hand, we form the two new numbers

$$\frac{a + b}{\sqrt{2}} = \frac{20}{1.4142} = 14.142 \qquad \text{its square} = 200$$

$$\frac{a - b}{\sqrt{2}} = \frac{8}{1.4142} = 5.657 \qquad \text{its square} = \underline{\phantom{0}32}$$

$$\text{total} = 232$$

The above identity may also be written

$$\frac{(a - b)^2}{2} = a^2 + b^2 - \frac{(a + b)^2}{2}$$

Further, the left-hand side may be written

$$\frac{(a - b)^2}{2} = \left(a - \frac{a + b}{2}\right)^2 + \left(b - \frac{a + b}{2}\right)^2$$

By equating the right-hand side of the two expressions above, we obtain another identity, and so on.   The identities we shall study in this course are merely extensions or generalizations of some basic identities with which we have been familiar since our high school days.   There is no reason to regard them as profound mathematics.

### 3. Definitions

A definition is simply an *arbitrary name* (that is, a technical term) for a quantity calculated in a certain way. For example, for four numbers 4, 1, 5, 2, we decide to call the quantity $(4 + 1 + 5 + 2)/4 = 12/4 = 3$ the *arithmetic mean* (a name) of the four numbers. In general, if there are $n$ numbers, $y_1, y_2, \ldots, y_n$, we shall call the quantity

$$\bar{y} = \frac{y_1 + y_2 + \cdots + y_n}{n}$$

the arithmetic mean of the $n$ numbers. We give this quantity a special name and a special symbol simply because we shall use it over and over again, and it needs a convenient handle. You can neither solve it the way you solve an equation nor prove it the way you prove an identity, but you can study the properties of the quantity so defined. Thus, it will be shown in the next chapter that $\Sigma(y - \bar{y}) = 0$ is a consequence of the definition.

In reading the following chapters you should always keep these distinctions between equations, identities, and definitions in mind. To make this easier, I have, whenever possible, treated the different kinds of expressions in separate chapters. For instance, Chap. 3 is devoted to the establishment of an identity, Chap. 6 deals with the solutions of a set of linear equations, and so on. An algebraic identity, stating a pure mathematical fact, cannot be considered as a part of statistical theory, and it is less confusing to treat it separately.

### 4. Function and minimum value

Consider the expression

$$Q = Q(u) = (6 - u)^2 + (2 - u)^2$$

For each given value of $u$ we may calculate a corresponding value for $Q$. We say that $Q$ is a function of $u$, and by that we simply mean that the value of $Q$ depends on the value of $u$. More formally, it is written as $Q(u)$, reminding us that its value is determined by $u$; or, in other words, the expression on the right involves $u$. A few of the numerical values of $u$ and the corresponding values of $Q$ are listed below. We note that when

| $u$ | 0 | 1 | 2 | 3 | 4 | 5 | 6 | 7 | 8 $\cdots$ |
|---|---|---|---|---|---|---|---|---|---|
| $Q$ | 40 | 26 | 16 | 10 | 8 | 10 | 16 | 26 | 40 $\cdots$ |

$u = 4$, $Q = 8$ is the smallest of all the $Q$ values. How can we be sure that this is actually the case? To see that it is by the elementary algebraic method of *completing the square,* we rewrite the original function as

$$Q = (u^2 - 12u + 36) + (u^2 - 4u + 4)$$
$$= 2u^2 - 16u + 40 = 2(u^2 - 8u + 20)$$
$$= 2[(u - 4)^2 + 4]$$

The smallest value of $Q$ is attained when $(u - 4)^2$ is zero, that is, when $u = 4$. For any other value of $u$, the corresponding value of $Q$ must be larger than 8.

The required value of $u$ that makes $Q$ a minimum may be more readily found by the method of differential calculus, a subject into which we shall not delve in this book. Suffice it to say that the method of differential calculus gives in this particular case the following condition (equation) for the required value of $u$:

$$(6 - u) + (2 - u) = 0$$
$$8 - 2u = 0 \qquad \therefore u = 4$$

More generally, if the function is $Q = Q(u) = (a - u)^2 + (b - u)^2$, where $a$ and $b$ are known numbers, the value of $u$ that minimizes the value of $Q$ is given by the condition

$$(a - u) + (b - u) = 0$$
$$a + b - 2u = 0 \qquad \therefore u = \tfrac{1}{2}(a + b)$$

That is, when $u$ is the arithmetic mean of $a$ and $b$, the value of $Q$ is the smallest. Again, the function $Q$ we shall study in statistics is merely an extension of the one considered here.

### 5. Mathematical model

Taking various measurements of an American quarter (a coin worth 25 cents) I find that its diameter is approximately 2.26 centimeters (cm). What is the area of the quarter? I have no direct way to measure its area, but it may be shown that for a *perfect* circle, the area is

$$A = \pi r^2$$

where $r$ is the radius and $\pi$ is the ratio of the circumference of a perfect circle to its diameter. So, I shall *assume* that the quarter is circular in shape with radius $r = 1.13$ cm. The value of $\pi$ is difficult to determine empirically. According to an ancient Chinese mathematics book, it is approximately equal to $355/113$. Hence the area $A$ of the quarter may be taken roughly as

$$A = 355/113(1.13)^2 = 4 \text{ cm}^2$$

Although we know that there is no such thing as a perfect circle anywhere and that the $\pi$ value is not $355/113$, the approximation presented above is probably good enough for all practical purposes.

On close examination of the coin, another disturbing feature emerges. It will be found that the surface is not smooth and even as implied by our calculations above, but quite rugged. One one side of the coin there is a *head;* on the other there is a bird (whose name must be *tail* according to probability textbooks). Hence the areas of the two sides are probably not equal. Faced with the infinite complexity of the surface, we begin to realize that the determination of the exact area is virtually impossible by any kind of simple measurements. Unless we give up completely, we shall for practical purposes simply measure the diameter and use the relation $A = \pi r^2$ to calculate the area, ignoring all the minute albeit real irregularities. Then we say $A = \pi r^2$ is the *mathematical model* for the purpose of determining the area of the surface of a coin.

When we define them in the sense we defined them above, all the mathematical relationships we learned in physics are mathematical models when applied to real objects. In physical sciences the models are amazingly accurate; in biological sciences they are far less so. A biological entity is very complicated indeed, much like the surface of a coin. In its full details it would be next to impossible to study. In order to be able to make any study at all, we are forced to introduce certain simplifying assumptions. In quantitative studies, we have to limit our attention to certain main features of the characteristic and ignore many other details, which is analogous to calculating the area of the surface of a coin without worrying too much about the furrows and ridges. In statistical study of a quantitative characteristic, we also have to adopt a mathematical model (although an extremely simple one) to describe and analyze the data. The concept will be introduced in Chap. 6.

## 6. Symbolic expressions

Consider the expression $ab - a - b + 1$, whatever it stands for. Instead of writing it as is, we can always write it in the factorial form $(a - 1)$ $(b - 1)$, because expansion of the latter by the formal rules of algebra gives the former:

$$(a - 1)(b - 1) = ab - a - b + 1$$

The appearance of this expression is no different from that of an ordinary algebraic identity except that the symbols $a$, $b$, 1 here do not stand for numbers. This is merely a device to replace the long form by the shorter factorial form of an expression. The reader may fail to see the advantage of using it. Right, there is very little advantage in a simple case like this one. But consider the expression

$$a_1b_1 - 2a_1b_2 + a_1b_3 - 2a_2b_1 + 4a_2b_2 - 2a_2b_3 + a_3b_1 - 2a_3b_2 + a_3b_3$$

It is long and difficult to remember. Symbolically, however, it may be written

$$(a_1 - 2a_2 + a_3)(b_1 - 2b_2 + b_3)$$

which is much easier to write as well as to remember. Whenever we need it, the original long expression can be readily obtained by expanding the compact factorial form. We shall have occasion to make use of these symbolic writings later (Chap. 21).

## 7. Simplifications

The reader is not expected to know the precise meaning of the following expression and may skip this section altogether if he wishes,

$$T_{xy} = \sum_{i=1}^{t} \frac{\left(\sum_{j=1}^{n_i} X_{ij}\right)\left(\sum_{j=1}^{n_i} Y_{ij}\right)}{n_i} - \frac{\left(\sum_{i=1}^{t}\sum_{j=1}^{n_i} X_{ij}\right)\left(\sum_{i=1}^{t}\sum_{j=1}^{n_i} Y_{ij}\right)}{\sum_{i=1}^{t} n_i}$$

because we have not yet explained the meanings of the symbols. All he has to do at present is to be impressed by the great complexity of its appearance and watch out for typographical errors. Yet, by very straightforward substitutions which arise naturally in actual tabulation of data and numerical calculation of the quantity $T_{xy}$, the expression may be reduced to a very simple form. Each sum may be denoted by a single letter such as $N = \Sigma n_i$. In our Chap. 14, where a similar quantity is required, it is chopped into two pieces corresponding to actual arithmetic procedure (symbols to be defined in Chap. 14) viz.,

$$B = \sum_{i} \frac{X_i Y_i}{n_i} \qquad C = \frac{XY}{N}$$

so that $T_{xy} = B - C$. All similar expressions in this book are given in the latter form rather than in the former.

## EXERCISES

1.1 Show that

a  $(n - 1)^2 - n(n - 2) = 1$

b  $\dfrac{2}{n(n - 2)} = \dfrac{1}{n(n - 1)} + \dfrac{1}{(n - 1)(n - 2)}$

c  $\dfrac{n_1 n_2}{n_1 + n_2} = \dfrac{1}{1/n_1 + 1/n_2}$

**1.2**  Expand

**a**  $(a + b + c + d)^2$
**b**  $(a_1 - 2a_2 + a_3)(b_1 - b_3)$
**c**  $(a - 1)(b - 1)(c - 1)$
**d**  $(3a_1 + a_2 - a_3 - 3a_4)(b_1 - b_2)$

**1.3**  Show that

**a**  $(b - c)^2 = 2(b^2 + c^2) - (b + c)^2$
**b**  $(a - b)^2 + (a - c)^2 + (b - c)^2 = 3(a^2 + b^2 + c^2) - (a + b + c)^2$
**c**  $2(ab + ac + bc) = (a + b + c)^2 - (a^2 + b^2 + c^2)$

**1.4**  Two linear equations in two unknowns $y_1$ and $y_2$:

$$fy_1 + 2y_2 = Q_1$$
$$2y_1 + fy_2 = Q_2$$

where $Q_1$, $Q_2$, and $f$ are known numbers.  To solve these equations, we may use the elementary method of eliminating one of the unknowns.  To eliminate $y_1$, we multiply the first equation by 2 and the second equation by $f$ and obtain

$$2fy_1 + 4y_2 = 2Q_1$$
$$2fy_1 + f^2y_2 = fQ_2$$

Subtracting $\qquad\qquad (f^2 - 4)y_2 = fQ_2 - 2Q_1$

Hence, $\qquad\qquad\qquad y_2 = \dfrac{fQ_2 - 2Q_1}{f^2 - 4}$

Obtain a similar solution for $y_1$ by first eliminating $y_2$ from the equations.

**1.5**  Given the function

$$S = S(a) = (1 - a)^2 + (4 - a)^2 + (10 - a)^2$$

Calculate the following paired values of $a$ and $S$.  For what value of $a$ is $S$ the small-

| $a$ | 1 | 2 | 3 | 4 | 5 | 6 | 7 | 8 | 9 |
|-----|---|---|---|---|---|---|---|---|---|
| $S$ | 90 | | | | | 54 | | | |

est?  If you have had a course in differential calculus, show that such a value of $a$ is given by the solution of the equation $dS/da = 0$, which, upon simplification, reduces to

$$(1 - a) + (4 - a) + (10 - a) = 0$$

If you have not, show that the function may be rewritten in the form

$$S = S(a) = 3[(a - 5)^2 + 14]$$

At what value of $a$ will $S$ assume its minimum value?

# 2

# Properties of addition

In the branch of statistics with which this book is concerned, namely, the analysis of variance, the great bulk of the arithmetic is that of calculating a quantity known as the sum of squares. The precise meaning of this term will be made clear in the next chapter. Here the reader is expected only to be familiar with the operation of squaring a number and with the addition of numbers, that is, how to find the sum of a series of numbers. Not that we always add a series of simple numbers. Frequently we have to add a series of numbers which are themselves made up of several other numbers. In order to be able to handle such situations with confidence, it is well to note certain properties of addition before going on to the sum of squares.

## 1. Adding a constant

First of all, we see that, if $c$ is a number (say, $c = 5$),

$$c + c + c + c = 4c$$
$$5 + 5 + 5 + 5 = 4 \times 5 = 20$$

In other words, adding a constant quantity is equivalent to multiplying that quantity by the number of times it appears. Indeed, this is the definition of multiplication in arithmetic.

## 2. The sigma notation

If the quantities to be added are all different, we do not have the simple result above, but we can use a shorthand notation for the sum. Let $y_1$ be one number, $y_2$ be another, and so on. The sum of four such numbers is

$$y_1 + y_2 + y_3 + y_4 = \Sigma y$$
$$6 + 2 + 9 + 3 = 20$$

The $\Sigma$ notation is very fundamental in our type of work. In words, the symbol $\Sigma$ is to be read "the sum of"; hence, the expression $\Sigma y$ means the sum of the $y$ values.

## 3. The running subscript

A more formal and rigid symbolism requires that each $y$ value should have a subscript for identification. Thus the symbol $y_\alpha$ is a general notation for any one of these $y$'s. When $\alpha = 1$, the number is $y_1$; when $\alpha = 2$, the number is $y_2$; and so on. Note that $\alpha$ is not a quantity in the sense that $y$ is, but is merely a tag or a serial number. The sum of these four $y$'s may be written as follows, in the order of decreasing formality and of increasing simplicity.

$$\sum_{\alpha=1}^{\alpha=4} y_\alpha = \sum_\alpha y_\alpha = \sum y_\alpha = \sum y = Y$$

Be sure that all these expressions denote the same thing. When there is no possibility of confusion, we prefer to use the simpler notations $\Sigma y$ or $Y$. In the general case when there are $n$ numbers to be added together, $\alpha$ takes the successive serial numbers $1, 2, 3, \ldots, n$ and the $\Sigma$ covers every number from the first to the last.

## 4. A variate and a constant

If a constant quantity $c$ is added to each of the four $y$ values, the new sum and the original sum have a very simple relationship, as every business-man knows, although he may be unable to express or write it out in the following manner:

$$(y_1 + c) + (y_2 + c) + (y_3 + c) + (y_4 + c)$$
$$= y_1 + y_2 + y_3 + y_4 + c + c + c + c = \Sigma y + 4c$$

In general,

$$\sum_{\alpha=1}^{\alpha=n} (y_\alpha + c) = \sum (y_\alpha + c) = \sum y + nc$$

### 5. A constant multiplier

If each $y$ value is multiplied by a constant number $b$, the new sum obviously will be $b$ times the original sum:

$$by_1 + by_2 + by_3 + by_4 = b(y_1 + y_2 + y_3 + y_4)$$

that is,

$$\Sigma by = b\Sigma y$$

In terms of practical calculation, this means that a constant factor after a $\Sigma$ sign can always be taken out and put in front of the $\Sigma$ sign.

### 6. Two-way addition

If a finite number of quantities are to be added, we are free to add them in any order whatsoever and we shall always end up with the same grand total. However, in many cases some sort of systematic and meaningful order is preferred. For example, you may prefer to keep your three-day traveling account in the following manner. Adding horizontally, you

|  | Monday | Tuesday | Wednesday | Total |
|---|---|---|---|---|
| Hotels | $x_1$ | $x_2$ | $x_3$ | $\Sigma x$ |
| Meals | $y_1$ | $y_2$ | $y_3$ | $\Sigma y$ |
| Total | $x_1 + y_1$ | $x_2 + y_2$ | $x_3 + y_3$ | $\Sigma(x + y)$ |

know that $\Sigma x$ is the total hotel bill and $\Sigma y$ is the total of the meals. Adding vertically, you know that $x_1 + y_1$ has been spent on the first day, $x_2 + y_2$ on the second day, and $x_3 + y_3$ on the third. The grand total of the day-to-day expenses is, of course, the same as the grand total for hotels and meals. Algebraically, we write

$$\Sigma(x + y) = \Sigma x + \Sigma y$$

This fact is so fundamental that it is sometimes called the "law of addition." In more formal language it sounds like this: the summation of sums is the sum of summations.

## 7. A summary

In dealing with algebra, the general symbol $y$ may denote a positive or a negative number, and all the expressions given so far still hold. For instance,

$$\Sigma(y - x) = \Sigma y - \Sigma x$$

Reviewing all the simple points just mentioned, we may take the following expression as a summary ($a$, $b$, $c$, constants):

$$\Sigma(ax \pm by \pm c) = a\Sigma x \pm b\Sigma y \pm nc$$

This expression may, of course, be extended to any number of items with positive or negative signs.

## 8. The arithmetic mean

So far we have been talking about the sum or total of a group of numbers. Now we are ready to *define* a very simple but important quantity which is closely associated with the sum. The *arithmetic mean* of a group of numbers is denoted by $\bar{y}$ and defined as, for the case $n = 4$,

$$\bar{y} = \frac{y_1 + y_2 + y_3 + y_4}{4} = \frac{\Sigma y}{4} = \frac{Y}{4}$$

or

$$\bar{y} = \tfrac{1}{4}y_1 + \tfrac{1}{4}y_2 + \tfrac{1}{4}y_3 + \tfrac{1}{4}y_4$$

From the above form of definition it follows that

$$4\bar{y} = \bar{y} + \bar{y} + \bar{y} + \bar{y} = y_1 + y_2 + y_3 + y_4$$

More generally,

$$\Sigma\bar{y} = n\bar{y} = \Sigma y = Y$$

This means that, if we replace every $y_\alpha$ by $\bar{y}$ and then add, we shall obtain the same total. For example, with $\bar{y} = 5$,

$$6 + 2 + 9 + 3 = 20$$
$$5 + 5 + 5 + 5 = 20$$

## 9. Sum of deviations

From the expression and example of the preceding paragraph, it follows that

$$(y_1 - \bar{y}) + (y_2 - \bar{y}) + (y_3 - \bar{y}) + (y_4 + \bar{y}) = 0$$
$$(6 - 5) + (2 - 5) + (9 - 5) + (3 - 5) = 0$$
$$1 - 3 + 4 - 2 = 0$$

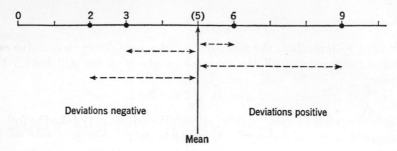

Fig. 2.1 Positive and negative deviations cancel each other so that the sum of all deviations is zero.

More generally,

$$\Sigma(y - \bar{y}) = \Sigma y - n\bar{y} = 0$$

The value $(y - \bar{y})$ is called the *deviation* of $y$ from the arithmetic mean. Some of the deviations are positive, others negative. The above expression says that the sum of deviations (from the mean) is zero. This is one of the most important properties of the arithmetic mean, and it is diagrammed in Fig. 2.1. Some other properties of the mean will be dealt with later.

**EXERCISES**

**2.1**  Find the arithmetic mean of the following 12 numbers:

$$2 \quad 6 \quad 4 \quad 10 \quad 4 \quad 9 \quad 12 \quad 8 \quad 11 \quad 6 \quad 9 \quad 3$$

**2.2**  Find the 12 deviations from the arithmetic mean, Exercise 2.1, and see if their sum is zero. Draw a diagram for the deviations similar to Fig. 2.1.

**2.3**  Find the grand total of the following array of numbers:

$$36 + 25 - 10 \times 6$$
$$4 + 25 - 10 \times 2$$
$$81 + 25 - 10 \times 9$$
$$9 + 25 - 10 \times 3$$

*Ans:* 30

**2.4**  Find the grand total of the following array of numbers:

$$36 + \phantom{0}4 - 2 \times 12$$
$$36 + 81 - 2 \times 54$$
$$36 + \phantom{0}9 - 2 \times 18$$
$$4 + 81 - 2 \times 18$$
$$4 + \phantom{0}9 - 2 \times \phantom{0}6$$
$$81 + \phantom{0}9 - 2 \times 27$$

*Ans:* 120

# 3

# The sum of squares

In squaring a number it is important to remember that the square of any real number, positive or negative, is always positive.   Thus: $(-3)^2 = +9$.

## 1. Square of a difference

In statistical work we frequently have to square the difference between two numbers.   To simplify algebraic expressions, the following *identity* has been employed over and over again:

$$(a - b)^2 = a^2 + b^2 - 2ab$$
$$(y_1 - M)^2 = y_1{}^2 + M^2 - 2y_1M$$
$$(y_1 - \bar{y})^2 = y_1{}^2 + \bar{y}^2 - 2y_1\bar{y}$$

Again, the thing to be remembered is that, whatever the values of $a$ and $b$, $y_1$ and $M$, $y_1$ and $\bar{y}$, the resulting square is a positive quantity.

## 2. Sum of squares of differences

Let $M$ be an arbitrary number (that is, any fixed number). For a group of four numbers, $y_1$, $y_2$, $y_3$, $y_4$, we may calculate the following quantity $Q$, which is the sum of squares of deviations of $y$ from the fixed number $M$:

$$Q = (y_1 - M)^2 + (y_2 - M)^2 + (y_3 - M)^2 + (y_4 - M)^2 = \Sigma(y_\alpha - M)^2$$

At this stage it seems that the calculation of $Q$ is quite pointless. Its usefulness to illustrate the meaning of another type of sum of squares and its application in dealing with several groups of numbers will be made clear in later chapters. Let the four $y$ values be 6, 2, 9, 3, as before. Then we see that the actual value of $Q$ depends on the value of $M$ inserted by us. The following numerical calculations (Table 3.1) are for the three different cases where the arbitrary number is taken as $M = 7$, $M = 5$, and $M = 2$.

*Table* 3.1   The sum of deviations and their squares from three different arbitrary numbers

|  | $M = 7$ | | $M = 5$ | | $M = 2$ | |
|---|---|---|---|---|---|---|
| $y$ | $(y - 7)$ | $(y - 7)^2$ | $(y - 5)$ | $(y - 5)^2$ | $(y - 2)$ | $(y - 2)^2$ |
| 6 | $-1$ | 1 | $+1$ | 1 | $+4$ | 16 |
| 2 | $-5$ | 25 | $-3$ | 9 | 0 | 0 |
| 9 | $+2$ | 4 | $+4$ | 16 | $+7$ | 49 |
| 3 | $-4$ | 16 | $-2$ | 4 | $+1$ | 1 |
| 20 | $-8$ | 46 | 0 | 30 | $+12$ | 66 |

## 3. The arbitrary $M$

Several observations may be made from Table 3.1. When $M = 5$, which happens to be also the value of the arithmetic mean $\bar{y} = 5$ of the four numbers, the sum of deviations is zero as it should be. In the other two cases, however, the sum of deviations is not zero; generally, it is equal to

$$\Sigma(y_\alpha - M) = \Sigma y_\alpha - nM = n\bar{y} - nM = n(\bar{y} - M)$$

Thus,

$$\Sigma(y_\alpha - 7) = 20 - 28 = 4(5 - 7) = -8$$
$$\Sigma(y_\alpha - 2) = 20 - 8 = 4(5 - 2) = +12$$

It is clear that if the arbitrary number $M$ is greater than the mean, the sum of deviations is negative; if $M$ is smaller than the mean, the sum of deviations is positive.   Briefly, the sign of the sum is determined by the sign of $(\bar{y} - M)$.

## 4. The smallest $Q$

Now if we examine the "square" columns of Table 3.1 we see that when $M = 5 = \bar{y}$, the sum of squares of the deviations is equal to $Q = 30$; and this quantity is smaller than the other two $Q$'s when $M$ does not coincide with the mean value.   This is not incidental; you will see in Sec. 8 that this is always the case.   The sum of squares of deviations from the arithmetic mean is of special interest in statistical analysis.   If the four numbers are widely different, the deviations will be large; and consequently, the sum of their squares also will be large.   On the other hand, if the four numbers are nearly equal, their deviations will be small, and the sum of their squares also will be small.

## 5. A fundamental identity

For various reasons it is desirable to express the sum of squares of deviations from the mean (*ssq* for brevity) in another form.   To do this, let us expand each squared deviation according to the identity

$$(a - b)^2 = a^2 + b^2 - 2ab$$

Thus when $n = 4$,

$$(y_1 - \bar{y})^2 = y_1{}^2 + \bar{y}^2 - 2y_1\bar{y}$$
$$(y_2 - \bar{y})^2 = y_2{}^2 + \bar{y}^2 - 2y_2\bar{y}$$
$$(y_3 - \bar{y})^2 = y_3{}^2 + \bar{y}^2 - 2y_3\bar{y}$$
$$(y_4 - \bar{y})^2 = y_4{}^2 + \bar{y}^2 - 2y_4\bar{y}$$

Adding   $ssq = \Sigma(y_\alpha - \bar{y})^2 = \Sigma y_\alpha{}^2 + n\bar{y}^2 - 2\bar{y}\Sigma y_\alpha$

The adding process used here follows from the law of addition given earlier.   Recalling that $\Sigma y = n\bar{y} = Y$ and substituting, the last two terms of the above expression may be simplified:

$$n\bar{y}^2 - 2\bar{y}\sum y = n\bar{y}^2 - 2\bar{y}n\bar{y} = -n\bar{y}^2 = -\frac{Y^2}{n}$$

The final form for the sum of squares is then

$$\sum (y_\alpha - \bar{y})^2 = \sum y_\alpha{}^2 - n\bar{y}^2 = \sum y_\alpha{}^2 - \frac{Y^2}{n}$$

The correctness of this identity may be readily verified by our numerical example in which $y = 6, 2, 9, 3$, with $Y = 20$ and $\bar{y} = 5$.    Thus:

$$
\begin{array}{llllll}
y^2: & 36 & 4 & 81 & 9 & \Sigma y^2 = 130
\end{array}
$$

$$ssq = 130 - 4(5)^2 = 130 - 100 = 30$$

or

$$ssq = 130 - \frac{20^2}{4} = 130 - 100 = 30$$

which is the same as that obtained in Table 3.1.    By using this identity, we avoid the labor of finding the value of each deviation.    In fact, even the calculation of $\bar{y}$ is unnecessary; the only two quantities we have to calculate are $\Sigma y^2$ and $Y$.    But the above identity has far more uses than serving merely as a convenient calculation form.    In some theoretical work, any property of the *ssq* may be obtained from that of $\Sigma y^2$ and $Y^2$. In the rest of the book the identity established in this paragraph will be used over and over again in deriving further and more complicated results.

### 6. A geometric representation

The identity in Sec. 5 is so fundamental in all types of statistical work that it is worthwhile to examine a visual aid to the algebraic expression.    The diagram in Fig. 3.1 is almost self-explanatory, so it needs only a very brief description.    Note that $y^2$ may be represented by a square area with sides of length $y$.    To construct the diagram, we first draw a base line of length $\Sigma y = Y$, upon which the individual squares are erected.    Then the total area of these $(n = 4)$ squares is

$$\text{(area) } A = y_1{}^2 + y_2{}^2 + y_3{}^2 + y_4{}^2 = \Sigma y^2$$

Directly beneath we draw another base line of the same length $Y$ and cut it into $n$ equal parts, each of which is then of length $\bar{y}$.    We now construct the $n$ squares, all of which are of the same size, and denote this area by

$$\text{(area) } C = \bar{y}^2 + \bar{y}^2 + \bar{y}^2 + \bar{y}^2 = n\bar{y}^2$$

It is a geometrical fact that the area $C$ is smaller than the area $A$.    That is, the area of $n$ squares of different sizes is always larger than the area of $n$ squares of equal size on a baseline of the same length.    Now we see that all the fundamental identity says is that the sum of squares of deviations from the mean is simply the difference between the two areas:

$$ssq = A - C$$

If a large square of side length $Y$ is constructed (dotted line in Fig. 3.1), we see that the area $C = n\bar{y}^2$ is only $1/n$ of the large $Y^2$; hence, $n\bar{y}^2 = Y^2/n$.

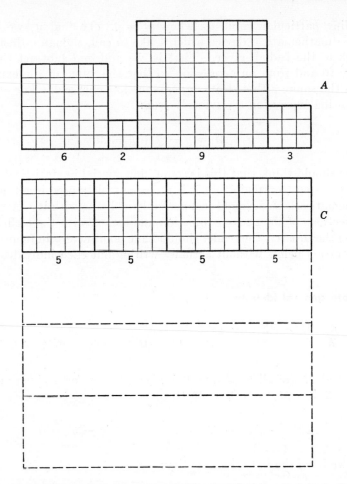

**Fig. 3.1** Visual aid to the calculation of the sum of squares of deviations from mean. The length of the base line is $6 + 2 + 9 + 3 = 20 = Y$. The area of the top row is $A = 6^2 + 2^2 + 9^2 + 3^2 = 130 = \Sigma y^2$. The area of the bottom row is $C = 5^2 + 5^2 + 5^2 + 5^2 = 100 = n\bar{y}^2$. The area of the big square indicated by dotted lines is $Y^2 = 20^2$. Hence, $C - n\bar{y}^2 = Y^2/n$. The difference between the two areas, $A - C$, is the desired $ssq$. [Modified from C. C. Li, A Diagrammatic Representation of the Sum of Squares and Products, *J. Am. Statist. Assoc.*, 50:1056–1063 (1955).]

## 7. The trick of the Indian giver

We are now prepared to study the quantity $Q = \Sigma(y - M)^2$ as introduced in Sec. 2 and exemplified in Table 3.1. We would like especially to find the relationship between the quantity $Q$ for an arbitrary $M$ and the corresponding particular value $Q_0$ when $M = \bar{y}$. To find it, we usually employ a mathematical device which I like to call, though unjustifiably, the trick of the Indian giver. That is, we add *and* subtract the same quantity to and from an expression so that the value of the expression remains the same, although its appearance has changed. For example, the following expression is obviously true:

$$y - M = y - \bar{y} + \bar{y} - M$$

Lest you should think that this is something special in statistical mathematics, I hasten to explain that it is a device widely used in all branches of mathematics. One should feel perfectly at home with it. Assured of the correctness of the above expression and armed with the technique of summing the squares, the reader should have no difficulty in following the demonstration below without actually writing out each individual term.

### 8. A more general identity

$$\begin{aligned}
\Sigma(y - M)^2 &= \Sigma[(y - \bar{y}) + (\bar{y} - M)]^2 \\
&= \Sigma[(y - \bar{y})^2 + (\bar{y} - M)^2 + 2(y - \bar{y})(\bar{y} - M)]
\end{aligned}$$

Since $\bar{y}$ and $M$ are all fixed quantities (constants) for a given group of numbers, $\Sigma(\bar{y} - M)^2 = n(\bar{y} - M)^2$ and

$$\Sigma 2(y - \bar{y})(\bar{y} - M) = 2(\bar{y} - M)\Sigma(y - \bar{y}) = 0$$

Hence, we have the following identity:

$$\Sigma(y - M)^2 = \Sigma(y - \bar{y})^2 + n(\bar{y} - M)^2$$

More briefly,

$$Q = Q_0 + n(\bar{y} - M)^2$$

The above identity brings out the important fact that the sum of squares of deviations from an arbitrary number (other than the mean value) is always larger than the sum of squares of deviations from the mean. Furthermore, we know that it is larger by the amount $n(\bar{y} - M)^2$. The more the $M$ diverges from $\bar{y}$, the larger the value of $Q$ as compared with $Q_0$. Returning to the numerical illustrations in Table 3.1, we see that when $M = 2$, $Q = Q_0 + n(\bar{y} - M)^2 = 30 + 4(5 - 2)^2 = 66$. This identity is a more general one than that of Sec. 5. In the former identity

$M$ can be any number at all. If we put $M = 0$, the identity reduces to

$$\Sigma y^2 = \Sigma(y - \bar{y})^2 + n\bar{y}^2$$

which is the fundamental identity of Sec. 5.

## 9. A product form

It may be pointed out that the sum of products of $y$ and its deviation $y - \bar{y}$ is also equal to the sum of squares, viz.,

$$\Sigma y(y - \bar{y}) = \Sigma y^2 - \bar{y}\Sigma y = \Sigma y^2 - n\bar{y}^2 = ssq$$

As an exercise, the reader should take out a few minutes to verify it with the numerical values $y = 6, 2, 9, 3$.

## 10. Constant, added or multiplied

If a constant $c$ is added to each of the $y$ values, their mean value must be increased by the same amount, $c$. Consequently, the deviations (from mean) remain the same, and the sum of squares of such deviations will also remain the same. Thus, the following groups of four numbers all have the same $ssq$:

$$
\begin{array}{ccccl}
6 & 2 & 9 & 3 & ssq = 30 \\
86 & 82 & 89 & 83 & ssq = 30 \\
+1 & -3 & +4 & -2 & ssq = 30
\end{array}
$$

If each of the $y$'s is multiplied by a constant $c$, the new mean value will be $c$ times the old mean. Likewise, the new deviation will be $c$ times the old deviation, and the square of the new deviation will be $c^2$ times the old squared deviation. Hence

$$\Sigma(cy - c\bar{y})^2 = c^2\Sigma(y - \bar{y})^2$$

For example,

$$
\begin{array}{lccccl}
y: & 6 & 2 & 9 & 3 & ssq = 30 \\
2y: & 12 & 4 & 18 & 6 & ssq = 30 \times 2^2 = 120 \\
cy: & 6c & 2c & 9c & 3c & ssq = 30c^2
\end{array}
$$

## 11. Two distinct values

A very special case is that in which there are only two numbers, $y_1$ and $y_2$. In that case the mean is $\bar{y} = \frac{1}{2}(y_1 + y_2)$, the midpoint between the two $y$ values. It follows that the two deviations must be equal in numerical value but opposite in sign, remembering that the sum of deviations is zero. That is, if one deviation is $d$, the other deviation is $-d$, so that the $ssq = d^2 + (-d)^2$. But $2d$ is the total length from $y_1$ to $y_2$; that is, $2d = y_1 - y_2$. Hence we obtain

$$ssq = d^2 + (-d)^2 = 2d^2 = \frac{(2d)^2}{2}$$

or

$$(y_1 - \bar{y})^2 + (y_2 - \bar{y})^2 = \frac{(y_1 - y_2)^2}{2}$$

which we leave to the reader to verify to his own satisfaction.

The above identity may be immediately extended to the case in which half of the numbers take one value and the other half another value, such as 8, 8, 8, 2, 2, 2. The mean of the entire group of six numbers is equal to the mean of the two distinct numbers, 8 and 2, which may be designated as $y_1$ and $y_2$. Since there are $n = 3$ of each, the sum of squares of the entire group is simply $n = 3$ times the sum of squares for the two distinct numbers. Hence, for the entire group,

$$ssq = n(y_1 - \bar{y})^2 + n(y_2 - \bar{y})^2 = \frac{n(y_1 - y_2)^2}{2} = \frac{(Y_1 - Y_2)^2}{2n}$$

where $Y_1 = ny_1$ and $Y_2 = ny_2$. In our numerical example,

$$ssq = \frac{3(8 - 2)^2}{2} = \frac{(24 - 6)^2}{6} = 54$$

The shortcut method described above may be further extended to a group in which there are unequal number of identical values, like the following

$$8 \quad 8 \quad 8 \quad 2 \quad 2 \quad 2 \quad 2 \quad 2 \quad 2$$

Here, $y_1 = 8$ and $y_2 = 2$ are the two distinct values, but there are $n_1 = 3$ of the former and $n_2 = 6$ of the latter. Then the sum of squares of deviations for the entire group of nine numbers is

$$ssq = n_1(y_1 - \bar{y})^2 + n_2(y_2 - \bar{y})^2$$

where

$$\bar{y} = \frac{n_1 y_1 + n_2 y_2}{n_1 + n_2} = \frac{24 + 12}{3 + 6} = 4$$

Substituting and simplifying, we obtain

$$ssq = \frac{n_1 n_2^2 (y_1 - y_2)^2}{(n_1 + n_2)^2} + \frac{n_2 n_1^2 (y_2 - y_1)^2}{(n_1 + n_2)^2}$$

$$= \frac{n_1 n_2 (y_1 - y_2)^2}{n_1 + n_2} = \frac{(y_1 - y_2)^2}{1/n_1 + 1/n_2}$$

In our example this value is

$$ssq = \frac{3 \times 6 \times (8 - 2)^2}{9} = 72$$

which may be verified by longhand calculation.

## 12. Two special values: 1 and 0

We have learned how to calculate the *ssq* of a group of $n$ numbers in general and then how to use a shortcut method when a group consists of only two distinct values, whether equal or unequal in number. Now we come to the very special case in which the two distinct values are 1 and 0. The reader naturally anticipates that the method of calculating *ssq* will be even simpler in such cases. Indeed it is.

The method is based on the fundamental identity established in Sec. 5. Being an identity, it applies to any group of numbers whatsoever, and it applies to a group of 1's and 0's. Consider the following group of nine ($n = 9$) very special numbers and apply our general method of calculating the *ssq*.

$$y_\alpha: \quad 1 \quad 0 \quad 1 \quad 1 \quad 0 \quad 1 \quad 0 \quad 1 \quad 1 \qquad Y = \Sigma y = 6$$

The mean of these nine numbers is

$$\bar{y} = \frac{\Sigma y}{n} = \frac{6}{9} = \frac{2}{3}$$

In a case such as this, the mean is simply the proportion of 1's in the group. Thus, two-thirds of the nine numbers are 1. A further simplification arises from the facts that $1^2 = 1$ and $0^2 = 0$. Hence, the quantity ("area") $A = \Sigma y^2 = \Sigma y = Y$. In view of the meaning of $\bar{y}$ in this particular situation, we may introduce a new symbol for it. If we write $p$ for $\bar{y}$, meaning *proportion*, then $Y = np$ and the sum of squares is

$$ssq = \Sigma y^2 - n\bar{y}^2 = np - np^2 = np(1 - p) = npq$$

where $q = 1 - p$. In our example, the $ssq = 9(\frac{2}{3})(\frac{1}{3}) = 2$. The reader should take out a few minutes to verify the answer by longhand calculation: find each single deviation, square it, and then adding the squares together.

This completes the essential algebra of the sum of squares. There are several other expressions for *ssq*—one of which is given in the exercises at the end of this chapter—but the preceding paragraphs have covered the most important and most frequently encountered ones.

## EXERCISES

**3.1** Calculate the following quantities for twelve numbers:

| $y$ | $y - \bar{y}$ | $(y - \bar{y})^2$ | $y(y - \bar{y})$ | $y^2$ |
|:---:|:---:|:---:|:---:|:---:|
| 2 | | | | |
| 6 | | | | |
| 4 | | | | |
| 10 | | | | |
| 4 | | | | |
| 9 | | | | |
| 12 | | | | |
| 8 | | | | |
| 11 | | | | |
| 6 | | | | |
| 9 | | | | |
| 3 | | | | |
| Total  $Y$ | | | | |

Verify

$$\sum (y - \bar{y})^2 = \sum y(y - \bar{y}) = \sum y^2 - \frac{Y^2}{12}$$

and draw a diagram to illustrate the last expression (right-hand side of the equation).

**3.2** Do the same thing as in Exercise 3.1 with the following nine numbers:

$$1 \quad 9 \quad 4 \quad 2 \quad 8 \quad 5 \quad 1 \quad 4 \quad 9$$

It is only through long calculations that one appreciates the identity in Sec. 5.
*Ans.:* $ssq = 83.55 = 83\frac{5}{9}$

**3.3** Let the total of $n$ numbers be $Y = y_1 + \cdots + y_n$. Expand

$$Y^2 = (y_1 + \cdots + y_n)^2$$

If you are not used to general expressions, try the case $n = 4$ and write out all the terms systematically of $(y_1 + y_2 + y_3 + y_4)^2$. This must be done before you proceed to the next exercise.

**3.4** Given $n = 4$ numbers, $y_1$, $y_2$, $y_3$, $y_4$. Take the difference between all possible pairs. The relationship to be established is that the sum of squares of such differences is a simple multiple of $ssq$ as calculated from deviations from the mean. Thus,

$$(y_1 - y_2)^2 = y_1{}^2 + y_2{}^2 - 2y_1y_2$$
$$(y_1 - y_3)^2 = y_1{}^2 + y_3{}^2 - 2y_1y_3$$
$$(y_1 - y_4)^2 = y_1{}^2 + y_4{}^2 - 2y_1y_4$$
$$(y_2 - y_3)^2 = y_2{}^2 + y_3{}^2 - 2y_2y_3$$
$$(y_2 - y_4)^2 = y_2{}^2 + y_4{}^2 - 2y_2y_4$$
$$(y_3 - y_4)^2 = y_3{}^2 + y_4{}^2 - 2y_3y_4$$

Adding,

$$\Sigma(y_\alpha - y_\beta)^2 = 3(y_1{}^2 + y_2{}^2 + y_3{}^2 + y_4{}^2) - 2\Sigma y_\alpha y_\beta \qquad \alpha < \beta$$

If there are $n$ numbers, $y_1, \ldots, y_n$, there will be $n - 1$ pairs involving $y_1$; and this is true for every number. Hence the general situation is

$$\begin{aligned}
\Sigma(y_\alpha - y_\beta)^2 &= (n - 1)\Sigma y^2 - 2\Sigma y_\alpha y_\beta \qquad \alpha < \beta \\
&= n\Sigma y^2 - (\Sigma y^2 + 2\Sigma y_\alpha y_\beta) \\
&= n\Sigma y^2 - Y^2 = n \times ssq
\end{aligned}$$

Verify this relationship, using the four numbers 6, 2, 9, 3.

3.5 Find the sum of squares of deviations from mean for the following nine numbers:

$$2 \quad 1 \quad 2 \quad 2 \quad 1 \quad 2 \quad 1 \quad 2 \quad 2$$

First use the method of Sec. 11 and then the method of Sec. 12 (by subtracting 1 from each number without affecting the sum of squares).

# Expected values

As we approach such topics as populations and samples, we can hardly avoid at least a cursory review of the theoretical background. In this chapter we shall need just enough to get by with the main problem of estimating the population variance based on a limited number of observations contained in a sample.

## 1. Population and sample

Suppose that there is a very large population from which a random sample of $n$ observations is taken, as illustrated in Fig. 4.1. We shall confine ourselves to only two of the *parameters* of the population: one is the mean of all the $y$'s in the population, denoted by $\mu$; the other is the mean value of all the $(y - \mu)^2$ in the population, denoted by $\sigma^2$. These two values, $\mu$ and $\sigma^2$, are considered to be *constants* as far as any one population is concerned. Conventionally, these population parameters are denoted by Greek letters.

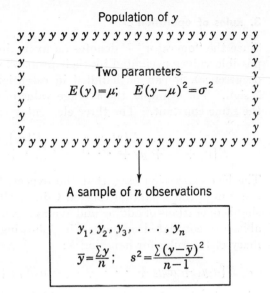

Fig. 4.1 A sample of $n$ random observations from a large $y$ population.

## 2. The operator $E$

A new symbol, $E$, that is very convenient and useful in theoretical work is here introduced (Fig. 4.1). Its meaning and the rules for handling it will be explained in some detail.

Suppose that we take only one $y$ from the population and assume that every $y$ in the population has the same chance of being selected by us. In other words, the particular $y$ we happen to observe could be, with equal probability, any one of the $y$'s in the population. This process is known as *random sampling;* and if random sampling is repeated many, many times, the long-range average value of the $y$'s will be the mean of the population. The symbol $E$ simply means "the long-range average value of," or, more formally, "the mathematical expectation of" or "the expected value of." Hence, we write $E\{y\} = \mu$. This is, in fact, the definition of the population mean. In practical mathematical manipulations, $E\{y\}$ may always be replaced by $\mu$.

Associated with each $y$ value there is a corresponding squared deviation $(y - \mu)^2$. The average value of all squared deviations in the population is the variance of the population, and hence we write $E\{y - \mu\}^2 = \sigma^2$. Again, this is the definition of $\sigma^2$; and in mathematical manipulations, $E\{y - \mu\}^2$ may always be replaced by $\sigma^2$. Note that in this and the preceding paragraph, the $y$ has no subscript attached, because we are not talking about any particular one of the $y$'s but are concerned with the long-term result of repeated random sampling. If a subscript is desired, we should use $y_\alpha$, where $\alpha$ may denote any one of the $y$'s.

### 3. Rules of operation

Since the "operator" $E$ denotes an averaging process (that is, adding all possible values, weighted by their probabilities or relative frequencies of occurrence), it can be handled in essentially the same manner as the $\Sigma$ sign. For example, the average value of a constant is still, obviously, the same constant. The three elementary rules are

$$E\{c\} = c \qquad E\{cy\} = cE\{y\}$$
$$E\{y_1 + y_2 + y_3 + \cdots\} = E\{y_1\} + E\{y_2\} + E\{y_3\} + \cdots$$

The last expression says that the expected value of a sum is equal to the sum of the separate expected values. This theorem results from the simple operations of adding and averaging, and it is true whether the variables are correlated or not. The following expression serves as a summary of the rules for handling $E$:

$$E\{a_1 y_1 + a_2 y_2 + \cdots + c\} = a_1 E\{y_1\} + a_2 E\{y_2\} + \cdots + c$$

where the $a$'s and $c$ are constants. If the expected (average) value of such quantities as $(ay)^2$ is desired, it follows from the above rules that

$$E\{ay\}^2 = E\{a^2 y^2\} = a^2 E\{y^2\}$$

### 4. Simple expected values

Our chief concern in this chapter is to find the expected value of quantities such as $\Sigma (y - \bar{y})^2/(n - 1)$ from repeated random samples of size $n$ drawn from a large population. Before we can do so, several elementary theorems must be clearly established.

1. First, suppose that we draw a random sample of three observations. It is immaterial which observation should be designated as $y_1$, which as $y_2$, and which as $y_3$. If the three observations are written down in random order, then each of the $y$'s could be any one from the population. Therefore, if samples of size 3 are repeated many times, the expected value of the so-called $y_1$'s is no different from that of the so-called $y_2$'s or $y_3$'s. In the long run they will all have an average value of $\mu$, the population mean. For a numerical illustration the reader may examine the first two columns of Table 4.1. Hence, in general

$$E\{y_1\} = \cdots = E\{y_n\} = \mu$$
$$E\{y_1 + y_2 + \cdots + y_n\} = n\mu$$

2. It should also be realized that in the type of random sampling under consideration, there is no correlation between any two of the observations in the sample. For example, in the sample of three observations, $y_1$ could

be any one from the population, but so could $y_2$ and $y_3$, and with the same probabilities. In such a case the $y$'s are said to be statistically independent. The expected value of the product of any two (say, the first two) observations in a large number of random samples is

$$E\{y_1 y_2\} = E\{y_1\}E\{y_2\} = \mu \cdot \mu = \mu^2$$

This theorem is true only with respect to statistically independent variables. It says that the expected value of a product is equal to the product of the separate expected values. This fact is demonstrated in all elementary textbooks on probability theory, but we shall have a numerical illustration in a later paragraph (Table 4.1).

3. The expected value of $y^2$ in repeated samples is also needed. It may be found directly from the definition of $\sigma^2$; thus

$$\begin{aligned}
\sigma^2 = E\{y - \mu\}^2 &= E\{y^2 - 2\mu y + \mu^2\} \\
&= E\{y^2\} - 2\mu E\{y\} + \mu^2 \\
&= E\{y^2\} - 2\mu \cdot \mu + \mu^2 = E\{y^2\} - \mu^2
\end{aligned}$$

Hence,

$$E\{y^2\} = \mu^2 + \sigma^2$$

This shows that the expected value of the square of a variable is always larger than that of the product of two statistically independent variables with the same mean value. It also follows that

$$E\{y_1^2 + \cdots + y_n^2\} = E\{\Sigma y^2\} = n(\mu^2 + \sigma^2)$$

4. With the aid of these preliminary theorems, we may find the expected value of $Y^2$ in repeated sampling. Thus,

$$E\{y_1 + \cdots + y_n\}^2 = E\{y_1^2 + \cdots + y_n^2 + 2y_1 y_2 + 2y_1 y_3 + \cdots\}$$

In the above expression there are $n$ square terms, each of which has an expected value of $\mu^2 + \sigma^2$, and $n(n - 1)$ product terms of the type $y_1 y_2$, each of which has an expected value of $\mu^2$. Hence,

$$E\{Y^2\} = n(\mu^2 + \sigma^2) + n(n - 1)\mu^2 = n^2\mu^2 + n\sigma^2$$

$$E\left\{\frac{Y^n}{n}\right\} = E\{n\bar{y}^n\} = n\mu^2 + \sigma^2$$

$$\frac{E\{Y^2\}}{n^2} = E\{\bar{y}^2\} = \mu^2 + \frac{\sigma^2}{n}$$

## 5. Numerical example

These theorems are simple, but it is important to grasp the physical meaning of what they say. For this purpose let us consider the following large population. The value 1 has a relative frequency (probability) of

$$
\begin{array}{ccccccc}
1 & 1 & 1 & 1 & 1 & \cdots & 1 \\
& 3 & 3 & 3 & 3 & 3 \cdots & 3 \\
5 & 5 & 5 & 5 & 5 \cdots & 5 \\
& 7 & 7 & 7 & 7 & 7 \cdots & 7
\end{array}
$$

$$
\mu = 4 \qquad\qquad \sigma^2 = 5
$$

25 per cent in the population and so does the value 3, etc.   The population mean is

$$\mu = \tfrac{1}{4}(1 + 3 + 5 + 7) = 4$$

Associated with the value 1 is the corresponding squared deviation $(1 - 4)^2 = (-3)^2 = 9$, which, of course, also has a relative frequency of 25 per cent in the population.   The average of all such squared deviations

*Table 4.1*   **Random samples of size $n = 2$ from the large population of Sec. 5**

| $y_1$ | $y_2$ | $y_1y_2$ | $y_1{}^2$ | $\bar{y}$ | $(\bar{y} - \mu)^2$ | $s^2$ |
|---|---|---|---|---|---|---|
| 1 | 1 | 1 | 1 | 1 | 9 | 0 |
| 1 | 3 | 3 | 1 | 2 | 4 | 2 |
| 1 | 5 | 5 | 1 | 3 | 1 | 8 |
| 1 | 7 | 7 | 1 | 4 | 0 | 18 |
| 3 | 1 | 3 | 9 | 2 | 4 | 2 |
| 3 | 3 | 9 | 9 | 3 | 1 | 0 |
| 3 | 5 | 15 | 9 | 4 | 0 | 2 |
| 3 | 7 | 21 | 9 | 5 | 1 | 8 |
| 5 | 1 | 5 | 25 | 3 | 1 | 8 |
| 5 | 3 | 15 | 25 | 4 | 0 | 2 |
| 5 | 5 | 25 | 25 | 5 | 1 | 0 |
| 5 | 7 | 35 | 25 | 6 | 4 | 2 |
| 7 | 1 | 7 | 49 | 4 | 0 | 18 |
| 7 | 3 | 21 | 49 | 5 | 1 | 8 |
| 7 | 5 | 35 | 49 | 6 | 4 | 2 |
| 7 | 7 | 49 | 49 | 7 | 9 | 0 |
| Total | 64 | 64 | 256 | 336 | 64 | 40 | 80 |
| Mean | 4 | 4 | 16 | 21 | 4 | 2·5 | 5 |
| Expected | $\mu$ | $\mu$ | $\mu^2$ | $\mu^2 + \sigma^2$ | $\mu$ | $\dfrac{\sigma^2}{n}$ | $\sigma^2$ |

in the population is

$$\sigma^2 = \tfrac{1}{4}[(-3)^2 + (-1)^2 + (1)^2 + (3)^2] = 5$$

Having specified the two parameters of the population, we now proceed to investigate the long-range results of repeated random sampling from this population. Let the $n$ random observations of the sample be $y_1$, $y_2$, . . . , $y_n$, which have been written down in random order, regardless of their actual values. The long-range average value of each of them will be $\mu = 4$, because each of them has a probability of $\tfrac{1}{4}$ of being 1, a probability of $\tfrac{1}{4}$ of being 3, etc. Next, let us form the products of any two random observations (say, $y_1$ and $y_2$). These possible products are listed in Table 4.1. Each of these products has a probability (or relative frequency) of $\tfrac{1}{16}$ to occur with statistical independence. Therefore the average value of these products is

$$E\{y_1 y_2\} = \tfrac{1}{16}(\text{sum of 16 products in Table 4.1})$$
$$= \tfrac{1}{16}(256) = 16 = \mu^2$$

The expected value of $y^2$ may also be verified numerically. Only the $y_1{}^2$ values are listed in Table 4.1, since $y_2{}^2$ assume the same values arranged in a different order. Each square has a probability of $\tfrac{1}{4}$ to occur. The average value of such squares is then

$$E\{y^2\} = \tfrac{1}{4}(1 + 9 + 25 + 49) = \tfrac{1}{4}(84)$$
$$= 21 = \mu^2 + \sigma^2 = 16 + 5$$

## 6. Expected value of $s^2$

Now we come to the main subject—that of finding the expected value of $s^2 = \Sigma(y - \bar{y})^2/(n - 1)$ for repeated samples of the fixed size $n$. In dealing with expectations, the constant denominator $(n - 1)$ may be ignored for the time being. Then we may concentrate on the numerator (that is, $s s y$), which can be broken into two parts, $\Sigma(y - \bar{y})^2 = \Sigma y^2 - Y^2/n$. The expectation of these two parts may be worked out separately by using the above preliminary theorems:

By item 3, Sec. 4: $\qquad\qquad E\{\Sigma y^2\} = n\mu^2 + n\sigma^2$
By item 4, Sec. 4: $\qquad\qquad E\{Y^2/n\} = n\mu^2 + \sigma^2$
Subtracting: $\qquad\qquad \overline{E\{\Sigma(y - \bar{y})^2\} = (n - 1)\sigma^2}$

Hence,

$$E\left\{\frac{\Sigma(y - \bar{y})^2}{n - 1}\right\} = E\{s^2\} = \sigma^2$$

where $s^2 = \Sigma(y - \bar{y})^2/(n - 1)$ is called the *sample variance*. For each sample of size $n$, there is a corresponding value of $s^2$. The above theorem says that if a large number of random samples of $n$ are taken, the long-range average value of $s^2$ is exactly equal to the population variance $\sigma^2$. For this reason, $s^2$ is called an *unbiased estimate* of $\sigma^2$. In the last column of Table 4.1 the value of $s^2$ is calculated for each sample, and it is seen that its average value is actually $\sigma^2 = 5$.

## 7. Expected value of $\bar{y}$

In establishing the theorem $E\{s^2\} = \sigma^2$, only the expectations $E\{y^2\}$ and $E\{Y^2\}$ have been used, and nothing has been said about the long-range properties of the sample mean $\bar{y}$. In the two preceding chapters, where only one group of numbers (which may be considered as one sample) is under consideration, $\bar{y}$ has been treated as a constant. For repeated sampling, however, $\bar{y}$ varies from sample to sample, as shown in Table 4.1. The long-range average value of these $\bar{y}$'s is

$$E\{\bar{y}\} = \frac{1}{n} E\{y_1 + \cdots + y_n\} = \frac{1}{n}(n\mu) = \mu$$

which is verified numerically in Table 4.1. For this reason, $\bar{y}$ is called an unbiased estimate of population mean $\mu$.

## 8. Variance of $\bar{y}$

Since $\bar{y}$ varies from sample to sample and since associated with each $\bar{y}$ there is a corresponding squared deviation $(\bar{y} - \mu)^2$, we define the variance of $\bar{y}$ as the average value of such squared deviations, viz.,

$$\sigma_{\bar{y}}^2 = E\{\bar{y} - \mu\}^2$$

To find this value, we may use the theorem that $E\{Y^2\} = n^2\mu^2 + n\sigma^2$ of part 4 of Sec. 4. Hence,

$$\sigma_{\bar{y}}^2 = E\{\bar{y} - \mu\}^2 = E\{\bar{y}^2\} - \mu^2$$
$$= \mu^2 + \frac{\sigma^2}{n} - \mu^2 = \frac{\sigma^2}{n}$$

Although $\bar{y}$ varies from sample to sample, its variance is only $1/n$ of the original population variance. By increasing the size of samples, the variance of $\bar{y}$ can be made small. In our numerical example, where $n = 2$, it is seen that $\sigma_{\bar{y}}^2 = \frac{1}{2}\sigma^2 = 2.5$, as shown in Table 4.1.

## 9. Another proof of $E\{s^2\} = \sigma^2$

This section gives an alternative proof of the result of Sec. 6 by using the relationship $\sigma_{\bar{y}}^2 = \sigma^2/n$. Let us recall the identity established in Chap. 3, Sec. 7, where $M$ is an arbitrary number. As far as any one sample is concerned, $\bar{y}$ is the mean from which the sum of squares is calculated, while $\mu$ plays the role of the arbitrary number $M$.  Thus we may rewrite that identity as follows:

$$\Sigma(y - \bar{y})^2 = \Sigma(y - \mu)^2 - n(\bar{y} - \mu)^2$$

Taking expectations of both sides, and remembering that the expected value of each $(y - \mu)^2$ is equal to $\sigma^2$ by definition and that of $(\bar{y} - \mu)^2$ is $\sigma^2/n$, we obtain

$$E\left\{\sum (y - \bar{y})^2\right\} = n\sigma^2 - n\left(\frac{\sigma^2}{n}\right) = (n - 1)\sigma^2$$

which is equivalent to saying $E\{s^2\} = \sigma^2$.

## 10. In terms of deviations

In preparation for later chapters, some of the results of this chapter may be restated by introducing another symbol.  Each $y$ value has a deviation from the population mean $\mu$.  Denote the deviation $(y - \mu)$ by $\epsilon$. Then each $y$ value may be regarded as consisting of two parts, viz.,

$$y_\alpha = \mu + \epsilon_\alpha$$

The statistical independence of two $y$ values implies the independence of two $\epsilon$ values, because the $\mu$ part is a constant.  Hence the following three expressions are equivalent:

$$E\{y\} = \mu \qquad E\{y - \mu\} = 0 \qquad E\{\epsilon\} = 0$$

and so are the following:

$$E\{y - \mu\}^2 = \sigma^2 \qquad E\{\epsilon^2\} = \sigma^2$$

The reader may show that the following are also equivalent:

$$E\{y_1y_2\} = \mu^2 \qquad E\{y_1 - \mu\}\{y_2 - \mu\} = 0 \qquad E\{\epsilon_1\epsilon_2\} = 0$$

For samples of size $n$, we have seen in Sec. 7 that $\bar{y}$ is an unbiased estimate of $\mu$.  Thus the sample deviation $(y - \bar{y}) = e$ is an estimate of $(y - \mu) - \epsilon$.  Then each number in the sample may be written as

$$y = u + e$$

where $u = \bar{y}$.  The sum of squares $\Sigma(y - \bar{y})^2$ then becomes $\Sigma e^2$ and the sample variance is $s^2 = \Sigma e^2/(n - 1)$.  Nothing in this section is new except for the two notations $\epsilon$ and $e$.

## 11. Degrees of freedom

At this stage it is convenient to introduce informally the concept of the *degrees of freedom*.  Since each of the $n$ observations in a random sample is independent of the others, we say that the quantity $\Sigma y^2$ has $n$ degrees of freedom. On the other hand, the quantity $n\bar{y}^2 = Y^2/n$ has only one degree of freedom, because, for any specified value of $n$, its value depends only on that of $\bar{y}$ or $Y$. For our purpose we shall accept as an axiom that the number of degrees of freedom of the difference between two quantities is equal to the difference between the two corresponding numbers of degrees of freedom. Thus the quantity

$$\Sigma y^2 - n\bar{y}^2 = \sum (y - \bar{y})^2 = \sum e^2$$

has $n - 1$ degrees of freedom. The degrees of freedom cannot be defined more precisely than this without touching upon the rank of a quadratic form. It should be pointed out that the $y$ values and the $\bar{y}$ involved in calculating the sum of squares are not independent but are related by the restriction $n\bar{y} = y_1 + \cdots + y_n$. This linear restriction decreases the degrees of freedom by 1, resulting in $n - 1$. Using this new terminology, the sample variance may be written as

$$s^2 = \frac{ssq}{df}$$

where $df$ denotes the number of degrees of freedom of $ssq$.

Referring to the geometric picture (Fig. 3.1), we may say that the number of degrees of freedom is simply equal to the number of *distinct* squares in an area. Thus the area $A = y_1{}^2 + \cdots + y_n{}^2$ has $n$ degrees of freedom, and the area $C = n\bar{y}^2 = \bar{y}^2 + \cdots + \bar{y}^2$ has only 1 degree of freedom, so that $ssq = A - C$ has $n - 1$ degrees of freedom. Note that by distinct squares we mean distinct in principle. In a large sample, two or more of the $y$'s may have the same numerical value but they are distinct in principle, and the quantity $A$ has $n$ degrees of freedom. On the other hand, the quantity $C$ necessarily involves only 1 distinct square.

## 12. Summary

The headings of the columns in Table 4.1 and their corresponding expected values shown at the bottom of that table should serve as a summary of this chapter. These theorems are very general, being true for any large population (or finite population with replacement in sampling). Note that throughout this chapter, no normality of the population has ever been assumed. In fact, the population used for our example consists of only four distinct values with uniform distribution. The additional assumption that the population should be a normal one is required only when we want to make use of the $F$ (variance-ratio) distribution. This will be dealt with in a later chapter.

## EXERCISES

**4.1** Although the example (Table 4.1) refers to a population of four distinct values with uniform distribution, the theorems established are applicable to any large population. Consider a population with three distinct values 3, 5, 9 with respective densities (relative frequencies) $\frac{1}{2}$, $\frac{1}{4}$, $\frac{1}{4}$. This population may simply be written as (3, 3′, 5, 9), each with relative frequency $\frac{1}{4}$. The prime of the second 3 distinguishes it from the first 3. Take from this population random samples of size $n = 2$ and construct a table similar to Table 4.1, enumerating all possible samples and finding the expected values of $y_1$, $y_1 y_2$, $y_2{}^2$, $\bar{y}$, $s^2$, and $(\bar{y} - \mu)^2$. Check your results with those predicted by theory.

NOTE: $\mu = 5$, $\sigma^2 = 6$.

**4.2** Random samples of size $n = 3$ from a large population (1, 4, 10) with mean $\mu = 5$ and variance $\sigma^2 = 14$. Verify the following results carefully.

| $y_1$ | $y_2$ | $y_3$ | $Y$ | $\bar{y}$ | $ssq$ | $(\bar{y} - \mu)^2$ |
|---|---|---|---|---|---|---|
| 1 | 1 | 1 | 3 | 1 | 0 | 16 |
| 1 | 1 | 4 | 6 | 2 | 6 | 9 |
| 1 | 1 | 10 | 12 | 4 | 54 | 1 |
| 1 | 4 | 1 | 6 | 2 | 6 | 9 |
| 1 | 4 | 4 | 9 | 3 | 6 | 4 |
| 1 | 4 | 10 | 15 | 5 | 42 | 0 |
| 1 | 10 | 1 | 12 | 4 | 54 | 1 |
| 1 | 10 | 4 | 15 | 5 | 42 | 0 |
| 1 | 10 | 10 | 21 | 7 | 54 | 4 |
| 4 | 1 | 1 | 6 | 2 | 6 | 9 |
| 4 | 1 | 4 | 9 | 3 | 6 | 4 |
| 4 | 1 | 10 | 15 | 5 | 42 | 0 |
| 4 | 4 | 1 | 9 | 3 | 6 | 4 |
| 4 | 4 | 4 | 12 | 4 | 0 | 1 |
| 4 | 4 | 10 | 18 | 6 | 24 | 1 |
| 4 | 10 | 1 | 15 | 5 | 42 | 0 |
| 4 | 10 | 4 | 18 | 6 | 24 | 1 |
| 4 | 10 | 10 | 24 | 8 | 24 | 9 |
| 10 | 1 | 1 | 12 | 4 | 54 | 1 |
| 10 | 1 | 4 | 15 | 5 | 42 | 0 |
| 10 | 1 | 10 | 21 | 7 | 54 | 4 |
| 10 | 4 | 1 | 15 | 5 | 42 | 0 |
| 10 | 4 | 4 | 18 | 6 | 24 | 1 |
| 10 | 4 | 10 | 24 | 8 | 24 | 9 |
| 10 | 10 | 1 | 21 | 7 | 54 | 4 |
| 10 | 10 | 4 | 24 | 8 | 24 | 9 |
| 10 | 10 | 10 | 30 | 10 | 0 | 25 |
| **Total** 135 | 135 | 135 | 405 | 135 | 756 | 126 |
| **Mean** 5 | 5 | 5 | 15 | 5 | 28 | 14/3 |
| **Expected** $\mu$ | $\mu$ | $\mu$ | $n\mu$ | $\mu$ | $(n-1)\sigma^2$ | $\sigma^2/n$ |

**4.3**  In presenting the various expected values we have proceeded step by step, first working out the expectation of simple quantities and then combining them to form more complicated ones.  Now, knowing that

$$Y^2 = (y_1 + \cdots + y_n)^2 = \Sigma y_i^2 + \Sigma y_i y_j \qquad i \neq j$$

and each

$$E\{y_i^2\} = \mu^2 + \sigma^2 \qquad E\{y_i y_j\} = \mu^2 \qquad i \neq j$$

prove that

$$\sigma_{\bar{y}}^2 = E\{\bar{y} - \mu\}^2 = \frac{\sigma^2}{n}$$

and

$$E\{s^2\} = E\left\{\frac{ssq}{n-1}\right\} = \sigma^2$$

or

$$E\{ssq\} = E\left\{\Sigma y^2 - \frac{Y^2}{n}\right\} = (n-1)\sigma^2$$

These two results are the most important ones, and the student must be familiar with them.

**4.4**  Prove that

$$E\{\bar{y}^2\} = \mu^2 + \frac{\sigma^2}{n}$$

and verify it numerically with the results of Exercise 4.2.

# 5

# The sum of squares of
# several groups

## 1. Groups of numbers

Before taking on the algebra and the new symbols of this chapter, let us consider a numerical example. Suppose that there are three ($k = 3$) groups of numbers, as shown in Table 5.1. For the time being, the

*Table* 5.1 The sum of squares of several groups

|  | Individual Item $y_{i\alpha}$ | Size $n_i$ | Total $Y_i$ | Mean $\bar{y}_i$ | Within Group $ssq_i$ |
|---|---|---|---|---|---|
| Group 1 | 6, 9, 3 | $n_1 = 3$ | $Y_1 = 18$ | $\bar{y}_1 = 6$ | 18 |
| Group 2 | 10, 4, 9, 12, 8, 11 | $n_2 = 6$ | $Y_2 = 54$ | $\bar{y}_2 = 9$ | 40 |
| Group 3 | 2, 6, 4 | $n_3 = 3$ | $Y_3 = 12$ | $\bar{y}_3 = 4$ | 8 |
| Pooled as a single group of twelve numbers |  | $N = 12$ | $Y = 84$ | $\bar{y} = 7$ | 120 |

method of calculating the *ssq* (as explained in Chap. 3) may be applied to each group separately. This means that the *ssq* for the first group is calculated from the mean ($\bar{y}_1 = 6$) of the first group, and so on. Then ignore the boundaries of the groups and regard the 12 numbers as constituting one single large group whose mean is $\bar{y} = 7$. The *ssq* of this one large group can, again, be calculated by the method of Chap. 3. Before proceeding further, the reader should do all this as an exercise and check with the answers given in Table 5.1. So far, there is nothing new.

## 2. Notation

When there is only one group of numbers, we use $y_\alpha$ as a general notation for any one of the $y$'s, and $\alpha$ can be regarded as a personal or individual name of the members of a single family. When there are several families, we also need a family name or surname, in addition to the "first name," to identify the individual completely. This is the basis for the new notation we are about to adopt. Each individual $y$ then has two subscripts, one to indicate the group to which it belongs (family name) and another to indicate the particular one in the group (given name). Thus, any number in the first group will be denoted by $y_{1\alpha}$, any one in the second group by $y_{2\alpha}$, and so on. For example, the fifth number in group 2 of the accompanying table is $y_{25} = 8$. Note that the subscripts here are 2 and 5, not 25. In general, any number in any group will be denoted by $y_{i\alpha}$, where $i$ identifies the group and $\alpha$ identifies the individual.

## 3. The general mean

Since $n$ has been used to denote the number of items in a single group, we shall now use $n_1$ to denote the number of items in the first group, as shown in Table 5.1. Generally, $n_i$ is the size of the $i$th group. Similarly, $Y_i$ and $\bar{y}_i$ are respectively the total and mean of the $i$th group. When all the groups are pooled together as a single large group, the total number of items and the grand total of all values are:

$$N = n_1 + n_2 + n_3 = \Sigma n_i$$
$$Y = Y_1 + Y_2 + Y_3 = \Sigma Y_i$$

while the general mean of the pooled large group is

$$\bar{y} = \frac{Y}{N} = \frac{Y_1 + Y_2 + Y_3}{N} = \frac{n_1\bar{y}_1 + n_2\bar{y}_2 + n_3\bar{y}_3}{n_1 + n_2 + n_3}$$

This shows that the general mean is not only the mean of the $N$ items but also the weighted (by size) mean of the group means.

## 4. Within-group *ssq*

With the help of these symbols, the arithmetic involved in Table 5.1 may now be stated in more general forms:

Group 1: $\quad ssq_1 = \sum_{\alpha} (y_{1\alpha} - \bar{y}_1)^2 = \sum_{\alpha} y_{1\alpha}^2 - \dfrac{Y_1^2}{n_1} = 18$

Group 2: $\quad ssq_2 = \sum_{\alpha} (y_{2\alpha} - \bar{y}_2)^2 = \sum_{\alpha} y_{2\alpha}^2 - \dfrac{Y_2^2}{n_2} = 40$

Group 3: $\quad ssq_3 = \sum_{\alpha} (y_{3\alpha} - \bar{y}_3)^2 = \sum_{\alpha} y_{3\alpha}^2 - \dfrac{Y_3^2}{n_3} = 8$

Adding,†

$$ssq_W = \sum_i (ssq_i) = \sum_i \sum_{\alpha} (y_{i\alpha} - \bar{y}_i)^2 = \sum_i \sum_{\alpha} y_{i\alpha}^2 - \sum \dfrac{Y_i^2}{n_i} = 66$$

This last quantity ($18 + 40 + 8 = 66$ in the example) is called the *within-group ssq*. The meaning of the double summation sign $\Sigma\Sigma$ should be clear from the above presentations. The second summation with respect to $i$ simply means adding such quantities for all the groups. On the other hand, if the boundaries between the groups are ignored, the *ssq* for the single large group of $N$ items with general mean $\bar{y} - 7$ will be

$$ssq_T = \sum \sum (y_{i\alpha} - \bar{y})^2 = \sum \sum y_{i\alpha}^2 - \dfrac{Y^2}{N} = 120$$

which is called the total *ssq*. As stated before, the above is merely an application of the fundamental identity (Chap. 3, Sec. 5) for each group separately.

## 5. Between-group *ssq*

The group mean $\bar{y}_i$ varies from group to group. The mean of the first group is $\bar{y}_1 = 6$, and it occurs three ($n_1 = 3$) times. Similarly, there are $n_2 = 6$ of $\bar{y}_2 = 9$, and so on. Since $\bar{y}$ is also the mean of these weighted group means, the sum of squares of the deviations of the group means from the general mean is

$$ssq_B = 3(6 - 7)^2 + 6(9 - 7)^2 + 3(4 - 7)^2 = 54$$

† The symbol *ssq* without subscript has been used in general as an abbreviation for the words "sum of squares." When a subscript is attached, as in this and the following paragraphs, it becomes a mathematical symbol to denote a particular sum of squares. This system of symbolism is easy to remember and economical of writing effort. It is borrowed from E. C. Fieller and C. A. B. Smith, Note on the Analysis of Variance and Intraclass Correlation, *Ann. Eugenics*, **16**(1):97–104 (1951).

that is,

$$ssq_B = n_1(\bar{y}_1 - \bar{y})^2 + n_2(\bar{y}_2 - \bar{y})^2 + n_3(\bar{y} - \bar{y})^2$$

Expanding each square term and simplifying, we have

$$
\begin{aligned}
ssq_B &= n_1(\bar{y}_1{}^2 + \bar{y}^2 - 2\bar{y}\bar{y}_1) + n_2(\bar{y}_2{}^2 + \bar{y}^2 - 2\bar{y}\bar{y}_2) + n_3(\bar{y}_3{}^2 + \bar{y}^2 - 2\bar{y}\bar{y}_3) \\
&= n_1\bar{y}_1{}^2 + n_2\bar{y}_2{}^2 + n_3\bar{y}_3{}^2 + N\bar{y}^2 - 2\bar{y}(N\bar{y}) \\
&= \frac{Y_1{}^2}{n_1} + \frac{Y_2{}^2}{n_2} + \frac{Y_3{}^2}{n_3} - \frac{Y^2}{N}
\end{aligned}
$$

Briefly, for any number of groups $(i = 1, 2, \ldots, k)$

$$ssq_B = \sum_i n_i(\bar{y}_i - \bar{y})^2 = \sum_{i=1}^{k} \frac{Y_i{}^2}{n_i} - \frac{Y^2}{N}$$

and this quantity is called the "between-group sum of squares." In our numerical example, this value is then

$$ssq_B = \frac{18^2}{3} + \frac{54^2}{6} + \frac{12^2}{3} - \frac{84^2}{12} = 54$$

in agreement with the result obtained previously.

## 6. An identity

Combining the results of the preceding two paragraphs, we obtain:

Between groups:

$$ssq_B = \sum_i n_i(\bar{y}_i - \bar{y})^2 = \sum_i \frac{Y_i{}^2}{n_i} - \frac{Y^2}{N} = 54$$

Within groups:

$$ssq_W = \sum_i \sum_\alpha (y_{i\alpha} - \bar{y}_i)^2 = \sum_i \sum_\alpha y_{i\alpha}{}^2 - \sum_i \frac{Y_i{}^2}{n_i} = 66$$

Total:    $$ssq_T = \sum_i \sum_\alpha (y_{i\alpha} - \bar{y})^2 = \sum_i \sum_\alpha y_{i\alpha}{}^2 - \frac{Y^2}{N} = 120$$

Thus, we see from the right-hand expressions that

$$ssq_B + ssq_W = ssq_T$$

In words, the sum of squares within the separate groups plus the sum of squares between the group means is equal to the total sum of squares of the single pooled group. This relationship has been established by a very slow procedure so that the reader can follow it step by step. For those who are used to algebraic manipulations the above relationship may be arrived at in one line by using the Indian giver's trick (Chap. 3,

Sec. 7):

$$\sum_i \sum_\alpha (y_{i\alpha} - \bar{y})^2 = \sum_i \sum_\alpha [(y_{i\alpha} - \bar{y}_i) + (\bar{y}_i - \bar{y})]^2$$

$$= \sum_i \sum_\alpha (y_{i\alpha} - \bar{y}_i)^2 + \sum_i n_i(\bar{y}_i - \bar{y})^2$$

The product term vanishes because the sum of deviations from $\bar{y}_i$ is zero for each group, that is, $\sum_\alpha (y_{i\alpha} - \bar{y}_i) = 0$, whereas the factor $(\bar{y}_i - \bar{y})$ is a constant for any fixed group.

## 7. Diagram and practical calculation

The above algebraic fact (identity) is very fundamental, and this type of calculation is very common in the analysis of variance. It is helpful to set up a routine arithmetic procedure by which the manual labor is minimized. For this purpose, a visual aid to practical calculation is provided in Fig. 5.1 (p. 44). For each group the construction of the squares is the same as in Fig. 3.1. When the three groups are connected together (the first and second rows of the diagram above) the total length of the base line is $Y = Y_1 + Y_2 + Y_3$. In the third row the total length $Y$ is divided into $N$ equal parts, on each of which a square is constructed. Then all one needs to do is to calculate the areas of the three series of squares:

$$(\text{area}) \ A = \sum \sum y_{i\alpha}^2 = 6^2 + 9^2 + \cdots + 6^2 + 4^2 = 708$$

$$(\text{area}) \ B = \sum n_i \bar{y}_i^2 = \sum \frac{Y_i^2}{n_i} = \frac{18^2}{3} + \frac{54^2}{6} + \frac{12^2}{3} = 642$$

$$(\text{area}) \ C = N\bar{y}^2 = \frac{Y^2}{N} = \frac{84^2}{12} = 588$$

Then we have:

| | | | |
|---|---|---|---|
| Between groups: | $ssq_B = B - C = 642 - 588 = $ | 54 |
| Within groups: | $ssq_W = A - B = 708 - 642 = $ | 66 |
| Total: | $ssq_T = A - C = 708 - 588 = $ | 120 |

This is the easiest way to obtain the components of the sum of squares. Note that the calculation of the various means and deviations is unnecessary. It is also clear that it is the calculation of the area $A = \Sigma\Sigma y_{i\alpha}^2$ that takes the largest amount of labor and for which there is no mathematical shortcut. However, a desk calculator which can yield the value of $Y$ and $A$ in one continuous operation renders the arithmetic perfectly painless.

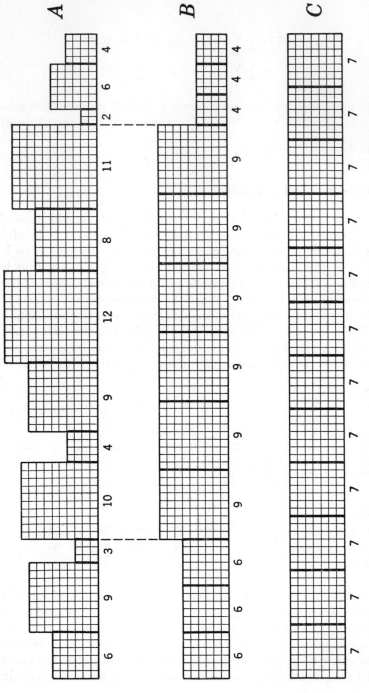

Fig. 5.1 Visual aid to the calculation of the sum of squares within and between groups. The first three squares of the top and middle row are those of group 1, and so on. The areas of the three rows are all that are necessary for the calculation. Detailed description is given in Sec. 7.

44

## 8. Groups of equal size

An important special case is that in which the groups are of equal size, say, $n_1 = n_2 = n_3 = n$. If there are $k$ groups, the total number of items for the entire group will be $N = nk$. The algebraic expressions of the preceding paragraphs, of course, still hold, except that one may wish to put $n$ in front of the $\Sigma$ sign. The only difference it makes in practical calculation is that now

$$(\text{area}) \ B = \frac{1}{n} \sum Y_i^2$$

Table 5.2 gives a numerical example for three groups of the same size and shows how easy the routine calculation is. (Note that these 12 numbers are the same 12 as used in Table 5.1 and, therefore, the areas $A$ and $C$ and the total $ssq$ must remain the same. The total $ssq$ has nothing to do with the grouping.)

*Table 5.2*   **Sum of squares for three equal groups**

| Group | $y_{i\alpha}$ | | | | $Y_i$ | $Y_i^2$ |
|---|---|---|---|---|---|---|
| 1 | 6 | 2 | 9 | 3 | 20 | 400 |
| 2 | 4 | 10 | 4 | 6 | 24 | 576 |
| 3 | 9 | 12 | 8 | 11 | 40 | 1,600 |
| Total | | | | | $84 = Y$ | $2{,}576 = \Sigma Y_i^2$ |
| Areas | $A = 6^2 + \cdots + 11^2 = 708$ | | | | $C = 84^2/12 = 588$ | $B = 2{,}576/4 = 644$ |
| $ssq$ | $ssq_T = A - C = 120$ | | | | $ssq_B = B - C = 56$ | $ssq_W = A - B = 64$ |

## 9. Concluding remark

This chapter deals only with the arithmetic and algebra of the subdivision of the total $ssq$ of an entire group into two main components: one is the $ssq$ within the separate groups and another is the $ssq$ between the groups. This is purely an algebraic fact. An algebraic identity per se may or may not have applications in statistics. In the next two chapters we shall consider a statistical model which will make use of this algebraic fact and explain why we should undertake such a subdivision of the sum of squares and what the meanings of the subdivisions are in the light of the statistical model proposed.

## EXERCISES

**5.1**   In Table 5.2 are listed 12 numbers divided into 3 groups, each of 4 numbers. Draw a diagram similar to Fig. 5.1 for these numbers, showing the areas $A$, $B$, $C$.

**5.2**   Calculate $ssq_1$, $ssq_2$, $ssq_3$, $ssq_W$, $ssq_B$, and $ssq_T$ for the following three groups of numbers.

$$
\begin{array}{lcccccl}
\text{Group 1:} & 3 & 2 & 2 & 5 & & Y_1 = 12 \\
\text{Group 2:} & 2 & 4 & 4 & 6 & & Y_2 = 16 \\
\text{Group 3:} & 2 & 1 & 1 & 4 & & Y_3 = 8
\end{array}
$$

Check your answer by the identity $ssq_B + ssq_W = ssq_T$.

**5.3**   The subdivision of the total $ssq$ of a group of numbers into two major components—one within the subgroupings and one between the subgroupings—is based on an algebraic identity. Hence the subdivision applies to any group of numbers whatsoever. That is, the $y_{i\alpha}$ can assume any finite values, including the special value 0. Apply the method of the present chapter of calculating the areas $A$, $B$, and $C$ and $ssq_B$, $ssq_W$, $ssq_T$ to the data given in the accompanying table. The last column of the table provides an independent check for $ssq_W$, for each of the entries is a $ssq_i$ for that group.

| Group | $y_{i\alpha}$ | | | | | | | | | | | $Y_i$ | $n_i$ | $Y_i^2/n_i$ | $\sum_{\alpha} y_{i\alpha}^2 - Y_i^2/n_i$ |
|---|---|---|---|---|---|---|---|---|---|---|---|---|---|---|---|
| 1 | 1 | 0 | 1 | 1 | 0 | 1 | 1 | 1 | | | | 6 | 8 | 4.50 | 1.50 |
| 2 | 1 | 1 | 0 | 1 | 1 | 0 | 1 | 1 | 1 | 0 | | 7 | 10 | 4.90 | 2.10 |
| 3 | 1 | 0 | 1 | 0 | 0 | 1 | | | | | | 3 | 6 | 1.50 | 1.50 |
| 4 | 0 | 0 | 1 | 0 | 0 | 1 | 0 | 1 | 0 | 1 | | 8 | 16 | 4.00 | 4.00 |
|   | 1 | 0 | 1 | 0 | 1 | 1 | | | | | | | | | |
| Total | | | | | | | | | | | | $Y = 24$    $N = 40$ | | 14.90 | 9.10 |
|   | | | | | | | | | | | | | | $\parallel$ | $\parallel$ |
| Area | $A = \displaystyle\sum_i \sum_\alpha y_{i\alpha}^2 = 24$ | | | | | | | | | | | $C = 24^2/40 = 14.40$ | | $B$ | $ssq_W$ |

$$
\begin{aligned}
ssq_B &= B - C = ? \\
ssq_W &= A - B = ? \\
\hline
ssq_T &= A - C = ?
\end{aligned}
$$

**5.4**   In going through the calculations of the preceding exercise, the student cannot help noticing that (Chap. 3) for any group of numbers which are either 0 or 1,

**a**   $\Sigma y^2 = \Sigma y = Y = \text{total}$

**b**   $\bar{y} = \dfrac{Y}{n} = p = \text{proportion of 1's}$

Thus, if we write group mean $\bar{y}_i = Y_i/n_i = p_i$ and the general mean $\bar{y} = Y/N = P$, then show that

$$
ssq_T = NP(1 - P) \qquad ssq_W = \Sigma n_i p_i (1 - p_i)
$$

and

$$ssq_B = ssq_T - ssq_W = \sum \frac{Y_i^2}{n_i} - \frac{Y^2}{N} = \Sigma n_i p_i^2 - NP^2$$

HINT: $Y = NP = \Sigma n_i p_i = \Sigma Y_i$.

5.5 Calculate the total, between-group, and within-group sum of squares for the following four groups, each consisting of six numbers.

| Group | $y_{i\alpha}$ | | | | | | $Y_i$ |
|-------|----|----|----|----|----|----|-----|
| I | 16 | 22 | 16 | 10 | 18 | 8 | 90 |
| II | 28 | 27 | 17 | 20 | 23 | 23 | 138 |
| III | 16 | 25 | 16 | 16 | 19 | 16 | 108 |
| IV | 28 | 30 | 19 | 18 | 24 | 25 | 144 |
| | | | | | | | $480 = Y$ |

CHECK: $A = 10{,}324$, $B = 9{,}924$, $C = 9{,}600$.

# 6

# Linear model and
# least-square estimates

To introduce the linear model in the analysis of variance, I like to tell the following story.

### 1. Grandma's birthday party

To celebrate her 84th birthday, Grandma decides to distribute $84 among her 12 grandchildren. Being impartial and loving them all, she gives $7 to each grandchild ($7 \times 12 = 84$). Grandpa observes that 4 of the children are graduate students, 4 are in college, and the remaining 4 are still in high school. He does not think that they should all have the same amount of money to spend; and he suggests that the older children should get more than the younger ones. Obligingly, each of the 4 high school students contributes $2, and each of the 4 college students contributes $1, to the group of graduate students. The total amount of money among the twelve children is still $84, but now the four graduates have a total of \$40, each having \$10 instead of the original \$7. After this redistribution of gifts, the children decide to play poker at three separate tables: the four graduates at one table, the four college students at another table, and the four high school youngsters at still another table. All agree to the house rule that they play with nickels as the betting unit. (If the younger ones were to play with pennies at one table and the older ones were to play with quarters at another table, that would be a different story.) At the end of the evening, each child counts his gain or loss at the game and how much is left to him of the original gift of \$7. The detailed account for each child is given in Table 6.1, which should be read from left to right. Note that the gain or loss at the poker game is purely a "table affair," which means that for each table, the gain or loss must cancel among the four players of that table.

*Table 6.1* **Happenings at grandma's birthday party**

| Group | (1) Original amount $\bar{y}$ | (2) Contribution $t_i$ | (3) At start of game $\bar{y}_i$ | (4) Gain or loss $e_{i\alpha}$ | (5) At end of party $y_{i\alpha}$ | (6) Net from original $d_{i\alpha} = y_{i\alpha} - \bar{y}$ |
|---|---|---|---|---|---|---|
| High school | 7 | −2 | 5 | +1 | 6 | −1 |
|  | 7 | −2 | 5 | −3 | 2 | −5 |
|  | 7 | −2 | 5 | +4 | 9 | +2 |
|  | 7 | −2 | 5 | −2 | 3 | −4 |
| College | 7 | −1 | 6 | −2 | 4 | −3 |
|  | 7 | −1 | 6 | +4 | 10 | +3 |
|  | 7 | −1 | 6 | −2 | 4 | −3 |
|  | 7 | −1 | 6 | 0 | 6 | −1 |
| Graduate | 7 | +3 | 10 | −1 | 9 | +2 |
|  | 7 | +3 | 10 | +2 | 12 | +5 |
|  | 7 | +3 | 10 | −2 | 8 | +1 |
|  | 7 | +3 | 10 | +1 | 11 | +4 |
| Totals | 84 | 0 | 84 | 0, 0, 0† | 84 | 0 |

† The three zero's emphasize the fact that the sum of "gain or loss" at each table is zero.

## 2. Lawyer reconstructing the events

Soon after the party is over, the family lawyer drops in to say "Happy birthday" to Grandma. He then hears about Grandma's handouts and the children's poker games, but without details. He learns how much each child has left at the party's end: for the youngest group, 6, 2, 9, 3; for the college group, 4, 10, 4, 6; for the graduate group, 9, 12, 8, 11.   In other words, all he learns are the end results entered in column 5 of Table 6.1.   These are his "observations," so to speak.   Can the lawyer, who has never had a course in statistics, reconstruct the whole series of events (that is, the entire Table 6.1) from his observations alone?   Yes, he can, if he is told of two general conditions:

1. At each of the three poker tables, every player had the same amount of money at the start of the game.
2. Originally each child received the same amount from Grandma.

Informed of these conditions, the lawyer can rapidly reconstruct the whole series of events as recorded in Table 6.1.   Since the total amount of money is $84, originally each child must have had $7.   Then, since the total amount is $20 for the high school group, each high school student must have had $5 at the beginning of the game.   Columns 1 and 3 are known immediately.   The first high school child ended up with $6; this means that he must have won $1 from the poker game.   But he contributed $2 before playing, so the net result is that he is still one dollar short of the original $7, and so on.

## 3. Three components of $y_{i\alpha}$

The experimentalist and the statistician are in a position quite like that of the family lawyer.   All the experiment yields is a set of observations that are the end result of the action of various factors.   The statistician's job is to help the experimenter to reconstruct from the observations something like our Table 6.1 and to try to interpret the observations from that point of view.   Now, let us examine Table 6.1 once more and this time describe it in statistical language.   The numbers in column 5, $y_{i\alpha}$, are observed first.   The number 7 in column 1 is called the *general mean* $\bar{y}$, and the numbers in column 3 are the *group mean* $\bar{y}_i$.   The numbers in column 2 are called the *treatment effects* in general and, in this particular case, may be called the *age effect*, because the redistribution is made on the basis of age.   The "gain or loss" in column 4 is called the *error* or *random variation* of an individual observation.   Finally, the "net" in the last column is the *total deviation*.   Having defined the terms, we may turn our attention to the very close and obvious relationships between the six columns of numbers.   For instance, the second high school child contributed $2 (age effect) at first and then lost another $3 at the game; so his total loss is $5.   Thus, column 6 = columns 2 + 4.   The age effect is $t_i = \bar{y}_i - \bar{y}$, and the error is $e_{i\alpha} = y_{i\alpha} - \bar{y}_i$. These relationships may be summarized as follows:

$$(y_{i\alpha} - \bar{y}) = (\bar{y}_i - \bar{y}) + (y_{i\alpha} - \bar{y}_i)$$

or

$$y_{i\alpha} = \bar{y} + t_i + e_{i\alpha}$$

for example

$$2 = 7 - 2 - 3 \qquad \text{for the second child}$$
$$11 = 7 + 3 + 1 \qquad \text{for the last child}$$

Generally speaking, each observed value $y_{i\alpha}$ may be split into three separate parts: the first part $\bar{y}$ is common to every observation; the second part $t_i$ is common to every member of any one group but varies from group to group; and the third part $e_{i\alpha}$ varies from individual to individual.

## 4. The linear model

The three groups ($k$ groups in general) of observations are visualized by statisticians as three independent samples from a large population whose general mean is $\mu$. Further, it is assumed that this population consists of three subpopulations whose respective mean values are $\mu_1 = (\mu + \tau_1)$, $\mu_2 = (\mu + \tau_2)$, $\mu_3 = (\mu + \tau_3)$ and that the samples are drawn from different subpopulations. In each of the latter, the deviation of a member from the subpopulation mean is $\epsilon_{i\alpha} = y_{i\alpha} - \mu_i$. Consequently, any member of the general population may be expressed in the form of a sum of separate parts:

$$y_{i\alpha} = \mu_i + \epsilon_{i\alpha} = \mu + \tau_i + \epsilon_{i\alpha}$$

This sort of conception is called the *linear model*, of which the above expression is the simplest. It is important to realize that this is *not* a law of nature like the laws established by physicists. It is merely a convenient way of conceiving the various values of a population. It is a model invented by man as a device for simplification. The reader may see that this model will work quite satisfactorily in the case of Grandma's birthday party, but not all things in nature occur that way. In most situations, the linear model is at best an approximation of the realities.

## 5. Least-square criterion

Since, on the one hand, we set up the above model and, on the other hand, we have a set of observations, the next question is what values we should adopt for the subpopulation means that would "best fit" our observations. In other words, how shall we estimate the population parameters $\mu$ and $\tau_i$ based on the sample observations? The term "best fit" is a technical one. If $u$ and $t_i$ are such sample estimates, the criterion for being the "best" is that the sum of squares of the errors $e_{i\alpha}$ is a minimum. This is the *principle of least squares*. The sample estimates $u$ and $t_i$ are then called the least-square estimates of the parameters. In symbols, the quantity to be minimized is

$$Q = \Sigma e_{1\alpha}{}^2 + \Sigma e_{2\alpha}{}^2 + \Sigma e_{3\alpha}{}^2$$
$$= \Sigma(y_{1\alpha} - u - t_1)^2 + \Sigma(y_{2\alpha} - u - t_2)^2 + \Sigma(y_{3\alpha} - u - t_3)^2$$

with the supporting restriction $\Sigma t = t_1 + t_2 + t_3 = 0$. The latter is necessary for obtaining explicit solutions, as shown in the next section.

## 6. Equations and solutions

Using the differential calculus, we would obtain four equations, one for $u$ and three for the $t$'s, as shown at the bottom of Table 6.2. Even without the benefit of differential calculus, I hope the reader will be able to see

*Table* 6.2   **Fitting constants by the method of least squares on the basis of $n = 4$ observed values in each group**

| Group 1 | Group 2 | Group 3 |
|---|---|---|
| $y_{11} - u - t_1 = e_{11}$ | $y_{21} - u - t_2 = e_{21}$ | $y_{31} - u - t_3 = e_{31}$ |
| $y_{12} - u - t_1 = e_{12}$ | $y_{22} - u - t_2 = e_{22}$ | $y_{32} - u - t_3 = e_{32}$ |
| $y_{13} - u - t_1 = e_{13}$ | $y_{23} - u - t_2 = e_{23}$ | $y_{33} - u - t_3 = e_{33}$ |
| $y_{14} - u - t_1 = e_{14}$ | $y_{24} - u - t_2 = e_{24}$ | $y_{34} - u - t_3 = e_{34}$ |
| $\Sigma y_{1\alpha} - nu - nt_1 = 0$ | $\Sigma y_{2\alpha} - nu - nt_2 = 0$ | $\Sigma y_{3\alpha} - nu - nt_3 = 0$ |
| | $\Sigma\Sigma y_{i\alpha} - Nu = 0$ | |

the reasonableness of the solutions. Note that each group of terms in $Q$ is concerned with a subpopulation and is equivalent to the $Q$ introduced early in Chap. 3. In that chapter it was shown that for a single group the value of $Q$ is a minimum when $M$ is equal to the mean of that group. Applying this result to each group separately in our present case, we see that $Q$ will be a minimum when $u + t_1$ is the mean of group 1, $u + t_2$ is the mean of group 2, and $u + t_3$ is the mean of group 3. These are the meanings of the three equations listed next to the bottom of Table 6.2. Finally, adding the three equations together and remembering that $\Sigma t = 0$, we obtain the last equation of Table 6.2. Now, writing $Y_1 = \Sigma y_{1a}$ for the total of the first group, etc., and $Y = Y_1 + Y_2 + Y_3 = \Sigma\Sigma y_{ia}$ for the grand total, the four equations and their solutions are as follows:

| *(Least-square) equations* | *Solutions (estimates)* |
|---|---|
| $Y - Nu = 0$ | $u = \dfrac{Y}{N} = \bar{y}$ |
| $Y_1 - nu - nt_1 = 0$ | $t_1 = \dfrac{Y_1}{n} - u = \bar{y}_1 - \bar{y}$ |
| $Y_2 - nu - nt_2 = 0$ | $t_2 = \dfrac{Y_2}{n} - u = \bar{y}_2 - \bar{y}$ |
| $Y_3 - nu - nt_3 = 0$ | $t_3 = \dfrac{Y_3}{n} - u = \bar{y}_3 - \bar{y}$ |

## 7. Summary

The results of the preceding sections may be summarized as follows:

Population model:     $y_{ia} = \mu + \tau_i + \epsilon_{ia}$

Sample estimates:     $y_{ia} = u + t_i + e_{ia}$

or     $y_{ia} = \bar{y} + (\bar{y}_i - \bar{y}) + (y_{ia} - \bar{y}_i)$

The last two expressions take us back to Table 6.1. It is now apparent that what we did before was to obtain the least-square estimates of the population parameters of a linear model based on the sample observations, assuming that the 12 numbers consist of 3 independent samples, each from a different subpopulation. When the actual values of the sample estimates $u$ and $t_i$ which, when combined, form $\bar{y}_i$, are substituted in the expression for $Q$ in Sec. 5, it becomes the *ssy* within groups (that is, *ssq*$_W$) of Chap. 5, Sec. 4. Hence, we conclude that the *least-square estimates make the within-group sum of squares a minimum*. The discussion in this chapter may be clearly generalized to any number of groups; we use three for the sake of concreteness and simplicity.

## EXERCISES

**6.1** Construct a table similar to Table 6.1 with the 20 numbers given in the table top of page 54. That is, play the role of the family lawyer.

| Group | Original amount $\bar{y}$ | Contri- bution $t_i$ | At start of game $\bar{y}_i$ | Gain or loss $e_{i\alpha}$ | At end of party $y_{i\alpha}$ | Net from original $d_{i\alpha} = y_{i\alpha} - \bar{y}$ |
|---|---|---|---|---|---|---|
| 1 | | | | | 11 18 24 15 | |
| 2 | | | | | 26 25 22 11 | |
| 3 | | | | | 13 19 22 10 | |
| 4 | | | | | 26 21 19 22 | |
| 5 | | | | | 19 27 28 22 | |
| Total | | | | | 400 | |

**6.2** Square each single entry of the preceding table and then calculate the total of the columns of squared entries.   Verify the totals shown in the accompanying table.

| | $\bar{y}^2$ | $t_i^2$ | $\bar{y}_i^2$ | $e_{i\alpha}^2$ | $y_{i\alpha}^2$ | $d_{i\alpha}^2 = (y_{i\alpha} - \bar{y})^2$ |
|---|---|---|---|---|---|---|
| | 400 | 9 | 289 | 36 | 121 | 81 |
| | . | . | . | . | . | . |
| | . | . | . | . | . | . |
| | . | . | . | . | . | . |
| | 400 | 16 | 576 | 4 | 484 | 4 |
| Total | 8,000 | 184 | 8,184 | 402 | 8,586 | 586 |

Review Chap. 5 on the sums of squares of several groups.   What are the relationships among the six column totals of squared entries?   Can you identify three of the six totals as areas *A, B, C*?   How would you obtain the values of the other three totals?

**6.3**   For those who have had differential calculus, it is clear that the quantity to be minimized is the sum of squares within the groups:

$$Q = Q(u, t_1, t_2, t_3) = \Sigma\Sigma(y_{i\alpha} - u - t_i)^2 = \Sigma\Sigma e_{i\alpha}^2$$

The four equations given in Sec. 6 are derived from the four conditions:

$$\frac{\partial Q}{\partial u} = 0 \qquad \frac{\partial Q}{\partial t_1} = 0 \qquad \frac{\partial Q}{\partial t_2} = 0 \qquad \frac{\partial Q}{\partial t_3} = 0$$

where the symbol $\partial$ denotes partial differentiation.   Solutions of these equations are the required values of $u, t_1, t_2, t_3$.

# 7

# The expectation of mean squares

A sum of squares divided by its number of degrees of freedom is called a *mean square* for brevity.

## 1. Mean square for a single group

Let us consider each subpopulation separately for a while. In the first subpopulation, whose mean is $\mu_1 = \mu + \tau_1$, the deviation of each member from the mean is $\epsilon_{1\alpha} = y_{1\alpha} - \mu_1$. The average value of $\epsilon_1{}^2$ is, by definition given in Chap. 4, the variance $\sigma_1{}^2$ of the first subpopulation. Also, by the theory developed in Chap. 4 this variance may be estimated from the $n_1$ observations in group 1. Thus, by virtue of Chap. 4, Sec. 6,

$$E\{ssq_1\} = E\{\Sigma(y_{1\alpha} - \bar{y}_1)^2\} = E\{\Sigma e_1{}^2\} = (n_1 - 1)\sigma_1{}^2$$

The mean square for the first group

$$s_1{}^2 = \frac{ssq_1}{n_1 - 1}$$

is an unbiased estimate of $\sigma_1^2$. That is, $E\{s_1^2\} = \sigma_1^2$. Similarly, $s_2^2$ and $s_3^2$ are the unbiased estimates of $\sigma_2^2$ and $\sigma_3^2$ of the second and third subpopulations, respectively.

## 2. Assumption of common variance

In the ordinary analysis of variance, one of the essential assumptions is that although the subpopulations may have different mean values, their variances are the same, viz.,

$$\sigma_1^2 = \sigma_2^2 = \sigma_3^2 = \sigma^2$$

That is to say, they have a common variance $\sigma^2$. The meaning of this assumption may again be illustrated by the story about Grandma's birthday party (Chap. 6, Sec. 1). You may recall that the three groups of poker players, although they did not have the same amount of capital to begin with, did agree to obey the house rule and play with the same standard of chips; so that the magnitude of gain or loss (variation; error) was about the same at all three tables. This is what we mean by a common variance. The assumption that all subpopulations, in spite of their differences in mean values, have the same variance is clearly open to question and is made purely for the sake of mathematical simplicity. Fortunately, experience shows that if the subpopulation variances do not differ widely, the usual procedure of analysis is applicable.

When all subpopulations have the same variance, they are said to be *homoscedastic*. A method of testing the homogeneity of group variances is given in Chap. 32. Except for that chapter, we assume a common variance throughout the book.

## 3. Pooled estimate of common variance

From now on we shall make no distinction between the various subpopulation variances and write the common variance as $\sigma_e^2$, where the subscript $e$ denotes *error*, or simply $\sigma^2$. In such a case, instead of making three estimates of $\sigma^2$ from the three groups separately, we may make a single estimate based on all three groups in the following manner.

$$E\{ssq_1\} = (n_1 - 1)\sigma^2$$
$$E\{ssq_2\} = (n_2 - 1)\sigma^2$$
$$E\{ssq_3\} = (n_3 - 1)\sigma^2$$

Adding
$$E\{ssq_W\} = (N - 3)\sigma^2 = (N - k)\sigma^2$$

In the very last expression, $k$ is the number of groups for the general case.    Thus,

$$s_W{}^2 = \frac{ssq_1 + \cdots + ssq_k}{(n_1 - 1) + \cdots + (n_k - 1)} = \frac{ssq_W}{N - k}$$

is the pooled unbiased estimate of $\sigma^2$ based on all $k$ groups.    The above expression gives the general rule that in making a pooled estimate of a common variance, the sums of squares are pooled and so are their corresponding degrees of freedom.

## 4. Expectations of basic quantities

To prepare for more complicated situations, it is desirable to introduce a systematic procedure for obtaining the expectations of the sums of squares.    One may recall that in calculating the $ssq$ of several groups, it is advisable to obtain the three basic quantities (areas) $A$, $B$, $C$ of Chap. 5, Sec. 7.    Similarly, to obtain the expected values of the various $ssq$, all we need are the three basic expectations, $E\{A\}$, $E\{B\}$, $E\{C\}$. The reader may wish to review Chap. 4 on expected values, on the basis of which the following results are obtained for each group:

$$E\{\Sigma y_{1\alpha}{}^2\} = n_1(\mu_1{}^2 + \sigma_e{}^2), \text{ etc.}$$
$$E\{n_1\bar{y}_1{}^2\} = n_1\mu_1{}^2 + \sigma_e{}^2, \text{ etc.}$$

Thus, for the entire group:

$$E\{A\} = E\{\Sigma\Sigma y_{i\alpha}{}^2\} = n_1\mu_1{}^2 + n_2\mu_2{}^2 + n_3\mu_3{}^2 + N\sigma_e{}^2$$
$$E\{B\} = E\{\Sigma n_i\bar{y}_i{}^2\} = n_1\mu_1{}^2 + n_2\mu_2{}^2 + n_3\mu_3{}^2 + k\sigma_e{}^2$$
$$E\{C\} = E\{N\bar{y}^2\} = N\mu^2 + \sigma_e{}^2$$

where $k = 3$ in this example.    The only necessary remark is that the population mean here is taken as the weighted mean of the subpopulation means; that is, $N\mu = n_1\mu_1 + n_2\mu_2 + n_3\mu_3$.

## 5. Expectation of mean squares

With the above basic results, the expected values of the $ssq$ between and within groups may be immediately obtained because of the fact that $E\{A - B\} = E\{A\} - E\{B\}$, etc.    Thus,

$$E\{ssq_B\} = E\{B\} - E\{C\} = (\Sigma n_i\mu_i{}^2 - N\mu^2) + (k - 1)\sigma_e{}^2$$
$$E\{ssq_W\} = E\{A\} - E\{B\} = (N - k)\sigma_e{}^2$$

It follows that the expectations of the two mean squares are

$$E\left\{\frac{ssq_B}{k-1}\right\} = E\{s_B{}^2\} = V(\mu) + \sigma^2$$

$$E\left\{\frac{ssq_W}{N-k}\right\} = E\{s_W{}^2\} = \sigma^2$$

where

$$V(\mu) = (\Sigma n_i\mu_i{}^2 - N\mu^2)/(k-1)$$

The quantity $V(\mu)$ arises from the fact that the subpopulation means vary. This quantity is always positive, because

$$\Sigma n_i\mu_i{}^2 - N\mu^2 = \Sigma n_i(\mu_i - \mu)^2$$

Therefore, the expected value of $s_B{}^2$ is always larger than that of $s_W{}^2$.

## 6. Fixed and random models

For brevity, in the preceding paragraphs, I have used $\mu_i$ directly instead of $\mu + \tau_i$. In order to prepare the reader for more generalizations in later chapters, I must now rewrite the previous results in terms of $\mu$ and $\tau_i$. In doing so I cannot avoid referring to the so-called model I and model II; a subject into which we shall not go here except to borrow some notations. Very briefly, model I (or the *fixed model*) supposes that $\tau_i$ of a given subpopulation is a fixed quantity, not a random variable; while model II (or the *random model*) supposes that the $\tau_i$ themselves are also random variables, the $k$ treatments being a random sample of many possible treatments. A few examples may be found to be more compatible with either the concept of model I or with that of model II. In the great majority of cases, however, the investigator may argue either way, depending on his mood and his handling of the subject matter. In other words, it is more a matter of assumption than a matter of reality. For our present purpose, it is essentially a matter of notation. The reader who is interested in this subject may examine Eisenhart [The Assumptions Underlying the Analysis of Variance, *Biometrics*, 3:1–21 (1947)] and some of the papers that have appeared since.

Adopting model I, in which the treatment (group) effects are fixed quantities, we adopt the restriction that $\Sigma n_i\tau_i = 0$, and then the quantity $V(\mu)$ in Sec. 5 may be written as

$$V(\mu) = \frac{\Sigma n_i(\mu_i - \mu)^2}{k-1} = \frac{\Sigma n_i\tau_i{}^2}{k-1}$$

which is a fixed positive quantity.

## 7. Expectation under random model

On the other hand, according to model II, the $\tau_i$ are assumed to be independently distributed variables with $E\{\tau_i\} = 0$ and $E\{\tau_i^2\} = \sigma_t^2$, so that for each term

$$E\{y_{i\alpha}^2\} = E\{(\mu + \tau_i + \epsilon_{i\alpha})^2\} = \mu^2 + \sigma_t^2 + \sigma_e^2$$

and for each group

$$E\{n_i\bar{y}_i^2\} = E\left\{\frac{(n_i\mu + n_i\tau_i + \Sigma e_{i\alpha})^2}{n_i}\right\} = n_i\mu^2 + n_i\sigma_t^2 + \sigma_e^2$$

Consequently, the three basic expectations of Sec. 4 will now take the following forms:

$$E\{A\} = N\mu^2 + N\sigma_t^2 + N\sigma_e^2$$
$$E\{B\} = N\mu^2 + N\sigma_t^2 + k\sigma_e^2$$
$$E\{C\} = N\mu^2 + \frac{\Sigma n_i^2\sigma_t^2}{N} + \sigma_e^2$$

The last expression is due to the fact that

$$E\{Y^2\} = E\{(N\mu + \Sigma n_i\tau_i + \Sigma\Sigma\epsilon_{i\alpha})^2\}$$
$$= N^2\mu^2 + \Sigma n_i^2\sigma_t^2 + N\sigma_e^2$$

which, when divided by $N$, is $E\{Y^2/N\} = E\{C\}$.

It is seen that $E\{ssq_W\} = E\{A\} - E\{B\} = (N - k)\sigma_e^2$, remaining the same as before, but

$$E\{ssq_B\} = E\{B\} - E\{C\} = \left(N - \frac{\Sigma n_i^2}{N}\right)\sigma_t^2 + (k - 1)\sigma_e^2$$

so that

$$E\left\{\frac{ssq_B}{k - 1}\right\} = E\{s_B^2\} = n_0\sigma_t^2 + \sigma_e^2$$

where

$$n_0 = \frac{N - \Sigma n_i^2/N}{k - 1}$$

which is a sort of average value of the $n_i$'s. For the numerical example of Table 5.1, in which $n_1 = 3$, $n_2 = 6$, $n_3 = 3$, we have

$$n_0 = \frac{12 - (3^2 + 6^2 + 3^2)/12}{3 - 1} = 3.75$$

The value of $n_0\sigma_t^2$ here takes the place of $V(\mu) = \Sigma n_i\tau_i^2/(k - 1)$ of model I. Further explanatory notes are given at the end of the chapter.

## 8. Expectation for equal groups

An important special case is that the groups are of equal size; that is, $n_1 = \cdots = n_k = n$, say, so that the total number of observations is $N = kn$, and $\Sigma n_i^2 = kn^2$. When this is true, substitution in $n_0$ yields

$$n_0 = \frac{kn - kn^2/kn}{k - 1} = \frac{kn - n}{k - 1} = n$$

The results for groups of equal size may be summarized as follows:

Model I: $\qquad E\{s_B^2\} = \dfrac{n\Sigma\tau^2}{k - 1} + \sigma_e^2$

Model II: $\qquad E\{s_B^2\} = n\sigma_t^2 + \sigma_e^2$

Thus, in this simple case, the two models may be reconciled by accepting the notation

$$\sigma_t^2 = \frac{\Sigma\tau^2}{k - 1}$$

In any case, the very fundamental fact is that the expected value of $s_B^2$ is larger than that of $s_W^2$ if the subpopulations have different mean values.

As a numerical example of equal groups, let us consider the data of Table 5.2 once more.

| Group | $y_{ia}$ | | | | $Y_i$ | $Y$ |
|---|---|---|---|---|---|---|
| 1 | 6 | 2 | 9 | 3 | 20 | |
| 2 | 4 | 10 | 4 | 6 | 24 | 84 |
| 3 | 9 | 12 | 8 | 11 | 40 | |

The basic quantity (area) $A$ depends on the 12 single values, the quantity $B$ depends on the 3 group totals; and the quantity $C$ depends on the grand total. Previously we have found that

$$A = 708 \qquad B = 644 \qquad C = 588$$

so that we have

| Source | df | ssq | Mean square | Expected value |
|---|---|---|---|---|
| Between groups | 2 | $B - C = 56$ | $s_B^2 = 28$ | $\sigma_e^2 + n\sigma_t^2$ |
| Within groups | 9 | $A - B = 64$ | $s_W^2 = 7\frac{1}{9}$ | $\sigma_e^2$ |

where $n = 4$ in this example.  Here we see that the mean square for between groups is larger than that for within groups, conforming with the prediction by theory.

## 9. Estimation of variance

The within-group mean square $s_W^2$ is a direct estimate of error variance $\sigma_e^2$ and needs no comment.  To estimate the value of $\sigma_t^2$, we take

$$\frac{s_B^2 - s_W^2}{n} = \frac{28.00 - 7.11}{4} = 5.22 = \hat{\sigma}_t^2$$

where the "hat" over $\sigma$ means "the estimate of."  In certain problems the investigator is concerned with the estimation of this parameter. We shall not pursue the subject any further, because we are mainly concerned with the question whether the quantity $\sigma_t^2$ is consistent with the value zero, as explained in the next chapter.

## EXERCISES

**7.1**    There are 20 observations distributed in three groups ($k = 3$) of unequal size. Calculate the value of $n_0$ for each of the cases in the accompanying table.  Which of the $n_0$ is closest to the arithmetic mean $\bar{n} = 6.67$?

| $n_1$ | $n_2$ | $n_3$ | $N$ | $n_0 = \dfrac{N - \Sigma n_i^2/N}{k-1}$ |
|---|---|---|---|---|
| 3 | 5 | 12 | 20 | 5.55 |
| 4 | 6 | 10 | 20 | |
| 6 | 7 | 7 | 20 | |

**7.2**    In the linear model $y_{i\alpha} = \mu + \tau_i + \epsilon_{i\alpha}$ the random component $\epsilon_{i\alpha}$ is the same in both models: $E\{e_{i\alpha}\} = 0$, $E\{e_{i\alpha}e_{i'\alpha'}\} = 0$, and $E\{e_{i\alpha}^2\} = \sigma_e^2$ or simply $\sigma^2$.  The difference is in the $\tau$'s.

In the model of fixed effects (model I) the $\tau$'s, being fixed values, are the parameters with the restriction $\Sigma \tau_i = 0$.  The expected values are $E\{t_i\} = \tau_i$ and $E\{t_i^2\} = \tau_i^2$. In the model of random effects (model II) $t$ is a random variable with mean zero and variance $\sigma_t^2$, which is the parameter.  The expected values are $E\{t_i\} = 0$ and $E\{t_i^2\} = \sigma_t^2$.

**7.3**    Find the expected values of the between-group mean square under the random model for the case of $k$ groups of equal size of $n$ each.  For the sake of concreteness we let $k = 3$ and $n = 4$, so that the total number of observations is $N = kn = 12$. Now, each single observation in the first group is

$$y_{1\alpha} = u + t_1 + e_{1\alpha}$$

Hence the total of the first group is

$$Y_1 = \Sigma y_{1\alpha} = nu + nt_1 + e_{11} + e_{12} + e_{13} + e_{14}$$
$$E\{Y_1{}^2\} = E\{nu + nt_1 + e_{11} + e_{12} + e_{13} + e_{14}\}^2$$
$$= n^2\mu^2 + n^2\sigma_t{}^2 + n\sigma_e{}^2$$
$$E\left\{\frac{Y_1{}^2}{n}\right\} = n\mu^2 + n\sigma_t{}^2 + \sigma_e{}^2$$

The expressions for $E\{Y_2{}^2/n\}$ and $E\{Y_3{}^2/n\}$ are the same. Hence the expected value of the basic quantity $B$ is $k$ times the expectation above:

$$E\{B\} = E\left\{\frac{Y_1{}^2 + Y_2{}^2 + Y_3{}^2}{n}\right\} = nk\mu^2 + nk\sigma_t{}^2 + k\sigma_e{}^2$$

Similarly, the grand total of all $nk$ observations is

$$Y = Nu + n(t_1 + \cdots + t_k) + e_{11} + \cdots + e_{kn}$$
$$E\{Y^2\} = E\{Nu + n(t_1 + \cdots + t_k) + e_{11} + \cdots + e_{kn}\}^2$$
$$= N^2\mu^2 + n^2(k\sigma_t{}^2) + N\sigma_e{}^2$$
$$E\{C\} = E\left\{\frac{Y^2}{N}\right\} = N\mu^2 + n\sigma_t{}^2 + \sigma_e{}^2$$

Hence the expected value of the between-group sum of squares and mean square are respectively

$$E\{B - C\} = E\{B\} - E\{C\} = (nk - n)\sigma_t{}^2 + (k - 1)\sigma_e{}^2$$
$$E\{s_B{}^2\} = \frac{E\{B\} - E\{C\}}{k - 1} = n\sigma_t{}^2 + \sigma_e{}^2$$

which is the required result.

**7.4** Rewrite the results of Sec. 7 for unequal groups.

# The variance-ratio test

In order to provide continuity with the preceding chapters and to avoid a sudden diversion to a different subject, we shall describe the variance-ratio test first and treat all the other tests as special cases.

## 1. Variation of group means

It is important to grasp the physical meaning of the results presented in the preceding chapter. The value of $s_W^2$ estimates the common variance $\sigma_e^2$ within groups. It varies from sample to sample for one reason only, and that is the process of random sampling, analogous to that shown in Table 4.1 for a single group. But $s_B^2$, which is based on the variation of group means $\bar{y}_i$, has a larger average value than that of $s_W^2$. This is because the group means have *two* reasons to vary from sample to sample. The first reason is that the true subpopulation mean values are actually different, so that $\bar{y}_1$ will vary around the value $\mu_1$, etc.   The second reason is that the usual random sampling causes the mean values to vary anyway, whether the subpopulation means are equal or different.

## 2. The null hypothesis; significance test

However, if the subpopulations have the same mean value, that is, $\mu_1 = \mu_2 = \mu_3 = \mu$ or $\tau_1 = \tau_2 = \tau_3 = 0$, then the quantity $\Sigma n_i \tau_i^2$, or $\sigma_t^2$, of the preceding chapter will be zero.   In such a case, the expectations of both $s_B^2$ and $s_W^2$ will be the same $\sigma_e^2$.   In actual experimentation, the three groups or three samples in our example will correspond to three different *treatments* of an otherwise homogeneous group of experimental units (see Chap. 15).   The investigator wishes to know first of all if these treatments have any effects on the variable under study.   In our terminology, he wishes to find out whether the subpopulation (treatment) means are the same or different.   To attack this problem, we first suppose that the treatments have *no effects* or, in other words, that the subpopulation means are all the same.   This supposition is called the *null hypothesis:*

$$H_0: \qquad \mu_i = \mu \qquad \text{or} \qquad \tau_i = 0$$

On the basis of $H_0$, the long-range average values of $s_B^2$ and $s_W^2$ should be the same, although each varies from sample to sample to a greater or lesser extent owing to the process of random sampling.   If, in our particular sample (experiment), the values of $s_B^2$ and $s_W^2$ turn out to be approximately the same, the finding is certainly not contradictory to the null hypothesis and therefore we lack evidence to assert that the treatments have effects.   In such a case, the null hypothesis will be retained and we say the treatment effects are *nonsignificant.*

On the other hand, if $s_B^2$ should turn out to be so much larger than $s_W^2$ that it cannot be explained away by the random sampling process, the natural alternative will be that the subpopulation means are actually different from each other or that the treatments indeed have effects. In such a case we discard the initial null hypothesis and conclude that the treatment effects are *significant.*   The problem is where to draw the line.   We need a statistical test of significance.   The line, wherever it is, must necessarily be an arbitrary one.   In other words, we need to agree to a predetermined rule by which we decide significance or nonsignificance.   Such a rule is called a test of significance.   The following sections explain the underlying assumptions and method of carrying out such a test.

### 3. Normality of error; the $F$ distribution

In devising a standard statistical test of the above problem, we make the basic assumption that the variable under study is *normally distributed*. Since each $y$ value may be regarded as consisting of three parts according to the linear model of Chap. 6 and the two parts $\mu$ and $\tau_i$ are supposed to be fixed values, this amounts to assuming that the error part $\epsilon$ is normally distributed with mean zero and variance $\sigma_e^2$. The shape of the normal distribution is shown in Fig. 33.2.

Note that this is the first time normal distribution is mentioned. Everything said so far is true for any population. The reason we assume normality here is that the sampling distribution of $F$ is known and has been partially (the part most frequently used in application) tabulated, where $F$ is the ratio of two independent estimates of what on the null hypothesis is the same variance. The $F$ distribution may be used, in general, to compare two independent samples to ascertain whether the presumably normal populations from which they are derived have equal or unequal variances.

In our present case, $s_B^2$ and $s_W^2$ are two independent estimates of the same variance on the null hypothesis. The mathematical proof of their independence is a long story and will not be given here, but the reader may see the reasonableness of their independence; for $s_B^2$ is based solely on the group (treatment) mean values and has nothing to do with the individual variations within any group, while $s_W^2$ is based solely on the individual variations within groups (that is, measured from their own group mean), whatever the group mean happens to be. Consequently, the ratio

$$F = \frac{s_B^2}{s_W^2} = \frac{ssq_B/(k-1)}{ssq_W/(N-k)}$$

has the $F$ distribution—the sampling distribution of the ratio due to random variations of both $s_B^2$ and $s_W^2$ from sample to sample.

### 4. The $F$ table and test

The theoretical background outlined in the preceding paragraphs leads to the simple practical procedure of testing the null hypothesis indicated in Table 8.1, which thus serves as a summary of the last few chapters. The remaining problem is to choose a *level of significance*. For the sake of concreteness, I have used the "5 per cent point" throughout the book, but it should be understood that this by no means prevents the reader from using other significance levels. The meaning of the level of significance may be found in all elementary textbooks of statistics, and only a few remarks are needed here.

*Table* 8.1  **The analysis of variance** $(t_i = \bar{y}_i - \bar{y},\ e_{i\alpha} = y_{i\alpha} - \bar{y}_i,$
$d_{i\alpha} = t_i + e_{i\alpha} = y_{i\alpha} - \bar{y})$

|  | | | Variation | df | ssq | msq | F |
|---|---|---|---|---|---|---|---|
| **Data (Table 5.1)** | | | | | | | |
| (1) | 6  9  3 | | Between groups | $k - 1 = 2$ | $\Sigma n_i t_i^2 = 54$ | $s_B^2 = 27$ | |
| (2) | 10  4  9  12  8  11 | | | | | | 3.68 |
| (3) | 2  6  4 | | Within groups | $N - k = 9$ | $\Sigma\Sigma e_{i\alpha}^2 = 66$ | $s_W^2 = 7.33$ | |
| | | | Total | $N - 1 = 11$ | $\Sigma\Sigma d_{i\alpha}^2 = 120$ | | |

|  | | | Variation | df | ssq | msq | F |
|---|---|---|---|---|---|---|---|
| **Data (Table 5.2)** | | | | | | | |
| (1) | 6  2  9  3 | | Between groups | $k - 1 = 2$ | $n\Sigma t_i^2 = 56$ | $s_B^2 = 28$ | |
| (2) | 4  10  4  6 | | | | | | 3.94 |
| (3) | 9  12  8  11 | | Within groups | $N - k = 9$ | $\Sigma\Sigma e_{i\alpha}^2 = 64$ | $s_W^2 = 7.11$ | |
| | | | Total | $N - 1 = 11$ | $\Sigma\Sigma d_{i\alpha}^2 = 120$ | | |

Since both $s_B^2$ and $s_W^2$ are estimates of the same $\sigma_e^2$ of a normal population when $H_0$ is true, they should be approximately of the same magnitude most of the time. Occasionally they could be very different because of the random sampling process. The $F$ table (Table 8.2) gives the values of $F$ such that 95 per cent of the time the observed value of $F$ from repeated sampling is smaller than the tabulated value, and 5 per cent of the time the observed value of $F$ is larger than the tabulated value— for the specified combinations of the number of degrees of freedom—when the null hypothesis is true. At the top of the table is the $df$ of $ssq_B$, and at the left margin is the $df$ of $ssq_W$. Looking at the column headed by $df = k - 1 = 2$ and at the row corresponding to $df = N - k = 9$, we find that the tabulated value of $F$ is 4.26. This means that in repeated samplings the relative frequency (probability) of $F$ *larger* than 4.26 is $P = 0.05$ under $H_0$, while the relative frequency of $F$ *smaller* than 4.26 is 0.95. Since our observed $F$ is 3.68, which is smaller than the tabulated value, it may be considered a usual event on the basis of random sampling, and we have no evidence to contradict the null hypothesis. Hence we say that the treatments have no significant effects.

On the other hand, if the observed $F$ were 5.68 (instead of 3.68), we would think otherwise. The value $F = 4.26$ should be exceeded only 5 per cent of the time on the basis of random sampling alone $(H_0)$, and yet our observed $F$ has actually exceeded 4.26. We must conclude that a rare event has occurred ("just one of those things") or that the treat-

*Table* 8.2   Values of variance ratio $F$ exceeded in 5 per cent of random samples

| $v_2 = df$ of denominator | $t$ | 1 | 2 | 3 | 4 | $v_1 = df$ of numerator<br>5 | 6 | 7 | 8 | 9 | $v_2$ |
|---|---|---|---|---|---|---|---|---|---|---|---|
| 5 | 2.57 | 6.61 | 5.79 | 5.41 | 5.19 | 5.05 | 4.95 | 4.88 | 4.82 | 4.78 | 5 |
| 6 | 2.45 | 5.99 | 5.14 | 4.76 | 4.53 | 4.39 | 4.28 | 4.21 | 4.15 | 4.10 | 6 |
| 7 | 2.365 | 5.59 | 4.74 | 4.35 | 4.12 | 3.97 | 3.87 | 3.79 | 3.73 | 3.68 | 7 |
| 8 | 2.31 | 5.32 | 4.46 | 4.07 | 3.84 | 3.69 | 3.58 | 3.50 | 3.44 | 3.39 | 8 |
| 9 | 2.26 | 5.12 | 4.26 | 3.86 | 3.63 | 3.48 | 3.37 | 3.29 | 3.23 | 3.18 | 9 |
| 10 | 2.23 | 4.96 | 4.10 | 3.71 | 3.48 | 3.33 | 3.22 | 3.14 | 3.07 | 3.02 | 10 |
| 11 | 2.20 | 4.84 | 3.98 | 3.59 | 3.36 | 3.20 | 3.09 | 3.01 | 2.95 | 2.90 | 11 |
| 12 | 2.18 | 4.75 | 3.88 | 3.49 | 3.26 | 3.11 | 3.00 | 2.91 | 2.85 | 2.80 | 12 |
| 13 | 2.16 | 4.67 | 3.80 | 3.41 | 3.18 | 3.02 | 2.92 | 2.83 | 2.77 | 2.72 | 13 |
| 14 | 2.145 | 4.60 | 3.74 | 3.34 | 3.11 | 2.96 | 2.85 | 2.76 | 2.70 | 2.65 | 14 |
| 15 | 2.13 | 4.54 | 3.68 | 3.29 | 3.06 | 2.90 | 2.79 | 2.71 | 2.64 | 2.59 | 15 |
| 16 | 2.12 | 4.49 | 3.63 | 3.24 | 3.01 | 2.85 | 2.74 | 2.66 | 2.59 | 2.54 | 16 |
| 17 | 2.11 | 4.45 | 3.59 | 3.20 | 2.96 | 2.81 | 2.70 | 2.61 | 2.55 | 2.50 | 17 |
| 18 | 2.10 | 4.41 | 3.55 | 3.16 | 2.93 | 2.77 | 2.66 | 2.58 | 2.51 | 2.46 | 18 |
| 19 | 2.09 | 4.38 | 3.52 | 3.13 | 2.90 | 2.74 | 2.63 | 2.54 | 2.48 | 2.43 | 19 |
| 20 | 2.086 | 4.35 | 3.49 | 3.10 | 2.87 | 2.71 | 2.60 | 2.51 | 2.45 | 2.40 | 20 |
| 25 | 2.06 | 4.24 | 3.38 | 2.99 | 2.76 | 2.60 | 2.49 | 2.40 | 2.34 | 2.28 | 25 |
| 30 | 2.04 | 4.17 | 3.32 | 2.92 | 2.69 | 2.53 | 2.42 | 2.34 | 2.27 | 2.21 | 30 |
| 40 | 2.02 | 4.08 | 3.23 | 2.84 | 2.61 | 2.45 | 2.34 | 2.25 | 2.18 | 2.12 | 40 |
| 60 | 2.00 | 4.00 | 3.15 | 2.76 | 2.52 | 2.37 | 2.25 | 2.17 | 2.10 | 2.04 | 60 |
| 100 | 1.98 | 3.94 | 3.09 | 2.70 | 2.46 | 2.30 | 2.19 | 2.10 | 2.03 | 1.97 | 100 |
| ∞ | 1.96 | 3.84 | 3.00 | 2.60 | 2.37 | 2.21 | 2.10 | 2.01 | 1.94 | 1.88 | ∞ |
| $\chi^2$ | | 3.84 | 5.99 | 7.81 | 9.49 | 11.07 | 12.59 | 14.07 | 15.51 | 16.92 | |
| $df$ | | 1 | 2 | 3 | 4 | 5 | 6 | 7 | 8 | 9 | |

ments did have effects which have boosted the value of $s_B{}^2$. When we face such a situation, we usually bend a little and do not stubbornly insist on the truth of the null hypothesis. In fact, we give up the null hypothesis and say that there are treatment effects (or that the treatment effects are significant). Pronouncing significance whenever an observed $F$ exceeds 4.26 makes us wrong 5 per cent of the time if $H_0$ is actually true. This is the calculated risk we take in testing significance.

## 5. $F$ test for equal groups

When the treatment groups are of equal size $n$, both theoretical consideration and practical calculation become easier and the same test, $F = s_B{}^2/s_W{}^2$, is used to test $H_0$. The expectation of $s_B{}^2$ may be obtained directly by elementary observations. The $ssq$ of the $k$ group means is $\Sigma(\bar{y}_i - \bar{y})^2$, and therefore, on the null hypothesis, the mean square $\Sigma(\bar{y}_i - \bar{y})^2/(k-1)$ is an unbiased estimate of the *variance of sample means* which is equal to $\sigma_e{}^2/n$ according to Chap. 4, Sec. 8. But $ssq_B = n$ times $\Sigma(\bar{y}_i - \bar{y})^2$; it follows that $s_B{}^2 = ssq_B/(k-1)$ estimates

$$n \times \frac{\sigma_e{}^2}{n} = \sigma_e{}^2$$

Hence, the $F$ test may be used to test the truth of $H_0$. The analysis of variance for three equal groups is given in the lower part of Table 8.1.

## 6. Two groups: $F = t^2$

The $F$ distribution, as we have learned, has two parameters: $\nu_1 = df$ of $ssq_B$ and $\nu_2 = df$ of $ssq_W$. When there are only two groups,

$$\nu_1 = k - 1 = 2 - 1 = 1$$

then the $F$ test presented above is equivalent to the $t$ test with $\nu_2$ degrees of freedom. When there are $n_1$ observations in the first group and $n_2$ observations in the second, the sum of squares within the two groups is

$$ssq_W = \sum (y_{1\alpha} - \bar{y}_1)^2 + \sum (y_{2\alpha} - \bar{y}_2)^2$$
$$= \sum y_{1\alpha}{}^2 + \sum y_{2\alpha}{}^2 - \frac{Y_1{}^2}{n_1} - \frac{Y_2{}^2}{n_2} = A - B$$

and the estimate of the error variance is

$$s^2 = s_W{}^2 = \frac{ssq_W}{(n_1 - 1) + (n_2 - 1)} = \frac{ssq_W}{N - 2}$$

Then, a statistic known as the Student's $t$ is defined as

$$t = \frac{\bar{y}_1 - \bar{y}_2}{\sqrt{s^2(1/n_1 + 1/n_2)}}$$

which is being employed to test the significance of the difference between two group means, using the $t$ table.  Now let us observe that

$$t^2 = \frac{(\bar{y}_1 - \bar{y}_2)^2}{s^2(1/n_1 + 1/n_2)}$$

where $s^2 = s_W^2$ is the denominator of $F$.  Now all we have to show is that the rest of the expression for $t^2$ is equal to the numerator of $F$.  Since $k = 2$, the numerator of $F$ is $s_B^2 = ssq_B/(k - 1) = ssq_B$.  Now, by virtue of Chap. 3, Sec. 11, we see that, remembering $Y = n_1\bar{y}_1 + n_2\bar{y}_2$,

$$\begin{aligned} ssq_B &= n_1(\bar{y}_1 - \bar{y})^2 + n_2(\bar{y}_2 - \bar{y})^2 \\ &= n_1\bar{y}_2^2 + n_2\bar{y}_2^2 - \frac{(n_1\bar{y}_1 + n_2\bar{y}_2)^2}{n_1 + n_2} = B - C \\ &= \frac{n_1 n_2(\bar{y}_1 - \bar{y}_2)^2}{n_1 + n_2} = \frac{(\bar{y}_1 - \bar{y}_2)^2}{1/n_1 + 1/n_2} \end{aligned}$$

Hence

$$t^2 = \frac{ssq_B}{s^2} = \frac{s_B^2}{s_W^2} = F$$

or $t = \sqrt{F}$ for two groups.  Hence the ordinary $t$ test using the $t$ table is entirely equivalent to the $F$ test using the $F$ table for two groups. Turning to Table 8.2, we see that the left-hand marginal entries are the $t$ values, which are the square root of the $F$ values of the first column ($\nu_1 = k - 1 = 2 - 1 = 1$).  Thus for $\nu_2 = 11$, $t = 2.20$, and $F = 4.84$, and so on.

### 7. Very large groups

There is in common use another statistic known as the chi square, denoted by $\chi^2$.  For our present purpose it may be defined as the sum of squares between groups divided by the variance of the population, viz.,

$$\chi^2 = \frac{ssq_B}{\sigma^2}$$

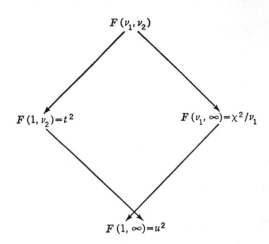

**Fig. 8.1** The relationship between the four fundamental sampling distributions based on normal parental populations. At the top is the general $F$ distribution specified by the degree of freedom of the numerator and the denominator. The arrows show the changes in the degrees of freedom. At the bottom, $u$ is the normal deviate and $u^2 = \chi^2$ with a single degree of freedom.

with $k - 1$ degrees of freedom. Now, in the analysis of variance when we have several ($k$) large groups, the number of degrees of freedom within the groups, $\nu_2 = N - k$, will be very large. Then the estimate $s_W^2 = ssq_W/(N - k)$ will be very nearly equal to the true population variance $\sigma^2$. Thus, in the limiting situation,

$$F = \frac{ssq_B/(k - 1)}{ssq_W/(N - k)} \doteq \frac{ssq_B/(k - 1)}{\sigma^2} = \frac{\chi^2}{k - 1} = \frac{\chi^2}{\nu_1}$$

or $(k - 1)F = \chi^2$ with $k - 1$ degrees of freedom. These values are given in the bottom margin of the $F$ table (Table 8.2). For instance, when there are eight ($k = 8$) large groups,

$$\nu_1 = 7 \qquad \text{and} \qquad \chi^2 = \nu_1 F = 7 \times 2.01 = 14.07$$

at the 5 per cent point.

Finally, when there are only two large groups, the $F$ test is equivalent to the *normal deviate $u$ test* using the normal table; for $F = t^2 = u^2 = \chi^2$ when $k = 2$ and $N$ is very large. At the 5 per cent point, $u = 1.96$; note that

$$u^2 = (1.96)^2 = 3.84 = \chi^2 = F(1, \infty)$$

The relations between the four statistics, all based on normal distribution, have been summarized in Fig. 8.1. A more detailed discussion may be found in Chapter 13 of the book by Jerome Li (1957).†

† When only author and date of a cited work are given, complete information will be found under References at the end of this book.

## 8. Fisher's z

The analysis of variance was originally developed by R. A. Fisher, who employed the statistic $z$ instead of the direct variance ratio $F$, where

$$z = \tfrac{1}{2}(\log_e s_B{}^2 - \log_e s_W{}^2) = \tfrac{1}{2}\log_e\left(\frac{s_B{}^2}{s_W{}^2}\right) = \tfrac{1}{2}\log_e F$$

Hence,

$$F = e^{2z}$$

The distribution of $z$ has been tabulated by Fisher, and later Snedecor converted these $z$ values to $F$ in order to save arithmetic. The letter $F$ is in honor of Fisher. (Otherwise, I would think that $F^2$ would be a better notation in view of its relationships with $t$ and $\chi^2$.) As a numerical example, let us take $\nu_1 = 3$ and $\nu_2 = 6$. The $F$ value is 4.76 at the 5 per cent point. The corresponding $z$ value is then

$$z = \tfrac{1}{2}\log_e 4.76 = \tfrac{1}{2}(1.560) = 0.780$$

Since the $F$ values have been extensively tabulated nowadays at various significance levels, we shall always use $F$ in practice.

## EXERCISES

**8.1**   Perform an analysis of variance and the $F$ test of significance for the following two groups of observations.

|        |   |   |    |   |    |    |            |             |
|--------|---|---|----|---|----|----|------------|-------------|
| $y_{1\alpha}$: | 9 | 4 | 12 | 8 | 11 | 10 | $Y_1 = 54$ | $\bar{y}_1 = 9$ |
| $y_{2\alpha}$: | 3 | 2 | 6  | 4 | 9  | 6  | $Y_2 = 30$ | $\bar{y}_2 = 5$ |

*Ans.:* $F = 6.67$.   Is the difference between the two means significant?

**8.2**   In analysis of variance the first quantity to be calculated is the so-called "area $A$." There is no shortcut to this value, although it may be obtained by a desk calculator quite easily. For two equal groups, each of size $n$, however, the $F$ value may be obtained by the following shortcut formula as soon as area $A$ is known.

$$t^2 = F = \frac{(n-1)(Y_1 - Y_2)^2}{nA - (Y_1{}^2 + Y_2{}^2)}$$

Write out the analysis-of-variance table in algebra and see that this formula for $F$ is correct. Verify it numerically, using the numbers in Exercise 8.1.

**8.3**   Since Chap. 5 we have been using the three groups (6, 2, 9, 3), (4, 10, 4, 6), and (9, 12, 8, 11) as numerical examples. Suppose that we doubled the number of observations without decreasing their variability; we would then have the data shown on top of page 73. What is the effect of having a larger number of observations?

| Group | Single values, $y_{i\alpha}$ | | | | | | | | Total $Y_i$ | $Y_i{}^2$ |
|---|---|---|---|---|---|---|---|---|---|---|
| (1) | 6 | 2 | 9 | 3 | 6 | 2 | 9 | 3 | 40 | 1,600 |
| (2) | 4 | 10 | 4 | 6 | 4 | 10 | 4 | 6 | 48 | 2,304 |
| (3) | 9 | 12 | 8 | 11 | 9 | 12 | 8 | 11 | 80 | 6,400 |
| Total | (Double of Table 5.2) | | | | | | | | 168 | 10,304 |

| Original set | New set |
|---|---|
| $A = 708$ | $A' = 2A = 1,416$ |
| $B = 644$ | $B' = 2B = 1,288$ |
| $C = 588$ | $C' = 2C = 1,176$ |

The analysis of variance

| Source | Original set | | | | New set | | | |
|---|---|---|---|---|---|---|---|---|
| | df | ssq | msq | F | df | ssq | msq | F |
| Treatments | 2 | 56 | 28 | | 2 | 112 | 56 | |
| | | | | 3.9 | | | | 9.2 |
| Error | 9 | 64 | 7.111 | | 21 | 128 | 6.095 | |
| Total | 11 | 120 | | | 23 | 240 | | |

**8.4** The results of two experiments are given in the accompanying table. The mean values of the treatments in these two experiments are the same. By inspection and without calculation, can you tell which experiment gives a larger value of $F$? Why? Then make an $F$ test for each experiment.

| | Experiment I | | | Experiment II | | |
|---|---|---|---|---|---|---|
| | Treatments | | | Treatments | | |
| | (a) | (b) | (c) | (a) | (b) | (c) |
| | 8 | 4 | 3 | 4 | 6 | 5 |
| | 3 | 8 | 2 | 5 | 7 | 4 |
| | 1 | 10 | 8 | 3 | 7 | 6 |
| | 4 | 6 | 7 | 4 | 8 | 5 |
| | 16 | 28 | 20 | 16 | 28 | 20 |

**8.5** The error variance between two means, based on $n_1$ and $n_2$ observations, is

$$s^2 \left( \frac{1}{n_1} + \frac{1}{n_2} \right)$$

Suppose that there are 20 experimental units available to test two treatments, how many would you assign to each treatment?   Calculate the values of

$$\tfrac{1}{2} + \tfrac{1}{18} \qquad \tfrac{1}{5} + \tfrac{1}{15} \qquad \tfrac{1}{8} + \tfrac{1}{12} \qquad \tfrac{1}{10} + \tfrac{1}{10}$$

Do you see the advantage of having equal groups for experimentation?

**8.6**   Verify the following:

| $\nu_2 = 6$ | $\nu_1 = 1$ | 2 | 3 | 4 | 5 |
|---|---|---|---|---|---|
| $F$ | 5.99 | 5.14 | 4.76 | 4.53 | 4.39 |
| $z$ | 0.895 | 0.819 | 0.780 | 0.756 | 0.739 |

# 9

# Hierarchical
# classifications

Hierarchical classification is a continued or repeated one-way classification into minor groups within each major group. It is a natural extension of what we have learned in the preceding three chapters.

## 1. A three-rank hierarchy

Notations and a numerical example are given in Table 9.1, in which the subscript $h$ specifies the *major* group, the subscript $i$ indicates a *minor* group within a major group, and $\alpha$, as usual, indicates the individual member in a minor group. A clear distinction should be made between the number of observations and the number of groups. The letters $n$ and $N$ are employed to denote the former and $K$ and $II$ to denote the latter. Thus, $n_{hi}$ is the number of observations in the $i$th minor group of the $h$th major group, so that $n_{hi}\bar{y}_{hi} = Y_{hi}$ is the total of that minor group. The "size" of a major group is $N_h = \sum_i n_{hi}$. The summation covers all the minor groups of the $h$th major group, so that $N_h \bar{y}_h = Y_h$ is the total of that major group. The grand total number of observations is

$$N = \sum_h N_h = \sum_h \sum_i n_{hi}$$

Let $H$ be the number of major groups. Then

$$N = N_1 + N_2 + \cdots + N_H$$

Also, let $k_1$ be the number of minor groups in the first major group, etc., so that $K = k_1 + k_2 + \cdots$ is the *total* number of minor groups in the entire set of data. An examination of Table 9.1 will make the meaning of the symbols clear.

The sizes ($n$ and $N$) appear explicitly in calculating the various sums of squares, and the number of groups ($K$ and $H$) appear in calculating the various degrees of freedom. The arithmetical procedure and the statistical model are such obvious extensions of the preceding chapters that some of the details are omitted (but understood) in the following paragraphs.

*Table 9.1* **Hierarchical classification with three ranks**

| Single values $y_{hi\alpha}$ | Minor groups | | | Major groups | | | Entire group | | |
|---|---|---|---|---|---|---|---|---|---|
| | Size $n_{hi}$ | Total $Y_{hi}$ | Mean $\bar{y}_{hi}$ | Size $N_h$ | Total $Y_h$ | Mean $\bar{y}_h$ | Size $N$ | Total $Y$ | Mean $\bar{y}$ |
| 6<br>2<br>9<br>3 | 4 | 20 | 5 | 6 | 36 | 6 | | | |
| 12<br>4 | 2 | 16 | 8 | | | | | | |
| 4<br>8 | 2 | 12 | 6 | | | | 12 | 84 | 7 |
| 9 | 1 | 9 | 9 | 6 | 48 | 8 | | | |
| 10<br>6<br>11 | 3 | 27 | 9 | | | | | | |
| | $K = 5$ minor groups | | | $H = 2$ major groups | | | One whole group | | |

## 2. Basic quantities and $ssq$

We may construct four rows of squares (Fig. 9.1) based on the data of Table 9.1 and similar to our Fig. 5.1 for simple classifications. The areas of the four series of squares are

$$A = \sum_h \sum_i \sum_\alpha y_{hi\alpha}^2 = 6^2 + 2^2 + 9^2 + \cdots + 6^2 + 11^2 = 708$$

$$B_1 = \sum_h \sum_i \left(\frac{Y_{hi}{}^2}{n_{hi}}\right) = \frac{20^2}{4} + \frac{16^2}{2} + \frac{12^2}{2} + \frac{9^2}{1} + \frac{27^2}{3} = 624$$

$$B_2 = \sum_h \left(\frac{Y_h{}^2}{N_h}\right) - \frac{36^2}{6} + \frac{48^2}{6} = 600$$

$$C = \frac{Y^2}{N} = \frac{84^2}{12} = 588$$

These are the four basic quantities from which the various sums of squares are obtained. Thus,

| | |
|---|---|
| Within minor groups: | $ssq_W = A - B_1 = 84$ |
| Between minor, within major: | $ssq_B = B_1 - B_2 = 24$ |
| Between major groups: | $ssq_H = B_2 - C = 12$ |

Total: $ssq_T = ssq_W + ssq_B + ssq_H = A - C = 120$

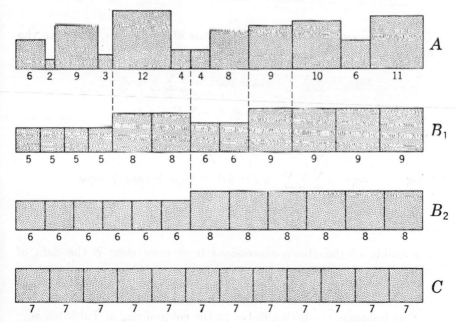

Fig. 9.1 The four areas $A$, $B_1$, $B_2$, $C$ corresponding to the groupings of Table 9.1.

*Table 9.2*   **The linear model of a three-rank classification**

| $\bar{y}$ | $m_h$ | $\bar{y}_h$ | $t_{hi}$ | $\bar{y}_{hi}$ | $e_{hi\alpha}$ | $y_{hi\alpha}$ | $d_{hi\alpha} = y_{hi\alpha} - \bar{y}$ |
|---|---|---|---|---|---|---|---|
| 7 | $-1$ | 6 | $-1$ | 5 | $+1$ | 6 | $-1$ |
| 7 | $-1$ | 6 | $-1$ | 5 | $-3$ | 2 | $-5$ |
| 7 | $-1$ | 6 | $-1$ | 5 | $+4$ | 9 | $+2$ |
| 7 | $-1$ | 6 | $-1$ | 5 | $-2$ | 3 | $-4$ |
| 7 | $-1$ | 6 | $+2$ | 8 | $+4$ | 12 | $+5$ |
| 7 | $-1$ | 6 | $+2$ | 8 | $-4$ | 4 | $-3$ |
| 7 | $+1$ | 8 | $-2$ | 6 | $-2$ | 4 | $-3$ |
| 7 | $+1$ | 8 | $-2$ | 6 | $+2$ | 8 | $+1$ |
| 7 | $+1$ | 8 | $+1$ | 9 | 0 | 9 | $+2$ |
| 7 | $+1$ | 8 | $+1$ | 9 | $+1$ | 10 | $+3$ |
| 7 | $+1$ | 8 | $+1$ | 9 | $-3$ | 6 | $-1$ |
| 7 | $+1$ | 8 | $+1$ | 9 | $+2$ | 11 | $+4$ |
| 84 | 0 | 84 | 0, 0 | 84 | 0, 0 | 84 | 0 |
|    |   |    |      |    | 0, 0, 0 |    |    |
| *ssq* | 12 | | 24 | | 84 | | 120 |

The various *ssq*, if written out in terms of deviations, will take the following forms:

$$ssq_W = \sum_h \sum_i \sum_\alpha (y_{hi\alpha} - \bar{y}_{hi})^2 = \sum_h \sum_i \sum_\alpha e_{hi\alpha}^2$$

$$ssq_B = \sum_h \sum_i n_{hi}(\bar{y}_{hi} - \bar{y}_h)^2 = \sum_h \sum_i n_{hi}t_{hi}^2$$

$$ssq_H = \sum_h N_h(\bar{y}_h - \bar{y})^2 = \sum_h N_h m_h^2$$

Adding   $$ssq_T = \sum_h \sum_i \sum_\alpha (y_{hi\alpha} - \bar{y})^2 = ssq_H + ssq_B + ssq_W$$

## 3. The linear model

The meaning of the above expressions is at once clear if the data of Table 9.1 are rewritten in the form of Table 9.2, which is analogous to our Table 6.1 for a simple one-way classification. Thus, we see that each observed number may be regarded as consisting of four component parts; for instance, the fifth number in the column $y_{hi\alpha}$ of Table 9.2 is

$$12 = 7 - 1 + 2 + 4$$

Generally,

$$y_{hia} = u + m_h + t_{hi} + e_{hia}$$

where $u = \bar{y}$ is the sample estimate of the general population mean, $m_h = \bar{y}_h - \bar{y}$ is the sample estimate of the major group effect, $t_{hi} = \bar{y}_{hi} - \bar{y}_h$ is the minor group effect, and $e_{hia} = y_{hia} - \bar{y}_{hi}$. Following the reasoning of Chap. 6, we see that these are the least-square estimates which minimize the quantity

$$Q = \Sigma\Sigma\Sigma(y_{hia} - u - m_h - t_{hi})^2 = \Sigma\Sigma\Sigma e_{hia}^2$$

In other words, the $ssq_W$ is a minimum when the estimates take on such values.

## 4. Expected value of mean square

This section may be omitted by those who do not care for the details of the expectations of the various mean squares. The numerical values of $ssq$'s given in Sec. 2, when divided by their corresponding degrees of freedom, may be subjected to the $F$ test in the usual manner. The four basic expectations to be found are those of $E\{A\}$, $E\{B_1\}$, $E\{B_2\}$, and $E\{C\}$, each single term of which is as follows:

(A single value)$^2$:

$$y_{hia}^2 = (u + m_h + t_{hi} + e_{hia})^2$$

(Minor group total)$^2$:

$$Y_{hi}^2 = \left(n_{hi}u + n_{hi}m_h + n_{hi}t_{hi} + \sum_a e_{hia}\right)^2$$

(Major group total)$^2$:

$$Y_h^2 = \left(N_h u + N_h m_h + \sum_i n_{hi}t_{hi} + \sum_i \sum_a e_{hia}\right)^2$$

(Grand total)$^2$:

$$Y^2 = \left(Nu + \sum_h N_h m_h + \sum_h \sum_i n_{hi}t_{hi} + \sum_h \sum_i \sum_a e_{hia}\right)^2$$

Some simplifying assumptions have to be made here as before, when a common variance was assumed for all the groups. By analogy, we assume that the variance of the groups of the same rank in the hierarchy is the same, so that $E\{e^2\} = \sigma_e^2$, no matter to which individual it belongs; $E\{t^2\} = \sigma_t^2$, no matter to which minor group it belongs; and $E\{m^2\} = \sigma_m^2$, no matter to which major group it belongs. These can be accepted as definitions of the expectations (or as a matter of notation). We shall not discuss the difference between the so-called models I and II, although the notation employed here is more consistent with the concept of model II. Another assumption is that the components $u$, $m$, $t$, $e$, are all statistically independent so that the expected value of any product

*Table 9.3*   The analysis of variance for hierarchical classifications (data taken from Table 9.1)

| Source of variation | df | ssq | msq | Expectation |
|---|---|---|---|---|
| Between major groups | $H - 1 = 1$ | $ssq_H = 12$ | $s_H{}^2 = 12$ | $\sigma_e{}^2 + c_1\sigma_t{}^2 + c_2\sigma_m{}^2$ |
| Between minor (within major) groups | $K - H = 3$ | $ssq_B = 24$ | $s_B{}^2 = 8$ | $\sigma_e{}^2 + c_3\sigma_t{}^2$ |
| Within minor groups (error) | $N - K = 7$ | $ssq_W = 84$ | $s_W{}^2 = 12$ | $\sigma_e{}^2$ |
| Total | $N - 1 = 11$ | $ssq_T = 120$ | | |

term is zero.   Now, squaring each term, taking the expectations, and summing, we obtain the following results:

$$E\{A\} = E\left\{\sum\sum\sum y_{hi\alpha}{}^2\right\} = N\mu^2 + N\sigma_m{}^2 + N\sigma_t{}^2 + N\sigma_e{}^2$$

$$E\{B_1\} = E\left\{\sum\sum\frac{Y_{hi}{}^2}{n_{hi}}\right\} = N\mu^2 + N\sigma_m{}^2 + N\sigma_t{}^2 + K\sigma_e{}^2$$

$$E\{B_2\} = E\left\{\sum\frac{Y_h{}^2}{N_h}\right\} = N\mu^2 + N\sigma_m{}^2 + \sum\frac{n_{hi}{}^2}{N_h}\sigma_t{}^2 + H\sigma_e{}^2$$

$$E\{C\} = E\left\{\frac{Y^2}{N}\right\} = N\mu^2 + \sum\frac{N_h{}^2}{N}\sigma_m{}^2 + \sum\frac{n_{hi}{}^2}{N}\sigma_t{}^2 + \sigma_e{}^2$$

The expected values of the various sums of squares may then be obtained by taking the appropriate differences of the above expectations.   For example,

$$E\{ssq_W\} = E\{A\} - E\{B_1\} = (N - K)\sigma_e{}^2$$

The complete analysis of variance is given in Table 9.3.   For brevity, the quantities like $(N - \Sigma N_h{}^2/N)$, obtained by taking differences, will simply be denoted by a constant symbol $c$.   The value of $F = s_B{}^2/s_W{}^2$ is used to test if $\sigma_t{}^2 = 0$, that is, if there are minor group effects.   The value of $F = s_H{}^2/s_W{}^2$ may be used to test if both $\sigma_t{}^2$ and $\sigma_m{}^2$ are zero. However, the use of $F = s_H{}^2/s_B{}^2$ for unequal groups is very doubtful, because in general $c_1 \neq c_3$.

## 5.  Hierarchy with equal groups

The last remark in the preceding section shows the importance of having groups of equal size in experimental work.   When all the groups of the same rank in the hierarchy are of the same size, the analysis is particularly

*Table 9.4*    **Hierarchical classification with groups of equal size**

| $y_{hia}$ | 4, 2 | 6, 4 | 8, 6 | 3, 9 | 12, 10 | 9, 11 | (area) $A$ = 708 | $N = 12$ |
|---|---|---|---|---|---|---|---|---|
| $Y_{hi}$ | 6 | 10 | 14 | 12 | 22 | 20 | (area) $B_1$ = 680 | $K = 6$ |
| $Y_h$ | | 30 | | | 54 | | (area) $B_2$ = 636 | $H = 2$ |
| $Y$ | | | 84 | | | | (area) $C$ = 588 | 1 |

The analysis of variance

| Source | df | ssq | msq | F |
|---|---|---|---|---|
| Major groups | $H - 1 = 1$ | $B_2 - C = 48$ | $_3H^2 - 48$ | |
| Minor groups | $K - H = 4$ | $B_1 - B_2 = 44$ | $s_R^2 = 11$ | $s_H{}^2/s_B{}^2 = 4.36$ |
| Error | $N - K = 6$ | $A - B_1 = 28$ | $_3w^2 = 4.67$ | $s_B{}^2/s_W{}^2 = 2.36$ |
| Total | $N - 1 = 11$ | $A - C = 120$ | | |

simple and the effect of each rank of classification can be tested separately. The data of Table 9.4 provide such an example, and the lower portion of that table presents the analysis. For the general case, suppose that there are $H$ major groups, each of which has $k$ minor groups, each of which in turn has $n$ observations. Then we have

| | Minor | Major | Total |
|---|---|---|---|
| Number of observations: | $n$ | $N_h = nk$ | $N = nkH$ |
| Number of groups: | $K = kH$ | $H$ | 1 |

When these values are substituted into the general expressions of $E\{A\}$, etc., of Sec. 4 to find the values of $E\{ssq\}$, the coefficients of the variances reduce to simple numbers. For instance, to find

$$E\{ssq_H\} = E\{B_2\} - E\{C\}$$

the coefficients of $\sigma_t{}^2$ and $\sigma_m{}^2$ are respectively

$$\sum \frac{n_{hi}{}^2}{N_h} - \sum \frac{n_{hi}{}^2}{N} = \frac{Hkn^2}{kn} - \frac{Hkn^2}{Hkn} = Hn - n = n(H - 1)$$

$$N - \frac{\Sigma N_h{}^2}{N} = Hkn - \frac{Hk^2n^2}{Hkn} = nk(H - 1)$$

so that

$$E\left\{\frac{ssq_H}{H - 1}\right\} = E\{s_H{}^2\} = \sigma_e{}^2 + n\sigma_t{}^2 + nk\sigma_m{}^2$$

Similarly,

$$E\left\{\frac{ssq_B}{K-H}\right\} = E\{s_B{}^2\} = \sigma_e{}^2 + n\sigma_t{}^2$$

$$E\left\{\frac{ssq_W}{N-K}\right\} = E\{s_W{}^2\} = \sigma_e{}^2$$

Hence, the value $F = s_H{}^2/s_B{}^2$ may be used to test if $\sigma_m{}^2 = 0$, that is, if the major groups have any significant effects *whatever the effects of the minor groups*. These relations also enable us to obtain an estimate of the variances. For example, the estimate of $\sigma_m{}^2$ may be obtained by setting $\sigma_m{}^2 = (s_H{}^2 - s_B{}^2)/nk$.

## 6. Extensions

The analysis of variance may be extended to any number of ranks in a hierarchical classification. The higher the number of ranks in the hierarchy, the more desirable it is to have groups of equal size so that the effect of each rank of classification may be tested separately and the magnitude of the variance of each factor may be estimated. When groups of each rank are of equal size, the expectations of the various mean squares given in the preceding paragraph may be generalized immediately by observing that the coefficient of each $\sigma^2$ is the number of items in the group of that rank. Realistic examples of hierarchical classification may be found in the more comprehensive textbooks.

### EXERCISES

9.1   As shown in the accompanying table, there are three major groups, each of which contains two minor groups, each of which in turn contains four observations.

| $y_{hi\alpha}$ | 16 | 28 | 22 | 25 | 16 | 17 | 10 | 20 | 18 | 23 | 8 | 23 | (area) $A =$ | $N = 24$ |
|---|---|---|---|---|---|---|---|---|---|---|---|---|---|---|
|  | 28 | 16 | 27 | 30 | 19 | 16 | 18 | 16 | 19 | 24 | 25 | 16 | | |
| $Y_{hi}$ | 88 | | 104 | | 68 | | 64 | | 84 | | 72 | | (area) $B_1 =$ | $K = 6$ |
| $Y_h$ | 192 | | | | 132 | | | | 156 | | | | (area) $B_2 =$ | $H = 3$ |
| $Y$ | 480 | | | | | | | | | | | | (area) $C =$ | 1 |

Analysis of variance

| Source | df | ssq | msq | F |
|---|---|---|---|---|
| Between major | 2 | $B_2 - C = 228$ | 114.00 | |
| Between minor (within major) | 3 | $B_1 - B_2 = 52$ | 17.33 | ... |
| Within minor | 18 | $A - B_1 = 144$ | 24.67 | |
| Total | 23 | $A - C = 724$ | | |

Verify the results in the analysis of variance and make an $F$ test for the major group effects.   Note that $s_B^2$ is smaller than $s_W^2$.   In such a case we take $\sigma_t^2 = 0$ for minor group effects.   The appropriate test for major group effects is then the ratio $114/24.67$.

**9.2**   Construct a table similar to Table 9.2 with the observations of Exercise 9.1, showing the four components of each observed value.   Check by longhand calculation that

$$\Sigma d^2 = \Sigma m^2 + \Sigma t^2 + \Sigma e^2$$

where each summation covers the entire column of 24 numbers.   Do these sums of squares agree with those given in Exercise 9.1?

**9.3**   Rewrite the expectations in Sec. 5 for the case of equal groups.   In particular, let, as in Exercise 9.2,

| *Number of observations* | *Number of groups* | |
|---|---|---|
| $n_{hi} = n = 4$ | $k = 2$ | |
| $N_h = nk = 8$ | $H = 3$ | $K = Hk = 6$ |
| $N = HN_h = 24$ | | |

Verify:

$$
\begin{aligned}
E\{B_2\} &= 24\mu^2 + 24\sigma_m^2 + 12\sigma_t^2 + 3\sigma_e^2 \\
E\{C\} &= 24\mu^2 + 8\sigma_m^2 + 4\sigma_t^2 + \sigma_e^2 \\
\hline
E\{B_2 - C\} &= 16\sigma_m^2 + 8\sigma_t^2 + 2\sigma_e^2 \\
E\{s_H^2\} &= 8\sigma_m^2 + 4\sigma_t^2 + \sigma_e^2
\end{aligned}
$$

# 10

# Two-way classifications

In the branch of statistics known as the design of experiments, the numbers (or observations) are very frequently subjected to a two-way, or double, classification. For example, the 12 grandchildren mentioned in Chap. 6 were classified into 3 groups according to their age. At the same time they could also have been classified according to sex into two groups—boys and girls—quite independently of age. Thus each child would be subjected to two systems of classification simultaneously. But if one of the systems of classification is ignored, then the classification becomes an ordinary one-way type like the classifications studied in early chapters. In dealing with two-way classification data, we have to apply the method of one-way classification twice—once for each system of classification. To facilitate discussion, I shall speak of the (horizontal) rows as the row classification and the (vertical) columns as the column classification instead of identifying the rows and columns specifically with age, breed, culture, drug, education, fertilizer, inbred line, marital status, occupation, race, replicate, sex, temperature, vaccine, or any other "treatment."

## 1. Types of data

There are several types of two-way classifications according to the number of observations in each "cell" of the classification table as illustrated in Fig. 10.1.   Type (i), in which the number of observations in each cell is rather irregular, is difficult to handle because the method involves the long procedure of fitting constants by the least-square principle.   In designed experiments the investigator always tries to avoid this unpleas-

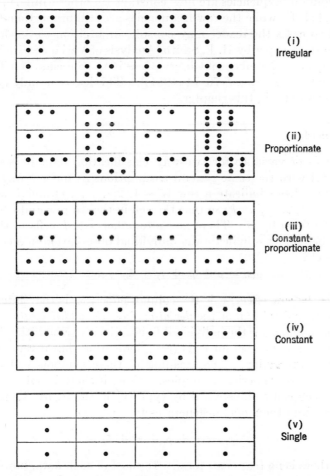

**Fig. 10.1   Types of two-way classifications according to the numbers of observations in a cell.**

ant situation in so far as he can.    I shall not give the method here; for details the reader is referred to Snedecor and other textbooks (see References) after he has mastered the methods for the simple cases presented in this chapter.    The cell frequencies of type (ii) in the figure are said to be *proportionate* since the row frequencies in the example are in the proportion of $3:2:4$ and the column frequencies are $1:2:1:3$.    For a table of this nature there is no difficulty in calculating the sum of squares but there is a problem in using the $F$ table to test the significance of certain effects or to estimate certain variance components, and this problem is quite analogous to that encountered in the unequal number of observations in a hierarchical classification.    Type (iii) is a simplification of type (ii) in that the column frequencies are now constant (in other words, in the proportion $1:1:1:1$), while the row frequencies are still proportionate.    For the last two cases the reader may consult, in addition to Snedecor and others, the short paper by H. F. Smith [Analysis of Variance with Unequal but Proportionate Numbers of Observations in the Subclass of a Two-way Classification, *Biometrics*, **7**:70–74 (1951)].    This leaves us only types (iv) and (v) to consider in this chapter.

## 2. Linear model

The analysis of variance for two-way classification data is based on a linear model with the same general assumptions as for one-way classification data.    Let $i$ indicate a row ($i = 1, 2, \ldots, r$) and $j$ indicate a column ($j = 1, 2, \ldots, k$) so that the cell in the $i$th row and $j$th column may be identified by $ij$.    There are, altogether, $r \times k$ cells in the table. Again, as before, let $\alpha$ indicate the individual observation in a cell.    The linear model is

$$y_{ij\alpha} = \mu + \rho_i + \tau_j + \delta_{ij} + \epsilon_{ij\alpha}$$

where $\rho_i$ is the row effect, $\tau_j$ is the column effect, $\delta_{ij}$ is the "interaction" between row and column or the *deviation from the additive value* of $\rho_i$ and $\tau_j$ for the cell, and $\epsilon_{ij\alpha}$ is the individual fluctuation or error within the cells. Another common symbol for $\delta_{ij}$ is $(\rho\tau)_{ij}$, indicating that it represents the row-column interaction.    The meaning of these symbols will be made clear later on by numerical examples.    Now, if there is only one observation in each cell as illustrated by type (v) in the figure, there will be no use for $\alpha$ and the model degenerates to

$$y_{ij} = \mu + \rho_i + \tau_j + \delta_{ij}$$

in which there is no true error term.    In most of such data, however, the rows or the columns themselves are merely replications of a set of treatments rather than a true classification according to another characteristic.

In this special model, $\delta_{ij}$ plays the role of an error ($\epsilon_{ij}$) on the assumption that there is no true interaction between row and column. That is,

$$y_{ij} = \mu + \rho_i + \tau_j + \epsilon_{ij}$$

We must, however, realize that this assumption is made here out of necessity rather than out of choice.

## 3. Notation and example

For the beginner who tries to follow this demonstration about half of the work is getting acquainted with the system of notations. To clarify the notations, they are given in the lower half of Table 10.1 in positions corresponding to those of the numerical values shown in the upper half of the table. Thus, $y_{ij}$ is the value in the cell of the $i$th row and $j$th column, and

$$Y_{i\cdot} = \sum_j y_{ij} = \text{total of } i\text{th row} \qquad \frac{Y_{i\cdot}}{k} = \bar{y}_{i\cdot} = \text{mean of } i\text{th row}$$

$$Y_{\cdot j} = \sum_i y_{ij} = \text{total of } j\text{th column} \qquad \frac{Y_{\cdot j}}{r} = \bar{y}_{\cdot j} = \text{mean of } j\text{th column}$$

*Table* 10.1 Two-way classification with single observation in each cell

| Three rows, $r = 3$ | Four columns, $k = 4$ | | | | Row total | Row mean | Row effect |
|---|---|---|---|---|---|---|---|
| | (1) | (2) | (3) | (4) | | | |
| (1) | 6 | 2 | 9 | 3 | 20 | 5 | −2 |
| (2) | 8 | 9 | 11 | 12 | 40 | 10 | +3 |
| (3) | 4 | 4 | 10 | 6 | 24 | 6 | −1 |
| Column total | 18 | 15 | 30 | 21 | 84 | | |
| Column mean | 6 | 5 | 10 | 7 | | 7 | |
| Column effect | −1 | −2 | +3 | 0 | | | 0 |

Corresponding notation

| | $j = 1$ | $j = 2$ | $j = 3$ | $j = 4$ | $Y_{i\cdot}$ | $\bar{y}_{i\cdot}$ | $r_i$ |
|---|---|---|---|---|---|---|---|
| $i = 1$ | $y_{11}$ | $y_{12}$ | $y_{13}$ | $y_{14}$ | $Y_{1\cdot}$ | $\bar{y}_{1\cdot}$ | $r_1$ |
| $i = 2$ | $y_{21}$ | $y_{22}$ | $y_{23}$ | $y_{24}$ | $Y_{2\cdot}$ | $\bar{y}_{2\cdot}$ | $r_2$ |
| $i = 3$ | $y_{31}$ | $y_{32}$ | $y_{33}$ | $y_{34}$ | $Y_{3\cdot}$ | $\bar{y}_{3\cdot}$ | $r_3$ |
| $Y_{\cdot j}$ | $Y_{\cdot 1}$ | $Y_{\cdot 2}$ | $Y_{\cdot 3}$ | $Y_{\cdot 4}$ | $Y_{\cdot\cdot}$ | | |
| $\bar{y}_{\cdot j}$ | $\bar{y}_{\cdot 1}$ | $\bar{y}_{\cdot 2}$ | $\bar{y}_{\cdot 3}$ | $\bar{y}_{\cdot 4}$ | | $\bar{y}_{\cdot\cdot}$ | |
| $t_j$ | $t_1$ | $t_2$ | $t_3$ | $t_4$ | | | 0 |

The grand total and the general mean are

$$Y = Y.. = \sum_i Y_{i\cdot} = \sum_j Y_{\cdot j} = \sum_i \sum_j y_{ij}$$

$$\bar{y} = \bar{y}.. = \frac{\sum_i \bar{y}_i}{r} = \frac{\sum_j \bar{y}_{\cdot j}}{k} = \frac{Y}{rk}$$

A row effect is equal to the row mean minus the general mean; a column effect is equal to the column mean minus the general mean:

$$r_i = \bar{y}_{i\cdot} - \bar{y}.. \qquad t_j = \bar{y}_{\cdot j} - \bar{y}..$$

so that $\Sigma r_i = 0$ and $\Sigma t_j = 0$. With this layout, each $y_{ij}$ may be expressed as the sum of four components as shown by the following identity:

$$y_{ij} = \bar{y} + (\bar{y}_{i\cdot} - \bar{y}) + (\bar{y}_{\cdot j} - \bar{y}) + (y_{ij} - \bar{y}_{i\cdot} - \bar{y}_{\cdot j} + \bar{y})$$
$$= u + r_i + t_j + e_{ij}$$

where $e_{ij}$ is the difference $y_{ij} - (u + r_i + t_j)$. Applying this method of partitioning to our numerical values of Table 10.1, we obtain the results of Table 10.2. The first component is $u = \bar{y} = 7$. The second component is $r_i$, so that in the first row it is $r_1 = -2$, and so on. The third component is $t_j$, so that in the first column it is $t_1 = -1$, and so on. The fourth component $e_{ij}$ is the balance of the observed number. Note that $e_{ij}$'s add up to zero for each row *and* for each column.

*Table* 10.2   Values of $y_{ij} = u + r_i + t_j + e_{ij}$ (data from Table 10.1)

| | (1) | (2) | (3) | (4) |
|---|---|---|---|---|
| **(1)** | 6 = 7<br>　-2<br>　　-1<br>　　　+2 | 2 = 7<br>　-2<br>　　-2<br>　　　-1 | 9 = 7<br>　-2<br>　　+3<br>　　　+1 | 3 = 7<br>　-2<br>　　+0<br>　　　-2 |
| **(2)** | 8 = 7<br>　+3<br>　　-1<br>　　　-1 | 9 = 7<br>　+3<br>　　-2<br>　　　+1 | 11 = 7<br>　+3<br>　　+3<br>　　　-2 | 12 = 7<br>　+3<br>　　+0<br>　　　+2 |
| **(3)** | 4 = 7<br>　-1<br>　　-1<br>　　　-1 | 4 = 7<br>　-1<br>　　-2<br>　　　+0 | 10 = 7<br>　-1<br>　　+3<br>　　　+1 | 6 = 7<br>　-1<br>　　+0<br>　　　+0 |

## 4. Least-square estimates

The above method of estimating the true row effects $\rho_i$ by $r_i = \bar{y}_{i\cdot} - \bar{y}$, and so on, is also based on the principle of least squares (Chap. 6). Here the sum of squares to be minimized by choosing appropriate values of $u$, $r_i$, $t_j$ is

$$Q = \Sigma\Sigma(y_{ij} - u - r_i - t_j)^2 = \Sigma\Sigma e_{ij}^2$$

Setting the partial derivatives of $Q$ with respect to $u$, $r_i$, $t_j$ equal to zero, we obtain a set of equations from which the required values of $u$, $r_i$, $t_j$ may be solved with the supporting restrictions $\Sigma r_i = 0$ and $\Sigma t_j = 0$. These equations simply turn out to be the sum of certain terms of the following array of expressions equating to zero:

$$
\begin{array}{llll}
y_{11} - u - r_1 - t_1 & y_{12} - u - r_1 - t_2 & y_{13} - u - r_1 - t_3 & y_{14} - u - r_1 - t_4 \\
y_{21} - u - r_2 - t_1 & y_{22} - u - r_2 - t_2 & y_{23} - u - r_2 - t_3 & y_{24} - u - r_2 - t_4 \\
y_{31} - u - r_3 - t_1 & y_{32} - u - r_3 - t_2 & y_{33} - u - r_3 - t_3 & y_{34} - u - r_3 - t_4
\end{array}
$$

To estimate $u$, we set the sum of all the twelve expressions to zero:

$$\sum\sum y_{ij} - rku = 0 \qquad \therefore u = \frac{Y}{rk} = \bar{y}$$

To estimate $r_1$, we ignore the expressions without $r_1$ and put the sum of the expressions involving $r_1$ (first row) equal to zero:

$$\sum_j y_{1j} - ku - kr_1 = 0 \qquad \therefore r_1 = \bar{y}_{1\cdot} - \bar{y}, \text{ etc.}$$

Similarly, to estimate the column effect $t_1$, we put the sum of the expressions in the first column equal to zero:

$$\sum_i y_{i1} - ru - rt_1 = 0 \qquad \therefore t_1 = \bar{y}_{\cdot 1} - \bar{y}, \text{ etc.}$$

Hence, the estimates we have adopted are least-square estimates and the value of $\Sigma\Sigma e_{ij}^2$ is a minimum.

## 5. Components of $ssq$

Having cleared the statistical model and its sample estimates, we come to the problem of subdividing the total sum of squares into appropriate components. Referring to the numerical values of the margins of Table 10.1 or the body of Table 10.2, we calculate the $ssq$ for $r_i$, $t_j$, $e_{ij}$ as follows:

$$ssq_R = 4[(-2)^2 + (3)^2 + (-1)^2] = 56 = k \sum_i r_i^2$$

$$ssq_K = 3[(-1)^2 + (-2)^2 + (3)^2 + (0)^2] = 42 = r \sum_j t_j^2$$

$$ssq_E = [2^2 + (-1)^2 + 1^2 + \cdots + 0^2] = 22 = \sum \sum e_{ij}^2$$

The sum of these three $ssq$'s is 120, which is the total $ssq$ for the 12 numbers, as we have seen in preceding chapters. The algebraic proof is easy. The *total*, or *gross*, *deviation* of each observed number from the general mean may be written as the sum of three deviations:

$$y_{ij} - \bar{y} = (\bar{y}_{i\cdot} - \bar{y}) + (\bar{y}_{\cdot j} - \bar{y}) + (y_{ij} - \bar{y}_{i\cdot} - \bar{y}_{\cdot j} + \bar{y})$$

Squaring each such term and then summing the cells of the $r$ rows *or* the $k$ columns, we obtain the required identity:

$$\sum_i \sum_j (y_{ij} - \bar{y})^2 = k \sum_i (\bar{y}_{i\cdot} - \bar{y})^2 + r \sum_j (\bar{y}_{\cdot j} - \bar{y})^2 + \sum_i \sum_j (y_{ij} - \bar{y}_{i\cdot} - \bar{y}_{\cdot j} + \bar{y})^2$$

$$= k \sum_i r_i^2 + r \sum_j t_j^2 + \sum_i \sum_j e_{ij}^2$$

all the product terms having vanished on summation. This last simplifying feature is due to the facts that for each fixed row the value of $(\bar{y}_{i\cdot} - \bar{y})$ is a constant while $\sum_j (\bar{y}_{\cdot j} - \bar{y}) = 0$, that a similar situation exists for each fixed column, and that the $e_{ij}$ add up to zero for each row as well as for each column. This last fact is one we have noticed for the numerical values in Table 10.2. For example, for a given row,

$$\sum_j (y_{ij} - \bar{y}_{i\cdot} - \bar{y}_{\cdot j} + \bar{y}) = \sum_j (y_{ij} - \bar{y}_{i\cdot}) - \sum_j (\bar{y}_{\cdot j} - \bar{y}) = 0 - 0 = 0$$

Thus, the total $ssq$ may be subdivided into three components—one due to the row effects, one due to the column effects, and one due to error as estimated by the deviation from the additive value of the preceding two.

### 6. Practical computation

The above demonstration reveals the meaning of the various sums of squares. For practical purposes, however, only four basic quantities (or areas if you visualize a diagram analogous to Fig. 5.1) need be calculated. The areas $A$ and $C$ are the same as before. Then regard the data as simply subdivided into $r$ rows (groups) in a one-way classification and calculate the area analogous to $B$ in Fig. 5.1. Finally, rearrange the data into $k$ columns (groups) in a one-way classification and again calculate the area analogous to $B$ in Fig. 5.1. Let us call these two areas $R$ and $K$, respectively. The arithmetic goes as follows (see Table 10.1):

$$\text{(area) } A = \sum_i \sum_j y_{ij}^2 = 6^2 + 2^2 + 9^2 + \cdots + 6^2 = 708$$

$$\text{(area) } R = \frac{1}{k} \sum_i Y_{i\cdot}^2 = \frac{1}{4}(20^2 + 40^2 + 24^2) = 644$$

$$\text{(area) } K = \frac{1}{r} \sum_j Y_{\cdot j}^2 = \frac{1}{3}(18^2 + 15^2 + 30^2 + 21^2) = 630$$

$$\text{(area) } C = \frac{1}{rk} Y^2 = \frac{1}{12}(84^2) = 588$$

The sum of squares may then be obtained as follows:

$$
\begin{array}{lll}
ssq_R = R - C & = & 56 \\
ssq_K = K - C & = & 42 \\
ssq_E = A - R - K + C & = & 22 \\
\hline
ssq_T = A - C & = & 120
\end{array}
$$

For numerical calculation, the value of $ssq_E$ is obtained by subtraction, since it is equal to $ssq_T - ssq_R - ssq_K$.

### 7. Expectation of $ssq$

The expectations of the sums of squares may be obtained in the same manner as in preceding chapters. As mentioned before, we have here assumed that there is no interaction between rows and columns. We thus treat $d_{ij}$ as a random error $e_{ij}$, which is independently and normally distributed with zero mean and variance $\sigma_e^2$. Furthermore, $r_i$ and $t_j$ are also assumed to be uncorrelated with each other or with the random error. Before giving the full results we should work out the expectation of one term as an example:

$$E\{Y_{i\cdot}^2\} = E\left\{\sum_j y_{ij}\right\}^2 = E\left\{ku + kr_i + \sum_j t_j + \sum_j e_{ij}\right\}^2$$
$$= k^2\mu^2 + k^2\sigma_r^2 + k\sigma_t^2 + k\sigma_e^2$$

In the above expression we write $\sigma_r^2$ for $E\{r_i^2\}$, etc. The expectations of all product terms are zero. Multiplying the above expression by $r$ (for there are $r$ rows) and dividing it by $k$ (for each row total is the sum of $k$ cells), we obtain the expectation of $R = \Sigma Y_{i\cdot}^2/k$. The complete results are as follows:

$$E\{A\} = rk\mu^2 + rk\sigma_r^2 + rk\sigma_t^2 + rk\sigma_e^2$$
$$E\{R\} = rk\mu^2 + rk\sigma_r^2 + r\sigma_t^2 + r\sigma_e^2$$
$$E\{K\} = rk\mu^2 + k\sigma_r^2 + rk\sigma_t^2 + k\sigma_e^2$$
$$E\{C\} = rk\mu^2 + k\sigma_r^2 + r\sigma_t^2 + \sigma_e^2$$

The expectations of the sums of squares may be obtained by taking the appropriate differences between the above expressions. Thus:

$$E\{ssq_R\} = E\{R - C\} = k(r - 1)\sigma_r^2 + (r - 1)\sigma_e^2$$
$$E\{ssq_K\} = E\{K - C\} = r(k - 1)\sigma_t^2 + (k - 1)\sigma_e^2$$
$$E\{ssq_E\} = E\{A - R - K + C\} = (rk - r - k + 1)\sigma_e^2$$
$$= (r - 1)(k - 1)\sigma_e^2$$

To obtain the expectations of the mean squares, we need only divide the above expressions by the number of degrees of freedom for each *ssq*. Since there are $r$ rows and $k$ columns, $ssq_R$ has $r - 1$ and $ssq_K$ has $k - 1$ degrees of freedom. Since the $e_{ij}$ add up to zero for each row as well as for each column, the $ssq_E$ has $(r - 1)(k - 1)$ degrees of freedom. These results are summarized in Table 10.3. It should be reiterated that we are not committed to either model I or model II, and in the table the symbol $\sigma_t^2$, for instance, may also be written as $\Sigma\tau_j^2/(k - 1)$, assuming $E\{t_j^2\} = \tau_j^2$ for each $j$.

*Table* 10.3   Analysis of variance for a two-way classification with a single observation in each cell

| Variation | *df* | Sum of squares | $E\{msq\}$ |
|---|---|---|---|
| Rows | $r - 1$ | $ssq_R = k \sum_i (\bar{y}_{i\cdot} - \bar{y})^2$ | $\sigma_e^2 + k\sigma_r^2$ |
| Columns | $k - 1$ | $ssq_K = r \sum_j (\bar{y}_{\cdot j} - \bar{y})^2$ | $\sigma_e^2 + r\sigma_t^2$ |
| Error | $(r - 1)(k - 1)$ | $ssq_E = \sum_i \sum_j (y_{ij} - \bar{y}_{i\cdot} - \bar{y}_{\cdot j} + \bar{y})^2$ | $\sigma_e^2$ |
| Total | $rk - 1$ | $ssq_T = \sum_i \sum_j (\bar{y}_{ij} - \bar{y})^2$ | |

Analysis of variance for data of Table 10.1

| Variation | *df* | *ssq* | *msq* | *F* | *F* at 5% point |
|---|---|---|---|---|---|
| Rows | $3 - 1 = 2$ | 56 | 28 | 7.64 | 5.14 |
| Columns | $4 - 1 = 3$ | 42 | 14 | 3.82 | 4.76 |
| Error | $2 \times 3 = 6$ | 22 | 3.67 | | |
| Total | $12 - 1 = 11$ | 120 | | | |

## 8. The $F$ test

Now we come to the final problem of the testing of significance of the row or column effects. From Table 10.3 it is clear that the row and column effects may be tested separately. Testing the null hypothesis that all columns have the same mean value, that is, $\sigma_t^2 = 0$, we calculate the value of $F = s_K^2/s_E^2$ and compare this value with that tabulated at the 5 per cent point as given on page 68.

Similarly, to test the row effects, we form the ratio $F = s_R^2/s_E^2$. From Table 10.3, which completes the numerical analysis of the data shown in Table 10.1, we conclude that the row effects are significant and the column effects are not. In many practical experiments, one of the classifications is "blocks," which are merely replicates of a set of treatments. Suppose that the rows here are blocks or replicates. Then we will be interested primarily in testing the significance of the column or treatment effects.

## 9. Eliminating row effects

In order to obtain further insight into the meaning of the various sums of squares, let us eliminate the effects of one of the classifications first and separately. If we assume that the rows in the above example are blocks and that we wish to eliminate the row effect without disturbing the column effect, we may do so simply by adjusting each observation according to the effect of the block in which the observation occurs. This has been done in Table 10.4, where the original data and the row-adjusted values are put side by side. Consider the first row, where the observations are 6, 2, 9, 3. The effect of the first row is that row mean minus the general mean, that is, $5 - 7 = -2$, as has been calculated in Table 10.1. This means that the values in this row are too small owing to the inferior row. Hence, we *add* 2 to each observation of this row. The adjusted values are then 8, 4, 11, 5. To adjust the second row, we subtract 3 from each observation, and so on (see Table 10.4).

Now, if we perform the same kind of calculation and analysis to these row-adjusted data, we shall discover that the *ssq* for rows is zero and the data should thus be analyzed as if they constituted a one-way classification, remembering that there is a loss of 2 *df* for adjusting the three rows. Then the new area $A$ is

$$A' = 8^2 + 4^2 + \cdots + 7^2 = 652$$

The areas due to columns and the grand total remain the same as before; that is, $K = 630$ and $C = 588$. The new "total" *ssq* is equal to $A' - C$, while the *ssq*'s due to columns and error remain the same as before.

*Table* 10.4    **Eliminating row effects from data**

| | Original data | | | | | | Row-adjusted data | | | | | |
|---|---|---|---|---|---|---|---|---|---|---|---|---|
| | (1) | (2) | (3) | (4) | $Y_i$ | $r_i$ | (1) | (2) | (3) | (4) | $Y_i$ | $r_i$ |
| (1) | 6 | 2 | 9 | 3 | 20 | −2 | 8 | 4 | 11 | 5 | 28 | 0 |
| (2) | 8 | 9 | 11 | 12 | 40 | +3 | 5 | 6 | 8 | 9 | 28 | 0 |
| (3) | 4 | 4 | 10 | 6 | 24 | −1 | 5 | 5 | 11 | 7 | 28 | 0 |
| | 18 | 15 | 30 | 21 | 84 | 0 | 18 | 15 | 30 | 21 | 84 | 0 |

Analysis of row-adjusted data

| Source | $df$ | $ssq$ | $msq$ | $F$ |
|---|---|---|---|---|
| Columns | 3 | $K - C = 42$ | 14.00 | 3.82 |
| Error | 6 | $A' - K = 22$ | 3.67 | |
| "Total" | 9 | $A' - C = 64$ | | |

The analysis is shown in the lower part of Table 10.4.  Comparing this with Table 10.3, we see that we have simply eliminated the rows from the picture and the results are those for columns and error and *their* total.

If we continue to adjust for the columns to the already row-adjusted data, we will discover that the "total" *ssq* is merely 22 for error, since both row and column effects have been eliminated.  Thus, the linear model permits both separate and successive elimination of certain effects. The chief purpose of our ordinary analysis is to isolate these various components in one comprehensive operation.

## 10. Student's paired method

Data frequently occur in pairs.  For instance, two treatments (1) and (2) may be applied to a pair of identical twins, one treatment to one twin. Table 10.5 gives the data on six ($n = 6$) pairs.  The variation from pair to pair is incidental, although probably unavoidable.  Our main interest here is to compare the effects of the two treatments and see if there is a difference between them.  To eliminate the effects of the various pairs and to determine the effects of treatments only, the method, described as *Student's paired method* in many textbooks, is to take the difference between the two treatments within each pair, as denoted by $D = (1) - (2)$ in Table 10.5.  Having done this, we shall regard $D$ as the variable to be studied.  In particular, we want to test the null hypothesis that the true mean value of the variable $D$ is zero.

Now, in our example, the observed total and mean of $D$ are respectively $\Sigma D = 12$ and $\bar{D} = 2.0$. The sum of squares of $D$ is calculated in the usual manner:

$$ssq_D = \Sigma D^2 - \frac{(\Sigma D)^2}{n} = 70 - \frac{12^2}{6} = 46$$

The estimated variance of $D$ is $ssq_D/(n - 1) = s_D^2$, and the estimated variance of the sample *mean* $\bar{D}$ is $s_D^2/n$. Finally, we obtain the value of $t = \bar{D}/s_{\bar{D}} = 1.615$. The details of the arithmetic have been given in the right-hand side of Table 10.5. This $t = 1.615$ has $n - 1 = 5$ degrees of freedom.

Alternatively, the data may be analyzed as a two-way classification case by the ordinary analysis-of-variance method. This has been done in the lower portion of Table 10.5. The procedure is routine and needs no explanation. Note that we now calculate the pair totals instead of the pair differences. Each pair is, in the language of analysis of variance, a block, a row in our table. The four basic quantities $A$, $R$, $K$, $C$ are calculated in the usual way.

There are two important relationships between the previous $t$ test and the present $F$ test. First we note that

$$ssq_D = \sum D^2 - \frac{(\Sigma D)^2}{n} = 46 \qquad s_D^2 = 9.2$$
$$ssq_E = A - R - K + C = 23 \qquad s_E^2 = 4.6$$

It may be shown (see Exercises) that the former is always twice as large as the latter. The physical meaning of this relationship is at once clear if we look at the linear model. The two observations of the $j$th pair are

$$y_{1j} = u + b_j + t_1 + e_{1j} \qquad y_{2j} = u + b_j + t_2 + e_{2j}$$

The difference between the two members of the pair is thus

$$D_j = y_{1j} - y_{2j} = (t_1 - t_2) + (e_{1j} - e_{2j})$$

which is a new (derived from the original) linear model; the first component is $t_1 - t_2$ and the error term is $e_{1j} - e_{2j}$. The estimated variance $s_D^2$ is the expected value of $(e_{1j} - e_{2j})^2$. Since the $e$'s are uncorrelated, and have equal variance, $s_D^2$ is simply the expected value of $e_{1j}^2 + e_{2j}^2$, or $2e^2$. Hence, $s_D^2 = 2s_E^2$, because $s_E^2$ is the expected value of one $e^2$. In other words, $s_D^2$ is the variance of the difference of two observations and $s_E^2$ is the variance of one observation. The second relationship to be noticed is that

$$t^2 = (1.615)^2 = 2.61 = F$$

This, corresponding to the case in Chap. 8, establishes the equivalence of these two tests, since there is only 1 *df* for treatments. In many

*Table* 10.5  **Analysis of** $n$ **paired observations**

Student's paired method; difference within pairs

| Treatments | | Difference, | $D^2$ | Calculations |
|---|---|---|---|---|
| (1) | (2) | $D = (1) - (2)$ | | |
| 6 | 2 | 4 | 16 | $ssq_D = 70 - 12^2/6 = 46$ |
| 8 | 9 | $-1$ | 1 | $s_D{}^2 = \dfrac{ssq_D}{n-1} = \dfrac{46}{5} = 9.2$ |
| 4 | 4 | 0 | 0 | |
| 9 | 3 | 6 | 36 | $s_{\bar{D}}{}^2 = \dfrac{s_D{}^2}{n} = \dfrac{46}{5 \times 6} = 1.533$ |
| 11 | 12 | $-1$ | 1 | |
| 10 | 6 | 4 | 16 | |
| Total  48 | 36 | 12 | 70 | $t = \dfrac{\bar{D}}{s_{\bar{D}}} = \dfrac{2}{\sqrt{1.533}} = 1.615$ |
| Mean  8 | 6 | $\bar{D} = 2$ | | |

Analysis of variance, two-way classification

| Treatments | | Pair total | Calculations |
|---|---|---|---|
| (1) | (2) | | |
| 6 | 2 | 8 | $A = 6^2 + 2^2 + \cdots + = 708$ |
| 8 | 9 | 17 | $R = \frac{1}{2}(8^2 + 17^2 + \cdots) = 673$ |
| 4 | 4 | 8 | $K = \frac{1}{6}(48^2 + 36^2) = 600$ |
| 9 | 3 | 12 | $C = \frac{1}{12}(84)^2 = 588$ |
| 11 | 12 | 23 | |
| 10 | 6 | 16 | |
| Total  48 | 36 | 84 | |

| | $df$ | $ssq$ | $msq$ | $F$ |
|---|---|---|---|---|
| Rows (pairs) | $n - 1 = 5$ | $R - C = 85$ | | |
| Treatments | $2 - 1 = 1$ | $K - C = 12$ | 12.0 | 2.61 |
| Error | $n - 1 = 5$ | $A - R - K + C = 23$ | $4.6 = s_E{}^2$ | |
| Total | $2n - 1 = 11$ | $A - C = 120$ | | |

textbooks, the Student's paired method is described as a special method to deal with paired data.  It is, as we have seen, merely a special case of the analysis of variance when there are only two treatments in a block. It is easy to write out the algebra showing that $t^2 = F$ as we have done for one-way classification in Chap. 8 (see Exercise 10.3).

## 11. *c* observations in each cell

Now we pass to the case in which there are *c* observations in each cell. Table 10.6 gives an example of two rows and three columns with two ($c = 2$) observations in each cell. The linear model is the one first introduced in Sec. 2. The calculation of the *ssq* may be conveniently executed in two stages. The first stage is to regard the $rk = 2 \times 3 = 6$ cells themselves as groups in a one-way classification so that we obtain an *ssq* within the cells and an *ssq* between the cells. The second stage, concerned with the latter only, is to concentrate on the *rk* cell totals and subdivide their *ssq* into three components—rows, columns, and "interaction"—by the method described in the preceding sections. The five basic quantities needed to accomplish these steps are given in the middle of Table 10.6, and the analysis of variance is given in the bottom portion.

At this point the procedure of testing significance needs some comment. If we adopt the concept of model II (p. 59) that $\rho_i$, $\tau_j$, $\delta_{ij}$ are independent random variables, the expectation of the various *mean squares* are then (the long algebraic write-up being omitted):

$$
\begin{array}{ll}
\text{Rows:} & \sigma_e^2 + c\sigma_d^2 + kc\sigma_r^2 \\
\text{Columns:} & \sigma_e^2 + c\sigma_d^2 + rc\sigma_t^2 \\
\text{Interaction:} & \sigma_e^2 + c\sigma_d^2 \\
\text{Within cells:} & \sigma_e^2
\end{array}
$$

To test the significance of the interaction, we use $F =$ interaction *msq*/within-cell *msq*. But to test the significance of the major row (or column) effects, we use $F =$ row (or column) *msq*/interaction *msq*. This is what was done in Table 10.6, and I believe that in most experimental situations this procedure is applicable. If the concept of model I ($\rho_i$, $\tau_j$, $\delta_{ij}$ being fixed quantities) is adopted, then the component $\sigma_d^2$ in the expectation of the row and the column mean squares should be deleted from the above expressions. Then all tests of significance should be made against the within-cell mean square. There are occasions when model I is clearly applicable. Suppose that one row represents boys and the other row represents girls. One can hardly claim that the two sexes are merely a random sample of all possible sexes! When one classification conforms with model I and the other classification conforms with model II (the mixed model), the *F* test with the interaction mean square as the denominator is still applicable. In many experiments the observations in a cell are repeated determinations of the same entity under the same conditions, and therefore the within-cell mean square measures the variability from determination to determination. This mean square is usually smaller, much smaller, than the interaction mean square. If the interaction is

*Table* 10.6 **Two-way classification with two observations in each cell**

| | Single values: $y_{ij\alpha}$ | | | Row total |
|---|---|---|---|---|
| | 9  11 | 2  4 | 4  6 | $Y_{1.}$ |
| | 10  12 | 6  8 | 9  3 | $Y_{2.}$ |
| Column total | $Y_{.1}$ | $Y_{.2}$ | $Y_{.3}$ | $Y_{..} = Y$ |

| | Cell totals: $Y_{ij}$ | | | Row total |
|---|---|---|---|---|
| | 20 | 6 | 10 | 36 |
| | 22 | 14 | 12 | 48 |
| Column total | 42 | 20 | 22 | 84 |

$$A = \sum_i \sum_j \sum_\alpha y_{ij\alpha}{}^2 = 9^2 + 11^2 + \cdots + 3^2 = 708$$

$$B = \frac{1}{c} \sum_i \sum_j Y_{ij}{}^2 = \frac{1}{2}(20^2 + \cdots + 12^2) = 680$$

$$R = \frac{1}{kc} \sum_i Y_{i.}{}^2 = \frac{1}{6}(36^2 + 48^2) = 600$$

$$K = \frac{1}{rc} \sum_j Y_{.j}{}^2 = \frac{1}{4}(42^2 + 20^2 + 22^2) = 662$$

$$C = \frac{1}{N} Y^2 = \frac{1}{12}(84)^2 = 588$$

Analysis of variance of a two-way classification with $c$ observations in each cell
(data above)

| Variation | df | ssq | | msq | F |
|---|---|---|---|---|---|
| Rows | $r - 1 =$  1 | $R - C =$ | 12 | 12 | 4.0 |
| Columns | $k - 1 =$  2 | $K - C =$ | 74 | 37 | 12.3 |
| Interaction | $(r - 1)(k - 1) =$  2 | $B - R - K + C =$ | 6 | 3 | |
| Between cells | $rk - 1 =$  5 | $B - C =$ | 92 | | |
| Within cells | $rk(c - 1) =$  6 | $A - B =$ | 28 | 4.67 | |
| Total | $rkc - 1 = 11$ | $A - C =$ | 120 | | |

absent between the two classification factors, these two mean squares are approximately of the same magnitude.

## EXERCISES

10.1 Given the two-way classification data in the accompanying table,

| | | \multicolumn{8}{c}{Columns (blocks = days)} |
|---|---|---|---|---|---|---|---|---|---|
| | | Mon (1) | Tues (2) | Wed (3) | Thurs (4) | Fri (5) | Sat (6) | Sun (7) | Row total |
| Rows | (1) | 11 | 15 | 10 | 19 | 17 | 12 | 21 | 105 |
| (treat- | (2) | 15 | 12 | 14 | 19 | 8 | 11 | 5 | 84 |
| ments) | (3) | 10 | 4 | 5 | 14 | 10 | 6 | 14 | 63 |
| Column total | | 36 | 31 | 29 | 52 | 35 | 29 | 40 | 252 |

calculate the following four basic quantities:

$$A = 11^2 + 15^2 + \cdots + 14^2 \qquad R = \tfrac{1}{7}(105^2 + 84^2 + 63^2)$$
$$K = \tfrac{1}{3}(36^2 + 31^2 + \cdots + 40^2) \qquad C = \tfrac{1}{21}(252^2)$$

Then complete the analysis-of-variance table. Are the three treatments significantly different?

| Variation | df | ssq | msq | F |
|---|---|---|---|---|
| Blocks | 6 | $K - C =$ | | |
| Treatments | 2 | $R - C =$ | | 3.706 |
| Error | 12 | $= 204$ | $s^2 =$ | |
| Total | 20 | $A - C =$ | | |

10.2 The accompanying table is a set of two-way classification data.

| 11 | 6 | 2 | 13 | 32 |
|---|---|---|---|---|
| 15 | 11 | 14 | 20 | 60 |
| 7 | 7 | 5 | 9 | 28 |
| 33 | 24 | 21 | 42 | 120 |

a Subdivide the total *ssq* into three components by first calculating the four basic quantities $A, R, K, C$, as shown in Exercise 10.1.
b Find the row effects $r_i = \bar{y}_i - \bar{y}$ and column effects $c_j = \bar{y}_j - \bar{y}$. Eliminate the row effects and the column effects separately from the original data. Then eliminate

both row and column effects in whatever order you prefer. The resulting numbers will have only the error component plus the general mean. Their *ssq* will therefore be the error *ssq*. Check your calculations with the results obtained in (*a*).

Column effects eliminated

| 10 | 8 | 5 | 9 | 32 |
|----|----|----|----|-----|
| 14 | 13 | 17 | 16 | 60 |
| 6 | 9 | 8 | 5 | 28 |
| 30 | 30 | 30 | 30 | 120 |

Row effects eliminated

| 13 | 8 | 4 | 15 | 40 |
|----|----|----|----|-----|
| 10 | 6 | 9 | 15 | 40 |
| 10 | 10 | 8 | 12 | 40 |
| 33 | 24 | 21 | 42 | 120 |

Both effects eliminated

| 12 | 10 | 7 | 11 | 40 |
|----|----|----|----|-----|
| 9 | 8 | 12 | 11 | 40 |
| 9 | 12 | 11 | 8 | 40 |
| 30 | 30 | 30 | 30 | 120 |

Error component

| +2 | 0 | −3 | +1 | 0 |
|----|----|----|----|-----|
| −1 | −2 | +2 | +1 | 0 |
| −1 | +2 | +1 | −2 | 0 |
| 0 | 0 | 0 | 0 | 0 |

**10.3** Suppose there are only two treatments in a block and there are $n$ blocks (pairs). Instead of using the formal $y_{ij}$ notation, we may use $a_i$ and $b_i$ to denote the two observations in the $i$th block. The data and some preliminary calculations

| | (1) | (2) | Total | Differ-ence $D$ | $D^2$ |
|---|-----|-----|-------|---------|-------|
| | $a_1$ | $b_1$ | $a_1 + b_1$ | $a_1 - b_1$ | $(a_1 - b_1)^2 = 2(a_1^2 + b_1^2) - (a_1 + b_1)^2$ |
| | $a_2$ | $b_2$ | $a_2 + b_2$ | $a_2 - b_2$ | $(a_2 - b_2)^2 = 2(a_2^2 + b_2^2) - (a_2 + b_2)^2$ |
| | $\cdots$ | $\cdots$ | $\cdots$ | $\cdots$ | $\cdots$ $\cdots$ $\cdots$ |
| | $a_n$ | $b_n$ | $a_n + b_n$ | $a_n - b_n$ | $(a_n - b_n)^2 = 2(a_n^2 + b_n^2) - (a_n + b_n)^2$ |
| Total | $T_1$ | $T_2$ | $T_1 + T_2$ | $T_1 - T_2$ | $\Sigma D^2 = 2A \qquad - 2R$ |

are given in the accompanying table. At this point we may well recall (see Chap. 3, Sec. 10) that

$$\frac{(T_1 - T_2)^2}{2n} = \frac{T_1^2 + T_2^2}{n} - \frac{(T_1 + T_2)^2}{2n} = K - C$$

This is, in fact, the same identity we have used to express $D^2$ above. The sum of squares for the $D$'s is then

$$ssq_D = \sum D^2 - \frac{(T_1 - T_2)^2}{n} = 2A - 2R - (2K - 2C) = 2ssq_E$$

where $A$, $R$, $K$, $C$ are the four basic quantities defined in the text. The relation $ssq_D = 2ssq_E$ leads immediately to $t^2 = F$. Now

$$t = \frac{\bar{D}}{\sqrt{ssq_D/(n-1)n}} = \frac{(T_1 - T_2)/n}{\sqrt{2ssq_E/(n-1)n}}$$

Squaring, substituting, and simplifying, we obtain

$$t^2 = \frac{(T_1 - T_2)^2/2n}{ssq_E/(n-1)} = \frac{K - C}{ssq_E/(n-1)} = F$$

**10.4** In the accompanying $3 \times 4$ (rows $\times$ columns) table there are two $(c = 2)$ single observations in each cell. Break down the total sum of squares in the manner shown in Table 10.6.

Single values: $y_{ij\alpha}$

| 2 4 | 1 1 | 4 5 | 1 2 |
|-----|-----|-----|-----|
| 5 3 | 3 6 | 7 4 | 4 8 |
| 3 1 | 1 3 | 4 6 | 4 2 |

Cell totals: $Y_{ij}$      $Y_{i\cdot}$

| 6 | 2 | 9 | 3 | 20 |
|---|---|---|---|----|
| 8 | 9 | 11 | 12 | 40 |
| 4 | 4 | 10 | 6 | 24 |

$Y_{\cdot j}$    18    15    30    21    84

Calculate the five basic quantities:

$$A = \Sigma y_{ij\alpha}^2 =$$
$$B = \tfrac{1}{2}\Sigma Y_{ij}^2 =$$
$$C = \tfrac{1}{24}(84^2) =$$

$$R = \tfrac{1}{8}\Sigma Y_{i\cdot}^2 =$$
$$K = \tfrac{1}{6}\Sigma Y_{\cdot j}^2 =$$

The breakdown of the sum of squares is given in the accompanying table. Note that the cell totals in this exercise are the same as the single observations of Table 10.1. Compare the $ssq$'s in this exercise with those shown in Table 10.3. What do you notice? What are the similarities and differences between these two cases?

| Variation | $df$ | Sum of squares |
|-----------|------|----------------|
| Rows | $r - 1 = 2$ | $R - C = 28$ |
| Columns | $k - 1 = 3$ | $K - C = 21$ |
| Interaction | $(r - 1)(k - 1) = 6$ | $B - R - K + C = 11$ |
| Between cells | $rk - 1 = 11$ | $B - C = 60$ |
| Within cells | $rk(c - 1) = 12$ | $A - B = 30$ |
| Total | $rkc - 1 = 23$ | $A - C = 90$ |

# 11

# Sum of squares
# as an invariant

## 1. Introductory remark

It is not possible here to formally introduce the algebra of matrices, because this subject cannot be disposed of in a few pages at the level set for this book.  To give a list of rules for handling matrices would be artificial and arbitrary and could have no meaning for the reader who lacks previous knowledge of the subject.  Instead, we shall take the sum of squares as our central subject and let the matrices appear as a natural consequence, examining the properties of the particular type of matrix as we encounter it.  The statistical applications of the method will be deferred to the next chapter.  Also, instead of studying the usual sum of squares of deviations from mean, we shall in this chapter give attention only to the sum of squares of the numbers themselves, that is

$$\Sigma y^2 = y_1{}^2 + y_2{}^2 + \cdots + y_n{}^2$$

## 2. New numbers built from old

To begin this new subject with a very simple example, let us consider only two numbers, $y_1 = 5$ and $y_2 = 10$, so that $y_1{}^2 + y_2{}^2 = 5^2 + 10^2 = 125$. Now let us build two new numbers, say, $z_1$ and $z_2$, based on the values of $y_1$ and $y_2$. The rules of building the new numbers are that $z_1$ should consist of 60 per cent of $y_1$ and 80 per cent of $y_2$ and that $z_2$ is obtained by subtracting 80 per cent of $y_1$ from 60 per cent of $y_2$. Briefly:

$$z_1 = 0.6y_1 + 0.8y_2 = 0.6(5) + 0.8(10) = 11$$
$$z_2 = -0.8y_1 + 0.6y_2 = -0.8(5) + 0.6(10) = 2$$

Lo and behold! The two new numbers, 11 and 2, have the same sum of squares as that of the two original numbers. That is, $11^2 + 2^2 = 125$ also. In other words, the numbers are changed (transformed) in such a way that their sum of squares remains the same:

$$z_1{}^2 + z_2{}^2 = y_1{}^2 + y_2{}^2$$

## 3. Linear transformation

Of course, we could build other values for $z_1$ and $z_2$, but in general their sum of squares would not remain the same as that of the original numbers. Whether the relationship $\Sigma z^2 = \Sigma y^2$ holds or not depends, evidently, on the particular way the new numbers are built. In our numerical example the method of building the new numbers may be characterized by the following four specifications:

$$\begin{pmatrix} 0.6 & 0.8 \\ -0.8 & 0.6 \end{pmatrix} \quad \text{or, more generally} \quad \begin{pmatrix} a_1 & a_2 \\ b_1 & b_2 \end{pmatrix}$$

Note that these four numbers are the four coefficients of the two linear equations of the preceding paragraph, arranged in corresponding positions. Such an arrangement of numbers, usually enclosed in brackets, is called a *matrix* (of numbers). A matrix is not a number; it is a collection of numbers arranged in a systematic manner. To study the properties of a change (transformation) of the $y$ values into $z$ values, we need only know the rules of changing (that is, those equations expressing the new values in terms of the old) as specified by the coefficients of those equations. This boils down to a study of the matrix in which the "elements" (that is, the individual numbers appearing in the matrix) are the coefficients of the equations. When the set of equations expressing $z$ in terms of $y$ is linear, as in our example, we say it is a *linear transformation*.

## 4. Required conditions

Now we proceed to investigate the conditions which the coefficients (elements of the matrix) must satisfy in order that $y_1^2 + y_2^2 = z_1^2 + z_2^2$, where

$$z_1 = a_1 y_1 + a_2 y_2 \qquad z_2 = b_1 y_1 + b_2 y_2$$

Substituting, we have

$$
\begin{aligned}
z_1^2 + z_2^2 &= (a_1 y_1 + a_2 y_2)^2 + (b_1 y_1 + b_2 y_2)^2 \\
&= a_1^2 y_1^2 + a_2^2 y_2^2 + 2 a_1 a_2 y_1 y_2 + b_1^2 y_1^2 + b_2^2 y_2^2 + 2 b_1 b_2 y_1 y_2 \\
&= (a_1^2 + b_1^2) y_1^2 + (a_2^2 + b_2^2) y_2^2 + 2(a_1 a_2 + b_1 b_2) y_1 y_2
\end{aligned}
$$

In order for this expression to be equal to $y_1^2 + y_2^2$, the coefficients must satisfy the following conditions:

$$a_1^2 + b_1^2 = 1 \qquad a_2^2 + b_2^2 = 1 \qquad a_1 a_2 + b_1 b_2 = 0$$

Looking at the positions of the elements in the matrix, we may state these conditions as follows: the sum of squares of the elements of each *column* in the matrix is unity, and the sum of products of corresponding elements of the two *columns* is zero. In our numerical example the coefficients do satisfy these conditions, for example, $(0.6)^2 + (-0.8)^2 = 1$.

## 5. Conditions for rows

Note that the last condition is $a_1 a_2 = -b_1 b_2$. One way of satisfying this condition is to make $a_1 = b_2$ and $a_2 = -b_1$, as hinted by our numerical example (see the comments in the next section). In such a case, the above three conditions for the coefficients may be rewritten as

$$a_1^2 + a_2^2 = 1 \qquad b_1^2 + b_2^2 = 1 \qquad a_1 b_1 + a_2 b_2 = 0$$

Referring once more to the matrix in Sec. 3, we see that the conditions are now as follows: the sum of squares of the elements in each *row* is unity, and the sum of products of corresponding elements of the two *rows* is zero. Our numerical example verifies these facts too. In other words, the rows and the columns of the matrix satisfy the same sort of conditions. A more general and formal treatment is given at the end of this chapter.

## 6. Normalized orthogonal matrix

In the preceding section we made $a_1 = b_2$, which seems an arbitrary decision. But this one seemingly arbitrary step enables us to bypass a long series of theorems in matrix algebra. A transformation matrix which makes $\Sigma y^2 = \Sigma z^2$ is known as a *normalized orthogonal matrix*. To explain systematically what it is in purely algebraic terms would take many pages. For our present purpose the conditions in the preceding paragraph may be accepted as the definition of an orthogonal (normalized) matrix, and we have seen that it does transform one set of values $(y_1,y_2)$ to another $(z_1,z_2)$ without changing the sum of squares. The sum of squares is then said to be an *invariant* (something that remains the same) under an orthogonal transformation. At this stage a geometric interpretation of an orthogonal transformation will help greatly to clarify the situation.

### 7. Rotation of axes

The values of $(y_1,y_2) = (5,10)$ may be regarded as a point $P$ relative to the $y_1$ and $y_2$ cartesian coordinate system as illustrated in Fig. 11.1. The square of the distance from the origin $O$ to point $P$ is $\overline{OP}^2 = y_1{}^2 + y_2{}^2$. Now turn (rotate) the axes to an angle of approximately $\theta - 53° 8'$, keeping the axes perpendicular to each other and at the same origin. Then the coordinates of the point $P$ relative to the new coordinate axes are seen to be $(z_1,z_2) = (11,2)$. The rotation of the axes in no way affects the distance from $O$ to $P$. Hence $\overline{OP}^2$ is also equal to $z_1{}^2 + z_2{}^2$. This explains the numerical results of our example.

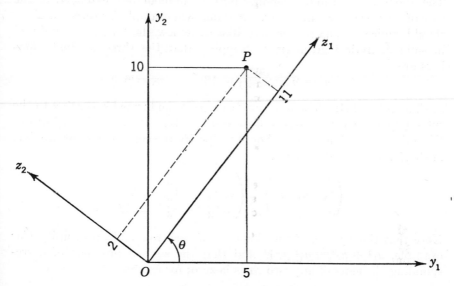

**Fig. 11.1   The rotation of axes through an angle $\theta$.**

## 8. General rotation

More generally, suppose that the axes are turned to an angle $\theta$. From plane analytical geometry we learn that the coordinate values of the point $P$ with respect to the new axes are related to those of the old axes in the following manner:

$$z_1 = (\cos \theta)y_1 + (\sin \theta)y_2$$
$$z_2 = -(\sin \theta)y_1 + (\cos \theta)y_2$$

In other words, the matrix

$$\begin{pmatrix} a_1 & a_2 \\ b_1 & b_2 \end{pmatrix} = \begin{pmatrix} \cos \theta & \sin \theta \\ -\sin \theta & \cos \theta \end{pmatrix}$$

is a normalized orthogonal matrix, for, whatever the value of $\theta$, the coefficients satisfy all the conditions stated in Sec. 5. We see now that there are many such orthogonal matrices which will make $\Sigma y^2 = \Sigma z^2$. In our previous numerical example, $\theta$ is taken to be $53° \, 8'$. The reader may try any other value of $\theta$ to satisfy himself. The important point is that an orthogonal transformation in algebra is equivalent to a rotation of axes in geometry.

## 9. Three numbers

The above method and reasoning may be immediately extended to the case of more than two numbers. A numerical example for three numbers should suffice. (The reader may like to review the "direction cosines" in solid analytical geometry.) Suppose that the three original numbers are

$$y_1 = 9 \qquad y_2 = 18 \qquad y_3 = 6$$

so that $\Sigma y^2 = 441$. The new numbers are to be formed according to the rules (equations) $z_1 = a_1 y_1 + a_2 y_2 + a_3 y_3$, etc. Instead of writing out these three equations, we need only to specify the coefficients, for brevity, in the following matrix form:

$$\begin{pmatrix} a_1 & a_2 & a_3 \\ b_1 & b_2 & b_3 \\ c_1 & c_2 & c_3 \end{pmatrix} = \begin{pmatrix} \tfrac{2}{3} & \tfrac{2}{3} & \tfrac{1}{3} \\ \tfrac{2}{3} & -\tfrac{1}{3} & -\tfrac{2}{3} \\ -\tfrac{1}{3} & \tfrac{2}{3} & -\tfrac{2}{3} \end{pmatrix}$$

Note that the squares of the elements of each row add up to unity (for example, $a_1^2 + a_2^2 + a_3^2 = 1$) and that the sum of products of corresponding elements of any two rows is zero; for example,

$$a_1 b_1 + a_2 b_2 + a_3 b_3 = 0$$

So it is a normalized orthogonal matrix. To calculate $z_3$, for instance, we obtain

$$z_3 = c_1 y_1 + c_2 y_2 + c_3 y_3 = -\tfrac{1}{3}(9) + \tfrac{2}{3}(18) - \tfrac{2}{3}(6) = 5$$

The three new numbers thus obtained are

$$z_1 = 20 \qquad z_2 = -4 \qquad z_3 = 5$$

so that $\Sigma z^2 = 441 = \Sigma y^2$. In geometrical language, the distance from origin $O$ to the point $(9,18,6)$ relative to the original axes is the same as the distance from $O$ to the point $(20, -4, 5)$ relative to the rotated axes. In this example the distance is $\sqrt{441} = 21$.

## 10. Another transform

As an introduction to a particular type of transformation, let us transform the same three numbers ($y_1 = 9$, $y_2 = 18$, $y_3 = 6$) by the following orthogonal matrix:

$$\begin{pmatrix} 1/\sqrt{2} & 0 & -1/\sqrt{2} \\ 1/\sqrt{6} & -2/\sqrt{6} & 1/\sqrt{6} \\ 1/\sqrt{3} & 1/\sqrt{3} & 1/\sqrt{3} \end{pmatrix} = \begin{pmatrix} 0.70711 & 0 & -0.70711 \\ 0.40825 & -0.81650 & 0.40825 \\ 0.57735 & 0.57735 & 0.57735 \end{pmatrix}$$

The $z$ values are calculated as follows:

$$z_1 = \frac{1}{\sqrt{2}}(9 + 0 - 6) \qquad = \frac{1}{\sqrt{2}}(3) \qquad = \quad 2.11132$$

$$z_2 = \frac{1}{\sqrt{6}}(9 - 2 \times 18 + 6) = \frac{1}{\sqrt{6}}(-21) = -8.57321$$

$$z_3 = \frac{1}{\sqrt{3}}(9 + 18 + 6) \qquad = \frac{1}{\sqrt{3}}(33) \qquad = \quad 19.05256$$

The sum of squares of these $z$'s is then

$$z_1^2 + z_2^2 + z_3^2 = 4.5 + 73.5 + 363.0 = 441$$

Again, we have $\Sigma z^2 = \Sigma y^2$.

The above treatment of orthogonal matrices must suffice for our present purpose. We shall accept as the definition of a normalized orthogonal matrix of *any size* that the sum of squares of the elements in each row is unity and that the sum of products of corresponding elements of any two rows is zero. Such a transformation matrix will always make $\Sigma y^2 = \Sigma z^2$. The essential point to be remembered is that not only is it possible to change one set of numbers into another set without changing the sum of squares but there are many ways to do so.

## EXERCISES

**11.1**  Let $y_1 = 5$ and $y_2 = 10$, so that $y_1{}^2 + y_2{}^2 = 125$.  Transform the $y$'s into $z$'s according to the following equations:

$$z_1 = 0.96y_1 + 0.28y_2$$
$$z_2 = -0.28y_1 + 0.96y_2$$

Is this a normalized orthogonal transformation?  Obtain the numerical values of the $z$'s and see if $z_1{}^2 + z_2{}^2 = 125$.

**11.2**  Transform the three numbers $y_1 = 9$, $y_2 = 18$, $y_3 = 6$ by the following matrix:

$$\begin{pmatrix} 1/\sqrt{2} & -1/\sqrt{2} & 0 \\ 1/\sqrt{6} & 1/\sqrt{6} & -2/\sqrt{6} \\ 1/\sqrt{3} & 1/\sqrt{3} & 1/\sqrt{3} \end{pmatrix}$$

Obtain the numerical values of the $z$'s and see if $\Sigma z^2 = \Sigma y^2 = 441$.

### 11.3  Orthogonal transformation matrix

We want to transform the three numbers $y_1$, $y_2$, $y_3$ into another three numbers $z_1$, $z_2$, $z_3$ in such a way that the sum of their squares remains the same, viz.,

$$y_1{}^2 + y_2{}^2 + y_3{}^2 = z_1{}^2 + z_2{}^2 + z_3{}^2$$

The equations (rules) relating the two sets of numbers are

$$z_1 = a_1y_1 + a_2y_2 + a_3y_3$$
$$z_2 = b_1y_1 + b_2y_2 + b_3y_3$$
$$z_3 = c_1y_1 + c_2y_2 + c_3y_3$$

The transformation matrix is thus

$$T \equiv \begin{pmatrix} a_1 & a_2 & a_3 \\ b_1 & b_2 & b_3 \\ c_1 & c_2 & c_3 \end{pmatrix}$$

In most mathematical writings the elements of the matrix (that is, the coefficients of the original linear equations) are denoted by a single letter with double subscripts such as $a_{ij}$ for perfect generalization.  Here we use different letters for different rows so that the rows and columns are more easily distinguished for nonmathematical readers.  Substituting and squaring, we have

$$z_1{}^2 = a_1{}^2y_1{}^2 + a_2{}^2y_2{}^2 + a_3{}^2y_3{}^2 + 2a_1a_2y_1y_2 + 2a_1a_3y_1y_3 + 2a_2a_3y_2y_3$$
$$z_2{}^2 = b_1{}^2y_1{}^2 + b_2{}^2y_2{}^2 + b_3{}^2y_3{}^2 + 2b_1b_2y_1y_2 + 2b_1b_3y_1y_3 + 2b_2b_3y_2y_3$$
$$z_3{}^2 = c_1{}^2y_1{}^2 + c_2{}^2y_2{}^2 + c_3{}^2y_3{}^2 + 2c_1c_2y_1y_2 + 2c_1c_3y_1y_3 + 2c_2c_3y_2y_3$$

Summing,

$$\Sigma z^2 = y_1{}^2 + y_2{}^2 + y_3{}^2 + 0 + 0 + 0$$

In order that $\Sigma z^2 = \Sigma y^2$, the coefficients of each $y^2$ must be unity and the coefficients of each product term $y_iy_j$ must be zero.  Hence the conditions for the transformation equations are

$$a_1{}^2 + b_1{}^2 + c_1{}^2 = 1, \text{ etc.}$$
$$a_1a_2 + b_1b_2 + c_1c_2 = 0, \text{ etc.}$$

That is, the sum of the squares of the elements of each *column* of the matrix $T$ should be unity, and the sum of the products of corresponding elements of any two *columns* should be zero. This completes the first part of the proof.

The second part is to prove that the same conditions hold for the *rows* of the transformation matrix, and this is what we are really after. Now, multiplying the first equation by $a_1$, the second by $b_1$, and the third by $c_1$, we obtain

$$a_1 z_1 = a_1{}^2 y_1 + a_1 a_2 y_2 + a_1 a_3 y_3$$
$$b_1 z_2 = b_1{}^2 y_1 + b_1 b_2 y_2 + b_1 b_3 y_3$$
$$c_1 z_3 = c_1{}^2 y_1 + c_1 c_2 y_2 + c_1 c_3 y_3$$

Adding,

$$a_1 z_1 + b_1 z_2 + c_1 z_3 = (a_1{}^2 + b_1{}^2 + c_1{}^2) y_1 = y_1$$

because $a_1 a_3 + b_1 b_3 + c_1 c_3 = 0$, etc. Multiplying the original three equations by $a_2$, $b_2$, $c_2$, respectively, we obtain an equation for $y_2$ in terms of the $z$'s. Finally, multiplication of the original equations by $a_3$, $b_3$, $c_3$ gives us $y_3$ in terms of the $z$'s. The three new equations thus obtained are

$$y_1 = a_1 z_1 + b_1 z_2 + c_1 z_3$$
$$y_2 = a_2 z_1 + b_2 z_2 + c_2 z_3$$
$$y_3 = a_3 z_1 + b_3 z_2 + c_3 z_3$$

and the new transformation matrix is thus

$$T' = \begin{pmatrix} a_1 & b_1 & c_1 \\ a_2 & b_2 & c_2 \\ a_3 & b_3 & c_3 \end{pmatrix}$$

The only difference between the original $T$ and the present $T'$ is that the rows of one become the columns of the other.

If we regard the three new equations as original equations and apply the same procedure of substituting, squaring, and summing, we reach the same conclusions regarding the *columns* of the new matrix $T'$, viz.,

$$a_1{}^2 + a_2{}^2 + a_3{}^2 = 1, \text{ etc.}$$
$$a_1 b_1 + a_2 b_2 + a_3 b_3 = 0, \text{ etc.}$$

But these refer to the elements of the *rows* of the original matrix $T$. This completes the proof. Although the illustration uses only three numbers, the procedure of demonstration is obviously general and applicable to orthogonal transformations of any size. In later chapters we shall concentrate our attention on the rows of the original $T$.

# 12

# Orthogonal contrasts

Usually we compare two groups at a time. In this chapter we shall enlarge the usual concept of comparisons.

## 1. Special orthogonal matrix

The orthogonal transformation as presented in the preceding chapter is not yet directly applicable in statistical work in which the main interest is in the quantity $\Sigma(y - \bar{y})^2 = \Sigma y^2 - Y^2/n$ rather than in $\Sigma y^2$ itself. Hence we further restrict ourselves to a particular type of orthogonal matrix. If there are $n$ numbers under consideration, the restriction is that the $n$ elements of one of the rows (say, the last row) of the matrix should always be taken as

$$\frac{1}{\sqrt{n}}, \frac{1}{\sqrt{n}}, \cdot \cdot \cdot , \frac{1}{\sqrt{n}}$$

The reason for this will be made clear in subsequent sections.  Note that the squares of such elements add up to unity:

$$\frac{1}{n} + \frac{1}{n} + \cdots + \frac{1}{n} = 1$$

satisfying the condition of an orthogonal matrix for that row.  This restriction on the elements of one row (henceforward referred to as the last row) leads to restrictions on the elements of the other $n - 1$ rows. For instance, consider the last and one of the other rows of the matrix:

$$\begin{pmatrix} a_1 & a_2 & \cdots & a_n \\ \cdots\cdots\cdots\cdots\cdots\cdots\cdots \\ \dfrac{1}{\sqrt{n}} & \dfrac{1}{\sqrt{n}} & \cdots & \dfrac{1}{\sqrt{n}} \end{pmatrix}$$

Orthogonality requires that the sum of products of corresponding elements of any two rows be zero.  Since the last row here has a constant element, this is equivalent to requiring that the sum of the elements of the other row be zero, in addition to the usual requirement that their squares add up to unity.  Hence, in the particular type of orthogonal matrix we shall use, each row of elements (except those of the last row) should meet the following two conditions:

$$a_1 + a_2 + \cdots + a_n = 0$$
$$a_1{}^2 + a_2{}^2 + \cdots + a_n{}^2 = 1$$

Note that the change of sign of all elements in a row does not affect these requirements.

## 2. An example with four numbers

With the last row fixed, it is easy to construct the first row.  All one has to do is to write down $n$ numbers, some positive and some negative, so that they add up to zero and then divide each number by the square root of the sum of squares of those numbers.  This procedure guarantees that the squares of the resulting elements add up to unity.  In choosing the second row of elements, additional attention should be directed to the requirement that the sum of products of corresponding elements of any two rows is zero, and so on.  To clarify the details, let us consider a numerical example with four ($n = 4$) $y$ values:

$$y_1 = 9 \qquad y_2 = 6 \qquad y_3 = 2 \qquad y_4 = 3$$

for which $\Sigma y = Y = 20$, $\Sigma y^2 = 130$, and

$$\Sigma(y - \bar{y})^2 = \Sigma y^2 - \frac{Y^2}{n} = 130 - 100 = 30$$

as we calculated in Chap. 3. Now suppose that the orthogonal matrix (of coefficients of the linear equations) employed to build the $z$'s from the $y$'s is as follows:

$$\begin{pmatrix} \dfrac{2}{\sqrt{10}} & \dfrac{1}{\sqrt{10}} & -\dfrac{1}{\sqrt{10}} & -\dfrac{2}{\sqrt{10}} \\[2ex] \dfrac{1}{\sqrt{4}} & -\dfrac{1}{\sqrt{4}} & -\dfrac{1}{\sqrt{4}} & \dfrac{1}{\sqrt{4}} \\[2ex] \dfrac{1}{\sqrt{10}} & -\dfrac{2}{\sqrt{10}} & \dfrac{2}{\sqrt{10}} & -\dfrac{1}{\sqrt{10}} \\[2ex] \dfrac{1}{\sqrt{4}} & \dfrac{1}{\sqrt{4}} & \dfrac{1}{\sqrt{4}} & \dfrac{1}{\sqrt{4}} \end{pmatrix}$$

Before proceeding further, the reader should check that this is an orthogonal matrix. Note that the denominator of each element is the square root of the sum of squares of the numerators of the row in which it appears; thus, for the first row, the denominator is

$$\sqrt{(2)^2 + (1)^2 + (-1)^2 + (-2)^2} = \sqrt{10}$$

The $z$'s and their squares are calculated as follows:

$$z_1 = \frac{1}{\sqrt{10}}(2y_1 + y_2 - y_3 - 2y_4) = \frac{16}{\sqrt{10}} \qquad z_1{}^2 = \frac{256}{10} = 25.6$$

$$z_2 = \frac{1}{\sqrt{4}}(y_1 - y_2 - y_3 + y_4) = \frac{4}{\sqrt{4}} \qquad z_2{}^2 = \frac{16}{4} = 4.0$$

$$z_3 = \frac{1}{\sqrt{10}}(y_1 - 2y_2 + 2y_3 - y_4) = \frac{-2}{\sqrt{10}} \qquad z_3{}^2 = \frac{4}{10} = 0.4$$

$$z_4 = \frac{1}{\sqrt{4}}(y_1 + y_2 + y_3 + y_4) = \frac{20}{\sqrt{4}} \qquad z_4{}^2 = \frac{400}{4} = 100.0$$

and, as expected from our previous theorem,

$$z_1{}^2 + z_2{}^2 + z_3{}^2 + z_4{}^2 = 130 = y_1{}^2 + y_2{}^2 + y_3{}^2 + y_4{}^2$$

### 3. Orthogonal subdivision of $ssq$

Now let us examine what $z_4{}^2$ really is. The elements of the last row of the orthogonal matrix are so chosen that, replacing 4 by $n$ for generality,

$$z_n = \frac{1}{\sqrt{n}}(y_1 + \cdots + y_n) = \frac{Y}{\sqrt{n}} \qquad z_n{}^2 = \frac{Y^2}{n}$$

It is due to this fact that we conclude that

$$z_1{}^2 + z_2{}^2 + z_3{}^2 = \sum y^2 - z_4{}^2 = \sum y^2 - \frac{Y^2}{n} = \sum (y - \bar{y})^2$$

In our numerical example it is seen that $z_1{}^2 + z_2{}^2 + z_3{}^2 = 30$, yielding the *ssq* for the four numbers directly. Evidently, this method may be applied to any value of $n$. The general conclusion is that any *ssq* with $n - 1$ degrees of freedom can always be subdivided into $n - 1$ components: $z_1{}^2$, $z_2{}^2$, . . . , $z_{n-1}{}^2$, each with a single degree of freedom. In fact, it may be shown that if the original $y$'s are independent normally distributed variables with the same mean and variance, then these $n - 1$ $z$'s are also independent variables, each being normally distributed with zero mean and the same variance $\sigma^2$ as that of $y$. This is one of the most useful theorems in experimental statistics.

## 4. Practical calculation

With this much background on the orthogonal subdivision of an *ssq*, we are now in a position to split the treatment *ssq* with $k - 1$ degrees of freedom into $k - 1$ components, each with a single degree of freedom. For practical calculations, many of the intermediate steps shown in the preceding example may be omitted, and a routine arithmetic procedure may easily be set up. For instance, the last row of the orthogonal matrix is always taken for granted and need not be written out except in theoretical work. Likewise, the denominator of each element need not be calculated as such, because, on squaring the $z$'s, the denominator of a $z^2$ is simply

$$D = a_1{}^2 + a_2{}^2 + a_3{}^2 + \cdots$$

for the first row, and so on. A routine arithmetic procedure of subdividing the treatment *ssq* is given in the following example taken from Chap. 10. Let us assume that the three rows of Table 10.1 are the three replications ($r = 3$) and that the four columns are four treatments ($k = 4$). Since we shall deal with treatment *ssq* only, the relevant data are as given in the accompanying table, the order of the columns of which has been changed to suit the present purpose. The sum of squares

| Row | Four treatments | | | |
|:---:|:---:|:---:|:---:|:---:|
| 1 | * | * | * | * |
| 2 | * | * | * | * |
| 3 | * | * | * | * |
| Total | 15 | 18 | 21 | 30 |

due to treatments, as we have found in Chap. 10, is

$$ssq_K = r \sum_j (\bar{y}_{\cdot j} - \bar{y})^2 = r \sum_j t_j^2 = 42$$

Since $ssq_K$ is $r$ times the $ssq$ of the $k$ treatment *means*, it is $1/r$ of the $ssq$ of the $k$ treatment *totals*. The practical computation is always based on totals rather than means.

### Example 1. Two drugs

For concreteness and simplicity we shall assume as the first example that two of the treatments are $S_1$ and $S_2$ (two sulfa drugs, say) and the remaining two are $A_1$ and $A_2$ (two antibiotics, say). Then the subdivision of the treatment $ssq$ given in Table 12.1 has a clear-cut physical meaning. The first comparison is $S_2 - S_1 = 18 - 15 = 3$, which is the total difference between the two sulfa treatments. This has been denoted by $Z$ in the table. The divisor is $D$, the sum of squares of the coefficients used in the comparison, multiplied by the number of replications $r$ in the experiment. Thus, for the first comparison, the divisor is $D \times r = (1^2 + 1^2) \times 3 = 6$. Then the quantity $Z^2/Dr = z^2$ is the $ssq$ for that comparison with a single degree of freedom. This may be tested against the error mean square ($s^2 = 3.67$ of Table 10.3), yielding a value of $F = t^2$. The meaning of the remaining two contrasts should be clear. As a check of the arithmetic involved, we may note that

$$z_1^2 + z_2^2 + z_3^2 = 1.50 + 13.50 + 27.00 = 42 = ssq_K$$

Such a set of three $(k - 1)$ comparisons is known as an *orthogonal set of contrasts*. The significance of each contrast may be tested separately by using the $F$ or $t$ table.

### Example 2. Placebo and drugs

Next, suppose that the first treatment is a placebo, the second and third are two sulfa drugs ($S_1$ and $S_2$), and the fourth is an antibiotic $A$. Then the meaning of the orthogonal contrasts given in the second part of Table 12.1 is obvious. The first contrast is placebo $P_0$ versus the other three drug treatments, and so on. For this contrast

$$\text{Divisor} = (3^2 + 1^2 + 1^2 + 1^2) \times 3 = 12 \times 3 = 36$$

It is seen that the greatest variation among the treatments is due to the difference between the antibiotic $A$ and the two $S$ drugs. The test of significance for each contrast is carried out as before.

*Table* 12.1   Four examples of an orthogonal set by which the treatment *ssq* is subdivided into three components, each with a single degree of freedom

| Example 1 | Treatment totals | | | | Effect $Z$ | Effect² $Z^2$ | Divisor $D \times r$ | $Z^2/Dr = z^2$ |
|---|---|---|---|---|---|---|---|---|
| | 15 $S_1$ | 18 $S_2$ | 21 $A_1$ | 30 $A_2$ | | | | |
| $S_2$ vs. $S_1$ | $-1$ | $+1$ | 0 | 0 | 3 | 9 | $2 \times 3$ | 1.50 |
| $A_2$ vs. $A_1$ | 0 | 0 | $-1$ | $+1$ | 9 | 81 | $2 \times 3$ | 13.50 |
| $A$ vs. $S$ | $-1$ | $-1$ | $+1$ | $+1$ | 18 | 324 | $4 \times 3$ | 27.00 |
| | | | | | | Total treatment *ssq* | | 42.00 |

| Example 2 | $P_0$ | $S_1$ | $S_2$ | $A$ | $Z$ | $Z^2$ | $D \times r$ | $z^2$ |
|---|---|---|---|---|---|---|---|---|
| $P_0$ vs. others | $-3$ | $+1$ | $+1$ | $+1$ | 24 | 576 | $12 \times 3$ | 16.00 |
| $A$ vs. $S$ | 0 | $-1$ | $-1$ | $+2$ | 21 | 441 | $6 \times 3$ | 24.50 |
| $S_2$ vs. $S_1$ | 0 | $-1$ | $+1$ | 0 | 3 | 9 | $2 \times 3$ | 1.50 |
| | | | | | | Total treatment *ssq* | | 42.00 |

| Example 3 | $a_1b_1$ | $a_1b_2$ | $a_2b_1$ | $a_2b_2$ | $Z$ | $Z^2$ | $D \times r$ | $z^2$ |
|---|---|---|---|---|---|---|---|---|
| $a_2$ vs. $a_1$ | $-1$ | $-1$ | $+1$ | $+1$ | 18 | 324 | $4 \times 3$ | 27.00 |
| $b_2$ vs. $b_1$ | $-1$ | $+1$ | $-1$ | $+1$ | 12 | 144 | $4 \times 3$ | 12.00 |
| Interaction | $+1$ | $-1$ | $-1$ | $+1$ | 6 | 36 | $4 \times 3$ | 3.00 |
| | | | | | | Total treatment *ssq* | | 42.00 |

| Example 4 | $a_0$ | $a_1$ | $a_2$ | $a_3$ | $Z$ | $Z^2$ | $D \times r$ | $z^2$ |
|---|---|---|---|---|---|---|---|---|
| Linear | $-3$ | $-1$ | $+1$ | $+3$ | 48 | 2,304 | $20 \times 3$ | 38.40 |
| Quadratic | $+1$ | $-1$ | $-1$ | $+1$ | 6 | 36 | $4 \times 3$ | 3.00 |
| Cubic | $-1$ | $+3$ | $-3$ | $+1$ | 6 | 36 | $20 \times 3$ | 0.60 |
| | | | | | | Total treatment *ssq* | | 42.00 |

### Example 3. Simplest factorial

Experimental designs in which the treatments involve various combinations of factors or ingredients will be dealt with in Part 2 of this book. Here we shall merely mention the simplest example of that type to illustrate still another way of subdividing the (treatment) sum of squares. Suppose that $a$ is a drug (think of aspirin, if you dislike generalities), administered at two levels $a_1$ and $a_2$ (for example, one tablet and two tablets) and that $b$ is another drug (Bufferin) also administered at two levels $b_1$ and $b_2$. The four treatments are then the four combinations of the two ingredients:

$a_1b_1$ = one tablet of aspirin and one tablet of Bufferin
$a_2b_1$ = two tablets of aspirin and one tablet of Bufferin
$a_1b_2$ = one tablet of aspirin and two tablets of Bufferin
$a_2b_2$ = two tablets of aspirin and two tablets of Bufferin

We say that this is an experiment involving two *factors*, each at two levels, abbreviated as a $2 \times 2$ experiment. The first contrast in Table 12.1 is treatment combinations containing $a_2$ versus those containing $a_1$. Similarly, the second contrast is those with $b_2$ versus those with $b_1$. These are called the *main effects*, or the average effects of factor $a$ and factor $b$, respectively. To see the meaning of the third contrast, we may reason as follows:

$$(a_2b_1 - a_1b_1) = \text{effect of } a, \text{ given } b_1 \text{ level}$$
$$(a_2b_2 - a_1b_2) = \text{effect of } a, \text{ given } b_2 \text{ level}$$

If the effect of factor $a$ is not influenced by the levels of factor $b$, the above two effects should be equal. Conversely, if the effect of $a$ varies with the levels of factor $b$, we say that there is "interaction" between the two factors. The interaction is defined as the difference between the above two effects:

$$\text{Interaction} = (a_2b_2 - a_1b_2) - (a_2b_1 - a_1b_1)$$
$$= a_2b_2 - a_1b_2 - a_2b_1 + a_1b_1$$

which is the third contrast in Table 12.1. If we were to proceed with the effects of $b$ at given $a$ levels, we would end up with the same expression for interaction, because the $2 \times 2$ experiment is symmetrical with respect to the two factors $a$ and $b$. Note that the three contrasts—two for main effects and one for interaction—are orthogonal to each other.

### Example 4. Four levels at equal intervals

Finally, let us suppose that the four treatments are four levels, at equal intervals, of a drug. For instance, you may take a pill once a day, twice a day, three times a day, or four times a day. Then the first contrast in the bottom portion of Table 12.1 gives us the "linear" component of the effect of the drug. The coefficients of the linear effect are (proportional to) the deviations of the levels from the mean level. Thus, writing the four levels as 0, 1, 2, 3 with mean level $(0 + 1 + 2 + 3)/4 = 1.5$, we see that the deviations of the levels are

|  | $(0 - 1.5)$ | $(1 - 1.5)$ | $(2 - 1.5)$ | $(3 - 1.5)$ |
|---|---|---|---|---|
| or | $-1.5$ | $-0.5$ | $0.5$ | $1.5$ |
| or | $-3$ | $-1$ | $1$ | $3$ |

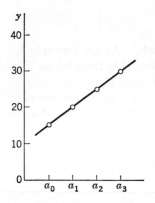

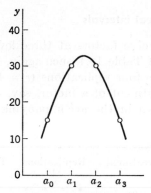

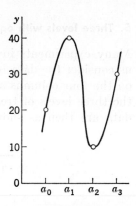

Fig. 12.1 Effects all linear. Other components zero.

Fig. 12.2 Effects all quadratic. Other components zero.

Fig. 12.3 Effects all cubic. Other components zero.

These are not only in arithmetic progression but also would add up to zero. If the treatment results were perfectly linear, for example,

$$15 \quad 20 \quad 25 \quad 30$$

for $\quad\quad a_0 \quad a_1 \quad a_2 \quad a_3$

(see Fig. 12.1) then the total *ssq* with three degrees of freedom for the four treatments would be given by this single linear component, and the other two components would be zero. On the other hand, if the four treatment results take a quadratic form for example,

$$15 \quad 30 \quad 30 \quad 15$$

for $\quad\quad a_0 \quad a_1 \quad a_2 \quad a_3$

(see Fig. 12.2) then the total *ssq* with three degrees of freedom for treatments will be concentrated into this single quadratic component, and the other two components will be zero. Finally, if the results for the four levels of dosage are of the type

$$20 \quad 40 \quad 10 \quad 30$$

(Fig. 12.3) the reader may verify that the treatment *ssq* will be equal to the single cubic component, and the other two components are zero. In our particular numerical example, with error mean square $s^2 = 3.67$, only the linear component is significant, while the quadratic and cubic effects are quite trivial. In many practical cases, especially when there are more than four levels, it is unnecessary to isolate all the single components, and it is sufficient to isolate out only the linear component and leave the rest as nonlinear component.

### 5. Three levels with equal intervals

Many experiments involve factors at three levels. As an example let us consider the data of Table 10.1 once again, but this time let us look on the four columns as four replications ($r = 4$) and the three rows as the three levels of treatment of a factor, say, $a_1$, $a_2$, $a_3$. The relevant data are then as shown in the accompanying table. The *ssq* due to

| Treatment | Replications | Total |
|-----------|:------------:|:-----:|
| $a_1$ | * * * * | 20 |
| $a_2$ | * * * * | 24 |
| $a_3$ | * * * * | 40 |

treatments is, as we have already seen in Chap. 10,

$$ssq_R = \frac{20^2 + 24^2 + 40^2}{4} - \frac{84^2}{12} = 56$$

with two degrees of freedom. If the three levels are at equal intervals, then the two orthogonal contrasts corresponding to the linear and quadratic effects and the subdivision of the treatment *ssq* into two single components are as shown in Table 12.2. The procedure of calculation is exactly the same as before and needs no explanation. It may be noted that if the results were perfectly linear (for example, 20, 30, 40 for the three levels), the treatment *ssq* would be equal to the linear component and the quadratic component would be zero. On the other hand, if the results were of the type 20, 40, 20, the linear component would be zero and the treatment *ssq* would be concentrated into the quadratic component alone. Further discussions will be found in Chap. 22.

*Table 12.2* The subdivision of treatment *ssq* into single components when the treatments are of three levels at equal intervals

| Treatment level: | (1) (2) (3) | $Z$ | $Z^2$ | $Dr$ | $z^2$ |
|---|---|---|---|---|---|
| Treatment total: | 20   24   40 | | | | |
| Linear | $-1$   $0$   $+1$ | 20 | 400 | $2 \times 4$ | 50 |
| Quadratic | $-1$   $+2$   $-1$ | $-12$ | 144 | $6 \times 4$ | 6 |
| | | | Treatment *ssq* with 2 *df* | | 56 |

*Table* 12.3 Orthogonal contrasts for three levels with unequal intervals

| | Levels | | | Levels | | | Levels | | |
|---|---|---|---|---|---|---|---|---|---|
| | (1) | (2) | (4) | (1) | (2) | (5) | (1) | (5) | (10) |
| Linear | −4 | −1 | +5 | −5 | −2 | +7 | −13 | −1 | +14 |
| Quadratic | −2 | +3 | −1 | −3 | +4 | −1 | −5 | +9 | −4 |

| | Levels | | | Levels | | | Levels | | |
|---|---|---|---|---|---|---|---|---|---|
| | (0) | (1) | (5) | (0) | (2) | (5) | (2) | (5) | (10) |
| Linear | −2 | −1 | +3 | −7 | −1 | +8 | −11 | −2 | +13 |
| Quadratic | −4 | +5 | −1 | −3 | +5 | −2 | −5 | +8 | −3 |

## 6. Unequal intervals

The levels of a factor, however, need not be at equal intervals. The three levels employed in many experiments are of the type 1, 2, 4; 1, 2, 5; or 1, 5, 10. The coefficients of the linear contrast for such cases may be found in the same manner as explained before. They are proportional to the deviation of the levels from the mean level. Once the linear contrast is written down, the remaining contrast is then determined. Some of the most frequently encountered unequal intervals of three levels are listed in Table 12.3.

Similarly, a factor administered at four levels is often not at equal intervals. In Table 12.4 are listed some of the common unequal intervals for four levels. In reading the levels, it should be remembered that 0, 1, 2, 4, for instance, means the same as 0, 5, 10, 20, or any other multiples of the same four numbers. It could also be taken to mean 1, 2, 3, 5, the shifting of origin being irrelevant. As mentioned earlier, for most practical purposes we need only isolate out the linear component and pool the rest as the nonlinear component.

## 7. Unequal number of replicates

The number of replications for each treatment likewise need not be equal. As an example of unequal numbers of replications, let us consider the data of Table 5.1 and assume that the three groups are the treatments of a factor at three levels with equal intervals. If $a_1$, $a_2$, $a_3$ are the coefficients of a contrast, then their weighted (by $r$, number of

*Table* 12.4   Orthogonal contrasts for four levels with unequal intervals

| Contrasts | Levels | | | | Levels | | | |
|---|---|---|---|---|---|---|---|---|
| | (0) | (1) | (2) | (4) | (1) | (2) | (4) | (5) |
| Linear | − 7 | − 3 | + 1 | + 9 | − 2 | − 1 | + 1 | + 2 |
| Quadratic | + 7 | − 4 | − 8 | + 5 | + 1 | − 1 | − 1 | + 1 |
| Cubic | − 3 | + 8 | − 6 | + 1 | − 1 | + 2 | − 2 | + 1 |
| | (0) | (1) | (2) | (5) | (1) | (2) | (4) | (8) |
| Linear | − 2 | − 1 | 0 | + 3 | −11 | − 7 | + 1 | +17 |
| Quadratic | +43 | −17 | −49 | +23 | +20 | − 4 | −29 | +13 |
| Cubic | − 6 | +15 | −10 | + 1 | − 8 | +14 | − 7 | + 1 |
| | (0) | (1) | (3) | (6) | (1) | (2) | (5) | (8) |
| Linear | − 5 | − 3 | + 1 | + 7 | − 3 | − 2 | + 1 | + 4 |
| Quadratic | + 9 | − 3 | −13 | + 7 | + 3 | − 1 | − 5 | + 3 |
| Cubic | − 5 | + 9 | − 5 | + 1 | − 9 | +14 | − 7 | + 2 |
| | (0) | (1) | (5) | (10) | (1) | (2) | (5) | (10) |
| Linear | − 4 | − 3 | + 1 | + 6 | − 7 | − 5 | + 1 | +11 |
| Quadratic | +107 | − 5 | −205 | +203 | +63 | − 4 | −107 | +48 |
| Cubic | − 18 | +25 | − 9 | + 2 | −10 | +15 | − 6 | + 1 |

replications) sum is zero:

$$a_1 r_1 + a_2 r_2 + a_3 r_3 = 0$$

Another contrast with coefficients $b_1$, $b_2$, $b_3$ is orthogonal to the previous one when the sum of the weighted products of the corresponding coefficients is zero:

$$a_1 b_1 r_1 + a_2 b_2 r_2 + a_3 b_3 r_3 = 0$$

The divisor is modified accordingly:

$$\text{Divisor} = a_1{}^2 r_1 + a_2{}^2 r_2 + a_3{}^2 r_3$$

The coefficients of the linear contrast are proportional to the deviations of the levels from the weighted mean level. The relevant data (from Table 5.1) are as shown in the accompanying table. The *ssq* due to the

|  | Replications | $r_i$ | $Y_i$ |
|---|---|---|---|
| Three treatments | * * * | 3 | 18 |
|  | * * * * * * | 6 | 54 |
|  | * * * | 3 | 12 |

three treatments is, as we have seen in Chap. 5,

$$ssq_B = \frac{18^2}{3} + \frac{54^2}{6} + \frac{12^2}{3} - \frac{84^2}{12} = 54$$

with two degrees of freedom. Now we wish to subdivide the $ssq_B = 54$ into a linear component and a quadratic component, each with a single degree of freedom. In Table 12.5 are given two examples: the first, assuming that the treatment totals 18, 54, 12 correspond to treatment levels (1), (2), (3,) respectively, shows that the nonlinear component is large. The second example, assuming that the treatment totals 12, 18, 54 correspond to treatment levels (1), (2), (3), respectively, shows that the linear component is predominant.

## 8. Variance of a contrast

If $s^2$ is the estimated variance of the single observations ($y$'s), then the variance of the contrast (difference) $y_1 - y_2$ is $s^2(1^2 + 1^2) - 2s^2$. Similarly, the variance of the contrast $y_1 + y_2 - 2y_3$ is $s^2(1^2 + 1^2 + 2^2) = 6s^2$.

**Table 12.5** Orthogonal contrasts with unequal number of replications

| Treatment level: | (1) | (2) | (3) | First example | | | |
|---|---|---|---|---|---|---|---|
| Treatment total: | 18 | 54 | 12 | | | | |
| Replication $r_i$: | 3 | 6 | 3 | $Z$ | $Z^2$ | Divisor | $z^2$ |
| Linear | $-1$ | 0 | 1 | $-6$ | 36 | 6 | 6.00 |
| Quadratic | $-1$ | 1 | $-1$ | 24 | 576 | 12 | 48.00 |
|  |  |  |  | Total treatment $ssq$ | | | 54.00 |

| Treatment level: | (1) | (2) | (3) | Second example | | | |
|---|---|---|---|---|---|---|---|
| Treatment total: | 12 | 18 | 54 | | | | |
| Replication $r_i$: | 3 | 3 | 6 | $Z$ | $Z^2$ | Divisor | $z^2$ |
| Linear | $-5$ | $-1$ | 3 | 84 | 7,056 | 132 | 54.45 |
| Quadratic | $-2$ | 4 | $-1$ | 6 | 36 | 66 | 0.54 |
|  |  |  |  | Total treatment $ssq$ | | | 54.00 |

Generally speaking, the variance of the contrast $l_1 y_1 + \cdots + l_n y_n$ among the single observations is

$$V(l_1 y_1 + \cdots + l_n y_n) = s^2(l_1{}^2 + \cdots + l_n{}^2)$$

If the contrast is among the $k$ group means $\bar{y}_i$ and each mean is determined by the same number $r$ of single observations, the variance is

$$V(l_1 \bar{y}_1 + \cdots + l_k \bar{y}_k) = \frac{s^2}{r}(l_1{}^2 + \cdots + l_k{}^2)$$

It follows that the variance of the difference between two group means is

$$V(\bar{y}_1 - \bar{y}_2) = \frac{2s^2}{r}$$

If the size of the group (number of replications) is $r_i$ for the $i$th group, the variance of the contrast among the group means is

$$V(l_1 \bar{y}_1 + \cdots + l_k \bar{y}_k) = s^2 \left( \frac{l_1{}^2}{r_1} + \cdots + \frac{l_k{}^2}{r_k} \right)$$

In particular, for the difference between two group means,

$$V(\bar{y}_1 - \bar{y}_2) = s^2 \left( \frac{1}{r_1} + \frac{1}{r_2} \right)$$

## 9. Hypotheses testing and orthogonal contrasts

In closing this chapter we shall attend briefly to the meaning of testing the significance of each single contrast. With four treatments, the general null hypothesis is $H_0$: $\mu_1 = \mu_2 = \mu_3 = \mu_4$. To use an orthogonal set of three comparisons is to subdivide the general null hypothesis into three independent and more specific sub-null hypotheses; for instance, in example 1 of Table 12.1 the three subhypotheses to be tested are

$$
\begin{aligned}
H_1&: & \mu_1 &= \mu_2 \\
H_2&: & \mu_3 &= \mu_4 \\
H_3&: & \mu_1 + \mu_2 &= \mu_3 + \mu_4
\end{aligned}
$$

It should be noted that only when all three subhypotheses are true, the general $H_0$ is true. Therefore, $H_1$, $H_2$, and $H_3$ may be considered as parts of $H_0$. In practical applications there frequently arises the question as to when the orthogonal comparisons should be tested. Some maintain that if the total treatment effects with three degrees of freedom are nonsignificant, the analysis should stop there and the orthogonal comparisons should not be employed, because the original null hypothesis $H_0$ has not been discarded. But one may argue that $H_1$, $H_2$, and $H_3$ *are* the original three hypotheses to be tested (and there has never been any general $H_0$ to begin with). Therefore, they should be tested, regardless of the total treatment effects with three degrees of freedom. The author agrees with the latter point of view on condition that the orthogonal comparisons to be tested are formulated beforehand—before the experimental results become known and not after the examination of the data tempts one to pick out the largest and smallest for comparison. Since many orthogonal sets of comparisons are available, the experimenter must choose the one which is most meaningful to him, and the meaningfulness depends largely on the physical setup of the treatments.

To summarize, the general $F$ test is for the hypothesis

$$H_0: \qquad \mu_1 = \mu_2 = \mu_3 = \mu_4$$

The individual $t$ test is for a particular comparison which is a part of $H_0$. The reader may refer to Chap. 31 for more discussions and for comparisons after examination of data.

## EXERCISES

**12.1** Given the four numbers $y_1 = 2$, $y_2 = 3$, $y_3 = 6$, $y_4 = 9$. Subdivide the $ssq = 30$ into three orthogonal components according to each of the following schemes:

| Scheme reference | $y_1$ | $y_2$ | $y_3$ | $y_4$ |
|---|---|---|---|---|
| Example 1 | $A_1$ | $A_2$ | $B_1$ | $B_2$ |
| Example 2† | $(0)$ | $A$ | $B_1$ | $B_2$ |
| Example 3 | $a_1b_1$ | $a_1b_2$ | $a_2b_1$ | $a_2b_2$ |
| Example 4 | $a_0$ | $a_1$ | $a_2$ | $a_3$ |

† Note carefully the order of arrangement.

**12.2** In Table 12.4 are listed eight examples of unequal intervals for four levels of a factor. For levels 1, 2, 4, 5, in which the middle level is missing, the various contrasts are actually those of the matrix in Sec. 2. Again, let 15, 18, 21, 30 be the treatment totals of $r = 3$ replications. Isolate the linear, quadratic, and cubic components of the treatment $ssq$ according to the scheme in the accompanying table.

| Treatment level: | (1) | (2) | (4) | (5) | $Z$ | $Z^2$ | $D \times r$ | $z^2$ |
|---|---|---|---|---|---|---|---|---|
| Treatment total: | 15 | 18 | 21 | 30 | | | | |
| Linear | $-2$ | $-1$ | $+1$ | $+2$ | | | | |
| Quadratic | $+1$ | $-1$ | $-1$ | $+1$ | | | | |
| Cubic | $+1$ | $-2$ | $+2$ | $-1$ | | | | |
| | | | | | | | $ssq$ due to treatment | $42.00$ |

**12.3** Suppose that 15, 18, 21, 30 are the treatment totals corresponding to levels (1), (2), (3), (4), with replication numbers 5, 3, 3, 5, respectively. Calculate the linear, quadratic, and cubic components of treatment *ssq* according to the contrasts in the accompanying table.

| Treatment levels: | (1) | (2) | (3) | (4) | $Z$ | $Z^2$ | *Div* | $z^2$ |
|---|---|---|---|---|---|---|---|---|
| Treatment totals: | 15 | 18 | 21 | 30 | | | | |
| Replication $r_i$: | 5 | 3 | 3 | 5 | | | | |
| Linear | −3 | −1 | +1 | +3 | | | | |
| Quadratic | +3 | −5 | −5 | +3 | | | | |
| Cubic | −1 | +5 | −5 | +1 | | | | |

Total treatment *ssq*

First check that each set of coefficients does provide a contrast. For instance, for the first row we have

$$(-3)5 + (-1)3 + (1)3 + (3)5 = 0$$

Next check that any two of the three rows are orthogonal; for the first two rows we have

$$(3)(-3)5 + (-5)(-1)3 + (-5)(1)3 + (3)(3)5 = 0$$

The $Z$ value for the first (linear) contrast is

$$Z_1 = (-3)15 + (-1)18 + (1)21 + (3)30 = 48$$

The divisor for the third row is

$$D_3 = (-1)^2 5 + (5)^2 3 + (-5)^2 3 + (1)^2 5 = 160$$

Finally, check your answer by observing that the sum of the three $z^2$ should be equal to the total treatment *ssq* with 3 *df*, viz.,

$$\frac{15^2}{5} + \frac{18^2}{3} + \frac{21^2}{3} + \frac{30^2}{5} - \frac{84^2}{16}$$

# 13

# Linear regression

Since every textbook of elementary statistics has a section on linear regression, we shall here review only those aspects of the subject that pave the way for the next chapter, which deals with the analysis of covariance.

## 1. An identity for the sum of products

Suppose that for each $y$ value we observe there is associated an $x$ value, so that we have pairs of values $(x,y)$. The total, mean, deviations, and the sum of squares of deviations may be calculated separately for $x$ and $y$ by the usual method. Now, a new quantity plays a cardinal role in linear regression and correlation analysis, and it is the sum of products of deviations of $x$ and $y$. Table 13.1 provides a numerical example for four pairs of $(x,y)$. Analogously to the identity concerning the sum of squares of deviations, the following identity enables us to avoid the

*Table* 13.1   The sum of products of deviations

| $x$ | $y$ | $(x - \bar{x})(y - \bar{y})$ | $xy$ |
|:---:|:---:|:---:|:---:|
| 3 | 2 | $(0)(-3) \ = \ 0$ | 6 |
| 2 | 3 | $(-1)(-2) \ = \ 2$ | 6 |
| 2 | 5 | $(-1)(0) \ = \ 0$ | 10 |
| 5 | 10 | $(2)(5) \ = \ 10$ | 50 |
| $X = 12$ | $Y = 20$ | $\Sigma(x - \bar{x})(y - \bar{y}) = 12$ | $\Sigma xy = 72$ |
| $\bar{x} = 3$ | $\bar{y} = 5$ | | $n\bar{x}\bar{y} = 60$ |

arithmetic of finding the deviations:

$$\sum (x - \bar{x})(y - \bar{y}) = \sum (xy - x\bar{y} - y\bar{x} + \bar{x}\bar{y})$$
$$= \sum xy - \bar{y}\sum x - \bar{x}\sum y + n\bar{x}\bar{y}$$
$$= \sum xy - n\bar{x}\bar{y} = \sum xy - \frac{XY}{n}$$

since $\bar{y}\Sigma x = \bar{x}\Sigma y = n\bar{x}\bar{y}$.   Note that if $x$ is replaced by $y$, it becomes the ordinary identity for the sum of squares.

## 2. Geometrical presentation

Each single product term $xy$ may be thought of as a rectangle, and $\Sigma xy$ is the total area of such rectangles (Fig. 13.1).   The "correction" term $n\bar{x}\bar{y}$ is the total area of $n$ rectangles, each of the size $\bar{x}\bar{y}$.   The diagrammatic representation of this identity is a generalization of that for the sum of squares.   The only—but important—difference is that the "area" $n\bar{x}\bar{y}$ may be larger than $\Sigma xy$, so that the sum of products of deviations may be a negative quantity, as shown by the bottom diagram of Fig. 13.1.   For practical calculation, the procedure is similar to that for calculating the sum of squares:

$$A = \Sigma xy \qquad C = \frac{XY}{n}$$

and the sum of products is then $A - C$.   For the numerical example in Table 13.1, the sum of products (of deviations of $x$ and $y$) is $72 - 60 = 12$.

## 3. Scatter diagram and regression line

Each pair of values $(x,y)$ may be plotted as a point on the $xy$ plane. Thus, the four pairs of values in Table 13.1 may be represented as four points shown in Fig. 13.2.   Generally, if there are $n$ pairs of $(x,y)$, there

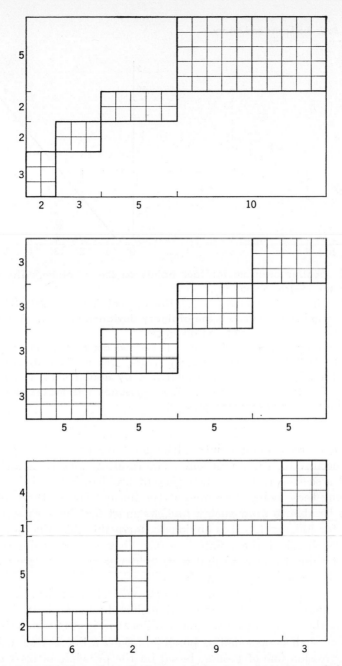

Fig. 13.1 Visual aid to the calculation of the sum of products of deviations of $x$ and $y$. The area at the top is $A = \Sigma xy$; the area in the middle is $C = n\bar{x}\bar{y} = XY/n$; and the required sum of products is the difference $A - C$. The bottom diagram shows the area of another four pairs of $(x,y)$ with the same $\bar{x}$ and $\bar{y}$. The sum of products for these four pairs is $A - C = 43 - 60 = -17$. [Modified from C. C. Li, *Journal of the American Statistical Association*, 50:1056–1063 (1955).]

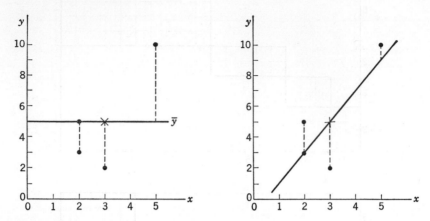

**Fig. 13.2**  Scatter diagram for four points on the $xy$ plane.  The cross $\times$ indicates the central point $(\bar{x}, \bar{y})$.

Left: The horizontal line is $y = \bar{y}$.  The vertical distance (dotted line) from a point to the horizontal line is the ordinary deviation $(y - \bar{y})$.  The ordinary *ssq* is the sum of squares of these distances.

Right: The regression line $y' = 2x - 1$ is drawn among the same four points. Note that this line also passes through $(\bar{x}, \bar{y})$.  The vertical distance from a point to the regression line, also indicated by dotted lines, is the residue deviation denoted by $e = y - y'$.  The regression line makes the quantity $\Sigma e^2$ smaller than any other straight line.

will be $n$ corresponding points, although some of them may overlap, that is, occupy the same position.  The resulting picture of such points is called a scatter diagram.  Our present problem is to "fit" a straight line among these points to represent the linear trend of the points.  In order to be able to draw such a line, we must first have some criterion or basis for judgment.  It is obviously impossible that the straight line pass through all of the points, because they are not collinear.  The next best thing to do is to find a straight line that is "closest" to the points.  But what do we mean by "closest"?  It is simply a matter of definition.  Largely for the sake of mathematical simplicity, the criterion we shall adopt is that the straight line should be such that the sum of squares of the *vertical* distances (dashed lines in Fig. 13.2) from the points to the line shall be as small as possible.  Such a line is then called the linear regression line of $y$ on $x$, based on the principle of least squares. The following sections explain how such a line (that is, the equation for the straight line) may be found.

From such a line, the expected value of $y$ on the basis of linearity, may be calculated for a given value of $x$.  This $y$ value, situated on the line, is usually denoted by $y'$ or $y_x$, or, more formally, $E(y|x)$.

In order to facilitate our appreciation of what the linear regression line does, we may first examine the meaning of the ordinary $ssq = \Sigma(y - \bar{y})^2$ in terms of the scatter diagram (Fig. 13.2). If we draw a horizontal line $y = \bar{y}$, then each of the ordinary deviations $y - \bar{y}$ represents a vertical distance from a point to the horizontal line. The sum of squares of such distances is the ordinary $ssq$ we have been using in preceding chapters.

If we also draw a vertical line at $x = \bar{x}$, it will cross the horizontal line at the point $(\bar{x}, \bar{y})$, the centroid of the $n$ points. Now, let us rotate the horizontal line in accordance with the linear trend of the points, using $(\bar{x}, \bar{y})$ as the center of rotation. For each position of the rotating line, there will be a corresponding sum of squares of the vertical distance from the points to the line. There will be a particular position for which the sum of squares of the vertical distances from the points to the line is the smallest. Such a position is shown in the right-hand diagram of Fig. 13.2 for the four points in our example. It is clear that the sum of squares of distances from the regression line is always smaller than the ordinary $ssq$, if there is any linear trend at all. In the following we shall not only outline the method of finding the slope of such a line but also show the relationship between the two sums of squares.

## 4. The linear model

Precisely as before, we shall assume that each $y$ value is made up of three components. The first component is a constant which represents the mean of the $y$ population. The second component is due to regression on $x$; that is, it represents a portion of $y$ which depends upon the value of its corresponding $x$. The third component is error, which is normally distributed with mean zero and variance $\sigma^2$. The explicit algebraic expression for this model is

$$y_\alpha = \mu + \beta(x_\alpha - \bar{x}) + \epsilon_\alpha$$

where $\beta$ is known as the regression coefficient of $y$ on $x$. In this model we consider $x$ as a given value, not a random variable. Our problem is to estimate the two parameters, $\mu$ and $\beta$, on the basis of a set of data which constitutes the sample.

## 5. Estimation of parameters

The method of estimation is also the same as before, namely, the method of least squares. The quantity to be minimized here is

$$Q = \Sigma(e_\alpha{}^2) = \Sigma[y_\alpha - u - b(x_\alpha - \bar{x})]^2$$

Equating the partial derivatives of $Q$ with respect to $u$ and $b$ to zero and dropping the subscript $\alpha$ for simplicity, we obtain the following two "normal equations":

$$\Sigma[y - u - b(x - \bar{x})] = 0$$
$$\Sigma[y - u - b(x - \bar{x})](x - \bar{x}) = 0$$

Since $\Sigma(x - \bar{x}) = 0$, the first equation yields $\Sigma y = nu$; that is, the estimate of $\mu$ is $u = \bar{y}$. Replacing $u$ by $\bar{y}$ in the second equation, we obtain

$$\Sigma(y - \bar{y})(x - \bar{x}) - b\Sigma(x - \bar{x})^2 = 0$$

and therefore

$$b = \frac{\Sigma(y - \bar{y})(x - \bar{x})}{\Sigma(x - \bar{x})^2}$$

where $b$ is the least-square estimate of $\beta$.

## 6. Components of $y$

Using the above solutions, we may write the value of each $y$ as the sum of three components. In our numerical example (Table 13.1),

$$\Sigma(y - \bar{y})^2 = 38 \qquad \Sigma(x - \bar{x})(y - \bar{y}) = 12 \qquad \Sigma(x - \bar{x})^2 = 6$$

so that

$$b = \frac{12}{6} = 2$$

The three components of $y$ are listed in the left half of Table 13.2. Note that the $e$ value (estimate of $\epsilon$) is taken as the difference between the actual value of $y$ and the first two components, that is, $\bar{y} + b(x - \bar{x})$.

Table 13.2   The components of $y$ and $\Sigma(y - \bar{y})^2$ based on the linear regression model (data from Table 13.1)

| $(x - \bar{x})$ | $y = \bar{y} + b(x - \bar{x}) + e$ | | | $b^2(x - \bar{x})^2$ | $e^2$ | $b(x - \bar{x})e$ |
|---|---|---|---|---|---|---|
| 0 | 2 = | 5 + 0 | −3 | 0 | 9 | 0 |
| −1 | 3 = | 5 − 2 | +0 | 4 | 0 | 0 |
| −1 | 5 = | 5 − 2 | +2 | 4 | 4 | −4 |
| 2 | 10 = | 5 + 4 | +1 | 16 | 1 | 4 |
| 0 | 20 = | 20 | 0    0 | 24 | 14 | 0 |

## 7. Components of $ssq$ of $y$

Next we come to the question of the sum of squares of $y$. The components of $y$ may be rewritten as

$$y - \bar{y} = b(x - \bar{x}) + e$$

Squaring both sides and summing the $n$ squares, we obtain

$$\Sigma(y - \bar{y})^2 = b^2\Sigma(x - \bar{x})^2 + \Sigma e^2$$

In our example,

$$38 = 24 + 14$$

That the sum of the cross product terms vanishes has been verified in the last column of Table 13.2. This fact follows from the second normal equation of a previous section by observing that

$$e = (y - \bar{y}) - b(x - \bar{x}) \qquad \text{and that} \qquad \Sigma e(x - \bar{x}) = 0$$

Thus the total $ssq$ of $y$ may be divided into two components. The component $b^2\Sigma(x - \bar{x})^2$ is due to the effect of $x$, and the component $\Sigma e^2 = \Sigma(y - y')^2$ is due to deviation of $y$ from the straight lines (see next section).

## 8. The equation of the regression line

The reason that the regression (of $y$ on $x$) problem is presented in the manner above is to show the close similarity of this case with the linear models of all the previous cases. Perhaps it is well at this stage to convert the above expressions into the more conventional form with which the reader is probably more familiar. Thus, we rewrite:

$$\begin{aligned} y &= \bar{y} + b(x - \bar{x}) + e \\ &= (\bar{y} - b\bar{x}) + bx + e \end{aligned}$$

Now, let

$$a = \bar{y} - b\bar{x}$$

then

$$y = a + bx + e$$

This is another way of writing the three components of each observed $y$ value. Now, we let the sum of the first two components be

$$y' = a + bx = \bar{y} + b(x - \bar{x})$$

This is the equation of the regression line fitted by the method of least squares. In our numerical example, $\bar{x} = 3$, $\bar{y} = 5$, $b = 2$, and

$$a = 5 - 2(3) = -1$$

The equation of the line shown in Fig. 13.2 is

$$y' = -1 + 2x = 2x - 1$$

The symbol $y'$ denotes the value of $y$ on the straight line, and $e = y - y'$ is the deviation of the actual $y$ from the "calculated" $y'$ and is usually known as the "residual." It is the vertical distance from a point to the regression line.   With respect to this straight line the sum of the squares of the "residuals" is smaller than that with respect to any other straight line.   This line passes through the point $(\bar{x}, \bar{y})$.   Also note that the component of $y$ that is due to regression is

$$y' - \bar{y} = b(x - \bar{x})$$

Consequently,

$$\Sigma(y' - \bar{y})^2 = b^2 \Sigma(x - \bar{x})^2$$

Hence, the identity in Sec. 7 concerning the subdivision of the sum of squares of $y$ may also be written as, remembering that $e = y - y'$,

$$\Sigma(y - \bar{y})^2 = \Sigma(y' - \bar{y})^2 + \Sigma(y - y')^2$$

In words, the total *ssq* for $y$ is equal to the *ssq*'s for $y'$ which lie on the regression line plus the *ssq*'s for the deviations of $y$ from the regression line.

## 9. Practical computation

Now we need to have a practical arithmetical procedure for calculating the components of the *ssq* of $y$.   In Table 13.2 are listed all the individual values of $y' = \bar{y} + b(x - \bar{x})$ and $e = y - y'$.   Calculating these individual values is very tedious even if there are only a moderate number of pairs of $x$ and $y$.   For practical purposes, all we need are the three basic quantities:

$$\Sigma(y - \bar{y})^2 \qquad \Sigma(x - \bar{x})(y - \bar{y}) \qquad \Sigma(x - \bar{x})^2$$

or abbreviated,

$$(yy) \qquad (xy) \qquad (xx)$$

Then $b = (xy)/(xx)$.   The component of $(yy)$ that is due to regression is, say,

$$\sum (y' - \bar{y})^2 = b^2(xx) = b(xy) = \frac{(xy)^2}{(xx)} = (yy)'$$

and the component due to residual deviations is obtained by subtraction, say,

$$\sum (y - y')^2 = \sum e^2 = (yy) - \frac{(xy)^2}{(xx)} = (yy)''$$

In our numerical example, $(xy)^2/(xx) = 12^2/6 = 24$, in agreement with the result of Table 13.2.   In this simplified notation, the identity of Secs. 7 and 8 may be summarized as

$$(yy) = (yy)' + (yy)''$$

where the first component is due to regression (that is, $y'$ on the line) and the second component is due to deviation from the line. This simplified notation will be adopted in the next chapter.

## 10. Test of significance of $b$

The total sum of squares of $y$ has $n - 1$ degrees of freedom. The component $b^2(xx) = (xy)^2/(xx)$ has one degree of freedom as its value is determined by $b$, remembering that the $x$'s are taken as given values in our model. The component $\Sigma e^2$ has $n - 2$ degrees of freedom (one lost to the estimate $\bar{y}$ and one lost to the estimate $b$). Then the mean square $\Sigma e^2/(n - 2)$ is an unbiased estimate of $\sigma^2$. The significance of $b$ may be tested by

$$F(1, n - 2) = t^2 = \frac{b^2(xx)}{\Sigma e^2/(n - 2)} = \frac{(yy)'}{(yy)''/(n - 2)}$$

Another explanation for this test is given in Sec. 12 of this chapter. The relationship between the regression coefficient $b$ and the correlation coefficient $r$ may be found in every textbook and will be omitted from our discussion, with only the comment that the test of significance for $b$ is equivalent to the test for $r$. If one is significant, so is the other.

## 11. Regression coefficient as an average slope

A few remarks may be made about the meaning of the regression coefficient as estimated by the method of least squares. The expression $b = (xy)/(xx)$ shows that the regression coefficient is the ratio of the covariance of $x$ and $y$ to the variance of $x$, but hardly anything else. To elucidate its meaning from another point of view, let us first consider the general method of finding the mean value of a series of numbers:

Value: $\quad\quad\quad\quad y_1 \quad y_2 \quad y_3 \quad y_4$
Weight: $\quad\quad\quad\quad w_1 \quad w_2 \quad w_3 \quad w_4$

Then the weighted mean of the $y$'s is

$$\bar{y} = \frac{y_1 w_1 + \cdots + y_4 w_4}{w_1 + \cdots + w_4} = \frac{\Sigma y_\alpha w_\alpha}{\Sigma w_\alpha}$$

In the particular case of equal weights ($w_\alpha = 1$), the expression becomes

$$\bar{y} = \frac{y_1 + \cdots + y_4}{1 + \cdots + 1} = \frac{\Sigma y}{\Sigma 1} = \frac{Y}{n}$$

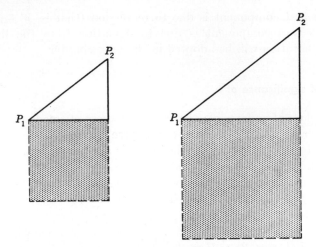

**Fig. 13.3** The concept of weight for the slope determined by two points. The slope from point $P_1(x_1,y_1)$ to $P_2(x_2,y_2)$ is $b_{12} = (y_2 - y_1)/(x_2 - x_1)$, and the associated weight is the square of the base, that is, $(x_2 - x_1)^2$. The longer the base, the greater the weight.

Now let us consider the following series of slopes (the $b$'s):

Value:

$$b_1 = \frac{y_1 - \bar{y}}{x_1 - \bar{x}} \qquad b_2 = \frac{y_2 - \bar{y}}{x_2 - \bar{x}} \qquad b_3 = \frac{y_3 - \bar{y}}{x_3 - \bar{x}} \qquad b_4 = \frac{y_4 - \bar{y}}{x_4 - \bar{x}}$$

Weight:

$$w_1 = (x_1 - \bar{x})^2 \qquad w_2 = (x_2 - \bar{x})^2 \qquad w_3 = (x_3 - \bar{x})^2 \qquad w_4 = (x_4 - \bar{x})^2$$

This concept of weight has been illustrated in Fig. 13.3. Note that $b_\alpha$ is the slope of the line joining the point $(x_\alpha, y_\alpha)$ with the central point $(\bar{x}, \bar{y})$. The weighted mean of these individual slopes is the least-square estimate $b$, which thus takes the form

$$b = \frac{\Sigma b_\alpha w_\alpha}{\Sigma w_\alpha} = \frac{\Sigma(y - \bar{y})(x - \bar{x})}{\Sigma(x - \bar{x})^2} = \frac{(xy)}{(xx)}$$

It is clear that $\bar{y}$ and $b$ are of the same form, both being linear functions, of the observed $y$'s. The main difference is in their weights. Hence the mathematical properties of $\bar{y}$ can be translated into those of $b$ by substituting the total weight $\Sigma w = (xx)$ for the weight $\Sigma w = n$. A more detailed discussion of the analogy between these two statistics may be found in chapter 19 of Jerome Li (1957). One of the more important properties is mentioned in the following section.

## 12. Variance of $b$

If the variance of $y$ is $\sigma^2$ and the mean $\bar{y}$ is determined from a sample of $n$ observations of equal weight, then the sampling variance of $\bar{y}$ is $\sigma_{\bar{y}}^2 = \sigma^2/\Sigma w = \sigma^2/n$, as we learned early in Chap. 4. When this true $\sigma^2$ is unknown, it may be estimated by $s^2 = \Sigma(y - \bar{y})^2/(n - 1)$, so that the estimate of $\sigma_{\bar{y}}^2$ may be taken as $s_{\bar{y}}^2 = s^2/n$. Similarly, if the residual variance of $y$ about the regression line is $\sigma^2$ and $b$ is determined from a sample of $n$ points, each with weight $(x - \bar{x})^2$, then the sampling variance of $b$ is

$$\sigma_b^2 = \frac{\sigma^2}{\Sigma w} = \frac{\sigma^2}{(xx)}$$

When the true residual variance is unknown, it may be estimated by $s^2 = \Sigma e^2/(n - 2)$, so that the estimate of $\sigma_b^2$ may be taken as $s_b^2 = s^2/(xx)$. Hence, $t = b/s_b$, the square of which, upon substitution, is

$$t^2 = \frac{b^2}{s_b^2} = \frac{b^2}{s^2/(xx)} = \frac{b^2(xx)}{\Sigma e^2/(n - 2)}$$

which is the $F$ test mentioned previously (Sec. 10).

From the relation $\sigma_b^2 = \sigma^2/(xx)$, it follows that for any given number of points, the wider the spread of the $x$ values, the smaller will be the sampling variance of $b$. Figure 13.4 shows two situations in which the left-hand estimate $b$ has a much smaller sampling variance than the one on the right-hand side, although the values of $b$ and $s^2$ are the same in both cases.

## 13. Regression coefficient as a contrast

There is yet another way of looking at the regression coefficient. Since $(xy) = \Sigma(x - \bar{x})(y - \bar{y}) = \Sigma(x - \bar{x})y$, we may write

$$b = \frac{(xy)}{(xx)} = \frac{(x_1 - \bar{x})}{(xx)}y_1 + \frac{(x_2 - \bar{x})}{(xx)}y_2 + \cdots + \frac{(x_n - \bar{x})}{(xx)}y_n$$

or, say,
$$b = l_1 y_1 + l_2 y_2 + \cdots + l_n y_n$$

But the coefficients $l_i$ add up to zero, that is, $\Sigma l_i = 0$. From Chap. 12 (Orthogonal contrasts) it follows that the expression for $b$ represents a contrast among the $y$ values. Furthermore, the $ssq$ for this contrast should be

$$\frac{b^2}{\Sigma l_i^2} = \frac{b^2}{(xx)/(xx)^2} = b^2(xx) = (yy)'$$

which is precisely what has been obtained in Secs. 7 to 9.

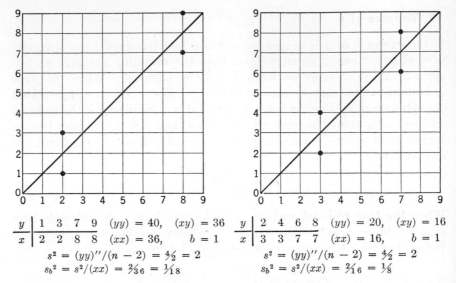

$$\begin{array}{c|cccc}
y & 1 & 3 & 7 & 9 \\
\hline
x & 2 & 2 & 8 & 8
\end{array}
\quad
\begin{array}{l}
(yy) = 40, \quad (xy) = 36 \\
(xx) = 36, \qquad b = 1
\end{array}
\qquad
\begin{array}{c|cccc}
y & 2 & 4 & 6 & 8 \\
\hline
x & 3 & 3 & 7 & 7
\end{array}
\quad
\begin{array}{l}
(yy) = 20, \quad (xy) = 16 \\
(xx) = 16, \qquad b = 1
\end{array}$$

$$s^2 = (yy)''/(n-2) = \tfrac{4}{2} = 2 \qquad\qquad s^2 = (yy)''/(n-2) = \tfrac{4}{2} = 2$$
$$s_b^2 = s^2/(xx) = \tfrac{2}{36} = \tfrac{1}{18} \qquad\qquad s_b^2 = s^2/(xx) = \tfrac{2}{16} = \tfrac{1}{8}$$

**Fig. 13.4** The sampling variance of $b$ is smaller when the $x$'s are farther apart than when the $x$'s are closer together. The residual variance of $y$ about the regression line is assumed to be the same in both cases.

Since $s^2 = \Sigma e^2/(n-2)$ is the estimated variance of the single $y$'s, it follows from the preceding chapter that the variance of the contrast $b$ is

$$V(b) = V(l_1 y_1 + \cdots + l_n y_n)$$
$$= s^2(l_1^2 + \cdots + l_n^2) = s^2 \left[ \frac{(xx)}{(xx)^2} \right] = \frac{s^2}{(xx)}$$

This is the result we have obtained in Sec. 12 through a different approach.

## 14. Regression and one-way classification

From the model of linear regression and the subsequent subdivision of the sum of squares shown in the foregoing sections, it is clear that linear regression analysis is also a form of analysis of variance. For a one-way classification with $k$ groups, the model is $y = u + t + e$, where $t$ varies from group to group and $\Sigma t = 0$. For linear regression, the model is $y = u + b(x - \bar{x}) + e$, where the term $b(x - \bar{x})$ varies from one $x$ to another and $\Sigma b(x - \bar{x}) = 0$. Thus, we see that the term $t$ in the former case plays the same role as $b(x - \bar{x})$ in the latter. The former is a group effect, and the latter is an $x$ effect. The term $u = \bar{y}$ is the same in both cases, and the term $e$ is a random variable with mean zero and variance $\sigma^2$ in both cases. The correspondence between the two cases is shown in Table 13.3.

*Table* 13.3  Correspondence between one-way classification and linear regression in the analysis of variance

| Point of comparison | One-way | | Regression | |
|---|---|---|---|---|
| Classification | Variation in treatments | | Variation in $x$ values | |
| Model | $y = u + t + e$ | | $y = u + b(x - \bar{x}) + e$ | |
| No. of parameters | $k$ | | 2 | |
| Effect *ssq*; *df* | $ssq_B = \Sigma n_i(\bar{y}_i - \bar{y})^2$ | $k - 1$ | $(yy)' = \Sigma(y' - \bar{y})^2$ | $2 - 1$ |
| Residual *ssq*; *df* | $ssq_W = \Sigma(y - \bar{y}_i)^2$ | $N - k$ | $(yy)'' = \Sigma(y - y')^2$ | $N - 2$ |
| Total *ssq*; *df* | $ssq_T = \Sigma(y - \bar{y})^2$ | $N - 1$ | $(yy) = \Sigma(y - \bar{y})^2$ | $N - 1$ |

The one-way classification model has $k$ parameters: 1 for $u$ and $k - 1$ for the $t$'s.  If we write $u_1 = u + t_1$, $u_2 = u + t_2$, etc., the $k$ parameters will be $u_1, \ldots, u_k$.  The regression model has two parameters: one for $u$ and one for $b$, remembering that the $x$'s are given values.  The $x$ effect is determined by $b$ for given $x$.  The chief difference between the two models is that the values of $u + b(x - \bar{x})$ are collinear while those of $u_1, \ldots, u_k$ are not necessarily so.

The correspondence between the two models in the subdivision of *ssq* has also been shown in Table 13.3.  The total *ssq* for a given set of $N$ values of $y$ remains the same in both cases.  For the same set of $y$'s, the number of degrees of freedom for the various *ssq* components is an indication of their magnitude.  Thus, in general,

$$\text{Between group:} \quad ssq_B > (yy)' \quad \text{due to regression}$$
$$\text{Within group:} \quad ssq_W < (yy)'' \quad \text{residue from line}$$

(See Exercises 13.4 to 13.6.)  These corresponding components are equal only when the group means $\bar{y}_i$ happen to be exactly collinear.  Since two points are always collinear, the one-way analysis of variance and linear regression analysis will yield identical results when there are only two groups ($k = 2$), whatever the two $x$ values are for the two groups of $y$.

## EXERCISES

**13.1**  Given two sets of paired observations:

| $x$: | 2 | 4 | 4 | 6 | $X = 16$ | | $x$: | 2 | 1 | 1 | 4 | $X = 8$ |
|---|---|---|---|---|---|---|---|---|---|---|---|---|
| $y$: | 4 | 9 | 11 | 16 | $Y = 40$ | | $y$: | 6 | 8 | 9 | 13 | $Y = 36$ |

For each set of data, calculate the regression coefficient $b$, the slope of the least-square fitted line.  Also, separately, draw a scatter diagram and the regression line which passes through the central point $(\bar{x}, \bar{y})$.  Let $y'$ be the value of $y$ on the straight line.

Calculate the following three kinds of sum of squares for each set of data.

$$(yy) = \Sigma(y - \bar{y})^2 = ssq \text{ of the } y \text{ deviations}$$
$$(yy)' = \Sigma(y' - \bar{y})^2 = ssq \text{ of the } y' \text{ deviations}$$
$$(yy)'' = \Sigma(y - y')^2 = ssq \text{ of vertical distances from point to line}$$

What relationship do these three quantities hold?
*Partial Ans.:* For the first set, $b = 3.0$; for the second, $b = 1.5$.

**13.2**   The slope from any point $(x_\alpha, y_\alpha)$ to the central point $(\bar{x}, \bar{y})$ is

$$b_\alpha = \frac{y_\alpha - \bar{y}}{x_\alpha - \bar{x}} \qquad \alpha = 1, 2, \ldots, n$$

We have shown that the slope of the regression line $b$ is the weighted mean of these $n$ individual slopes where the weight is $w_\alpha = (x_\alpha - \bar{x})^2$. It is to be recalled that $ssq$ for $n$ numbers (Exercise 3.4) may also be expressed in terms of the sum of squares of differences between all possible pairs of numbers. Similarly, instead of taking the slope from $(x,y)$ to $(\bar{x},\bar{y})$, we may take the slope between all possible pairs of numbers; for instance, for the first two points, the slope is

$$b_{12} = \frac{y_1 - y_2}{x_1 - x_2}$$

with weight $(x_1 - x_2)^2$. Then the regression coefficient $b$ is also the weighted mean of these slopes because

$$\Sigma(y_\alpha - y_\beta)(x_\alpha - x_\beta) = n\Sigma(y - \bar{y})(x - \bar{x}) \qquad \alpha < \beta$$

In the example

$$x: \quad 3 \quad 2 \quad 2 \quad 5$$
$$y: \quad 2 \quad 3 \quad 5 \quad 10$$

calculate the six slopes for the six possible pairs of points and verify that the weighted mean slope is $b = 2$, as obtained before.

**13.3**   It is instructive to place side by side the analogous expressions for the ordinary arithmetic mean $\bar{y}$ and the weighted average slope $b$. A few of them are as follows, and the reader may add others.

| | | |
|---|---|---|
| Single value: | $y_\alpha$ with weight 1 | $b_\alpha$ with weight $w_\alpha$ |
| Total value: | $\Sigma y_\alpha = Y$ | $\Sigma w_\alpha b_\alpha = (xy)$ |
| Total weight: | $\Sigma 1 = n$ | $\Sigma w_\alpha = (xx)$ |
| Mean: | $\bar{y} = \dfrac{Y}{n}$ | $b = \dfrac{(xy)}{(xx)}$ |
| Sum of squares: | $n\bar{y}^2 = \dfrac{Y^2}{n}$ | $(xx)b^2 = \dfrac{(xy)^2}{(xx)}$ |
| Degree of freedom: | 1 | 1 |
| Variance: | $V(\bar{y}) = \dfrac{\sigma_y{}^2}{n}$ | $V(b) = \dfrac{\sigma_e{}^2}{(xx)}$ |
| Test if it is 0: | $t^2 = \dfrac{\bar{y}^2}{s_{\bar{y}}{}^2} = \dfrac{n\bar{u}^2}{s_y{}^2}$ | $t^2 = \dfrac{b^2}{s_b{}^2} = \dfrac{(xx)b^2}{s_e{}^2}$ |

**13.4**   It often happens in practice that several observations on $y$ are obtained for the same value of $x$.   The following is an example of this situation.

$$x:\quad 1\ \ 1\ \ 1\ \ 1\ \ 2\ \ 2\ \ 4\ \ \ 4\ \ \ 5\ \ \ 5\ \ \ 5\ \ \ 5$$
$$y:\quad 2\ \ 4\ \ 3\ \ 5\ \ 8\ \ 6\ \ 9\ \ 13\ \ 11\ \ 10\ \ 16\ \ 9$$

The data on these twelve points (pairs of $x$, $y$) may be more succinctly written as in the accompanying table.   The columns under $n_i$, $Y_i$, and $\bar{y}_i$ are not needed for the

| $x$ | $y$ | | | | $n_i$ | $Y_i$ | $\bar{y}_i$ |
|----|----|----|----|----|----|----|----|
| 1 | 2 | 4 | 3 | 5 | 4 | 14 | 3.5 |
| 2 | 8 | 6 | | | 2 | 14 | 7.0 |
| 4 | 9 | 13 | | | 2 | 22 | 11.0 |
| 5 | 11 | 10 | 16 | 9 | 4 | 46 | 11.5 |
| | | | | | $12 = N$ | $96 = Y$ | $8.0 = \bar{y}$ |

time being.   Concentrating on the twelve individual points, fit a straight line.   The calculations are as follows:

$$(yy) = \Sigma(y - \bar{y})^2 = 194 \qquad (xy) = \Sigma(x - \bar{x})(y - \bar{y}) = 72 \qquad (xx) = \Sigma(x - \bar{x})^2 = 36$$

$$\text{Slope:}\qquad b = \frac{(xy)}{(xx)} = \frac{72}{36} = 2.0$$
$$\text{Equation:}\qquad y' = \bar{y} - b(x - \bar{x}) = 2 + 2x$$

The scatter diagram and the line are shown in Fig. 13.5.   Calculate each $y'$ value from the straight-line equation and verify by longhand the following:

$$(yy)' = \sum (y' - \bar{y})^2 = \frac{(xy)^2}{(xx)} = 144$$
$$(yy)'' = \sum (y - y')^2 = 194 - 144 = 50$$

**13.5**   Ignore the $x$ values of Exercise 13.4 and simply regard the data as divided into four groups with size $n_i$.   Using the method of one-way classification, verify the following subdivision of sum of squares:

| | |
|---|---|
| Between groups: | $ssq_B = 150$ |
| Within groups: | $ssq_W = \ \ 44$ |
| Total: | $ssq_T = 194$ |

**13.6**   Now let us consider the four mean points:

$$x:\quad 1\quad 2\quad 4\quad 5$$
$$\bar{y}_i:\quad 3.5\quad 7.0\quad 11.0\quad 11.5$$

where each mean point is weighted by $n_i$, the number of single observations that determine the mean for a fixed $x$.   For example, we consider that there are four points at the position (1, 3.5), etc.   Fit a straight line to these mean points.

It is obvious from Fig. 13.5 that the sum of squared vertical distance from mean point to straight line is much smaller than that for the original twelve points.   How much smaller?

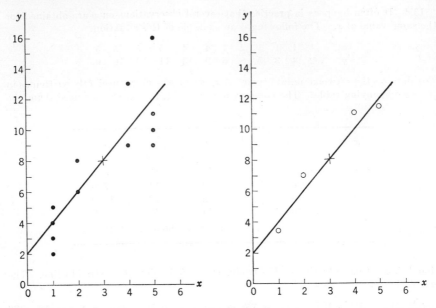

**Fig. 13.5** Left (Exercise 13.4): Scatter diagram for the twelve individual points. The cross $\times$ is the central point $(\bar{x},\bar{y})$. The equation of the regression line is $y' = 2x + 2$.
Right (Exercise 13.6): The points are $(x,\bar{y}_x)$, where the new symbol $\bar{y}_x$ denotes the mean value of the several $y$'s for a given $x$. Each such point has a weight proportional to the number of $y$'s at the given $x$. The regression line for these mean points is the same as that for the individual points, but the sum of the squared vertical distances from point to line is much smaller.

For these mean points, not only $\bar{x}$ and $(xx)$ but also $\bar{y}$ and $(xy)$ remain the same as those for the 12 original points, because $\Sigma x \cdot n_i\bar{y}_i = \Sigma xy$ for the 12 points. It follows that $b$ and the equation of the straight line all remain the same as before. The conclusion is this: Whether the regression line is fitted to the original individual points or fitted to the weighted mean points, it makes no difference. The line is the same.

Furthermore, since $(xx)$ and $(xy)$ remain the same for both cases, it follows that the *ssq* due to regression, that is, the *ssq* of the collinear $y'$ values, also remains the same, being $(xy)^2/(xx) = 144$.

The remaining question is what the residual *ssq* is for the mean points. Longhand calculation of $y'$ for each $x$ and then summing the squares of the differences $(\bar{y}_i - y')$ shows that $\Sigma n_i(\bar{y}_i - y')^2 = 6$, which is much smaller than the corresponding residue *ssq* $= 50$ for the 12 individual points. The *ssq* among the four weighted means is, $\Sigma n_i(\bar{y}_i - \bar{y})^2 = 150$. Hence, the residue *ssq* for the mean points is $150 - 144 = 6$, in agreement with that obtained by longhand calculation.

To summarize: for the 12 individual points the residue *ssq* $= 50$; for the mean points the residue *ssq* $= 6$. The difference is $50 - 6 = 44$, which is precisely the within-group *ssq* obtained in the one-way analysis of variance. The results of the last three exercises are summarized in the accompanying table. The degrees of freedom are shown to identify the components. The regression analysis on the mean points (right-hand portion of the table) is a subdivision of the between-group *ssq* = 150 with 3 *df*. The second line of the table shows the relationship of the within-group *ssq* and the two kinds of residue *ssq*. Thus, 44 = 50 − 6, and the degrees of freedom have the same relationship, that is, 8 = 10 − 2. Although the demonstration is based on a simple numerical example, the relationship is perfectly general.

| One-way analysis of variance | | | Linear regression | | | | |
| Source | *df* | *ssq* | Source | On individual points | | On mean points | |
| | | | | *df* | *ssq* | *df* | *ssq* |
| Between groups | 3 | 150 | Collinear | 1 | 144 | 1 | 144 |
| Within groups | 8 | 44 | Residue | 10 | 50 | 2 | 6 |
| Total | 11 | 194 | Total | 11 | 194 | 3 | 150 |

# 14

# Regression and
# analysis of variance

This chapter is a simple account of what is generally known as the *analysis of covariance*, which is a combined application of linear regression and the analysis of variance. The many ways in which the analysis of covariance can be applied in practical research have been given by the excellent review article by Cochran [Analysis of Covariance: Its Nature and Uses, *Biometrics*, **13**:261–281 (1957)]. The applications will be deferred to Chap. 20. In this chapter, we shall examine some of the underlying principles by which the analysis is made. If $y$ is the variable under investigation, the main idea is to correct or adjust the values of $y$, basing the correction on the value of another (concomitant) variate $x$. One very important assumption in using the covariance method is that the variation of the $x$ values is not due to the treatments themselves.

## 1. Subdivision of sum of products

First of all we need to establish the algebraic fact that the sum of products of deviations of $x$ and $y$, like the sum of squares of deviations, may be subdivided into two components: one for within the groups and one for between the groups. Let $x_{i\alpha}$, $y_{i\alpha}$ be the $\alpha$th pair of observations in the $i$th group ($i = 1, 2, \ldots, k$). This is a case of one-way classification arising from a completely randomized experiment. Using the same notation system as before, we may subdivide the sum of products of all the pairs in the following way:

$$\sum_i \sum_\alpha (x_{i\alpha} - \bar{x})(y_{i\alpha} - \bar{y}) = \sum_i \sum_\alpha [(x_{i\alpha} - \bar{x}_i) + (\bar{x}_i - \bar{x})][(y_{i\alpha} - \bar{y}_i) + (\bar{y}_i - \bar{y})]$$

$$= \sum_i \sum_\alpha (x_{i\alpha} - \bar{x}_i)(y_{i\alpha} - \bar{y}_i) + \sum_i n_i(\bar{x}_i - \bar{x})(\bar{y}_i - \bar{y})$$

or

$$(xy)_T = (xy)_W + (xy)_B$$
$$\text{(total)} \quad \text{(within)} \quad \text{(between)}$$

The other two terms vanish on summation, since $\Sigma(x_{i\alpha} - \bar{x}_i) = 0$ for each fixed $i$. Again, if $x$ is replaced by $y$, the identity becomes that for the subdivision of the sum of squares. For practical calculation we adopt a procedure analogous to that used in Chap. 5. For example,

*Table 14.1* Sum of products of three groups of equal size ($n = 4$, $N = 12$)

| | I | | | II | | | III | | | Grand total | | |
|---|---|---|---|---|---|---|---|---|---|---|---|---|
| | $x$ | $y$ | $xy$ | $x$ | $y$ | $xy$ | $x$ | $y$ | $xy$ | | | |
| | 3 | 2 | 6 | 2 | 4 | 8 | 2 | 6 | 12 | | | |
| | 2 | 3 | 6 | 4 | 9 | 36 | 1 | 8 | 8 | | | |
| | 2 | 5 | 10 | 4 | 11 | 44 | 1 | 9 | 9 | | | |
| | 5 | 10 | 50 | 6 | 16 | 96 | 4 | 13 | 52 | | | |
| $X_i$ | 12 | | | 16 | | | 8 | | | $36 = X$ | | |
| $Y_i$ | | 20 | | | 40 | | | 36 | | $96 = Y$ | | |
| | | | | | | | | | | $3,456 = XY = NC$ | | |
| $X_iY_i$ | 240 | | | 640 | | | 288 | | | $1,168 = nB$ | | |
| $\Sigma xy$ | | | 72 | | | 184 | | | 81 | $337 = A$ | | |

with the data of Table 14.1 we first calculate the following three quantities (that is, areas of three series of rectangles, diagrams omitted):

$$(\text{area}) \; A = \sum_i \sum_\alpha x_{i\alpha} y_{i\alpha} = 337$$

$$(\text{area}) \; B = n \sum_i \bar{x}_i \bar{y}_i = \frac{\Sigma X_i Y_i}{n} = \frac{1168}{4} = 292$$

$$(\text{area } C) = N \bar{x} \bar{y} = \frac{XY}{N} = \frac{3456}{12} = 288$$

so that the sums of products of deviations are

| | | |
|---|---|---|
| Within groups: | $(xy)_W = A - B =$ | 45 |
| Between groups: | $(xy)_B = B - C =$ | 4 |
| Total: | $(xy)_T = A - C =$ | 49 |

In the rest of this chapter it is understood that we talk about these three groups, each consisting of four pairs of observations.

## 2. Linear model and estimates

The linear model is a direct extension of that presented in the preceding chapter. If $\tau_i$ is the effect of the $i$th group (treatment), then the model is

$$y_{i\alpha} = \mu + \tau_i + \beta(x_{i\alpha} - \bar{x}) + \epsilon_{i\alpha}$$

The reader sees immediately that the expression above is merely a combination of the regression model and the one-way classification model summarized in Table 13.3. The estimates of the parameters should be so chosen that the quantity

$$Q = \sum_i \sum_\alpha [y_{i\alpha} - u - t_i - b(x_{i\alpha} - \bar{x})]^2$$

is a minimum. From the condition $\partial Q/\partial u = 0$, we obtain the equation

$$\sum_i \sum_\alpha [y_{i\alpha} - u - t_i - b(x_{i\alpha} - \bar{x})] = 0$$

Imposing the restriction $\Sigma t_i = 0$, we obtain the solution

$$Y = Nu \qquad \text{or} \qquad u = \bar{y}$$

Taking the partial derivative of $Q$ with respect to $t_1$, we may ignore the terms not involving $t_1$ and concentrate only on the $n$ terms of the first group involving $t_1$. Equating $\partial Q/\partial t_1$ to zero gives the following equation:

$$\sum_\alpha [y_{1\alpha} - u - t_1 - b(x_{1\alpha} - \bar{x})] = 0$$

Substituting the previous solution $u = \bar{y}$, and noting that

$$\Sigma(x_{1\alpha} - \bar{x}) = n(\bar{x}_1 - \bar{x})$$

and $\Sigma(y_{1\alpha} - \bar{y}) = n(\bar{y}_1 - \bar{y})$, we see that the solution is

$$t_1 = (\bar{y}_1 - \bar{y}) - b(\bar{x}_1 - \bar{x})$$

In general,

$$t_i = (\bar{y}_i - \bar{y}) - b(\bar{x}_i - \bar{x})$$

These are called the corrected or adjusted treatment effects, for, without the regression terms, $t_i = \bar{y}_i - \bar{y}$ as estimated in Chap. 6. Finally, setting $\partial Q / \partial b = 0$, we obtain the equation

$$\sum_i \sum_\alpha [y_{i\alpha} - u - t_i - b(x_{i\alpha} - \bar{x})](x_{i\alpha} - \bar{x}) = 0$$

By substituting $u = \bar{y}$ and the above solution for $t_i$ and adopting the simplified notation, the above equation may be written

$$(xy)_T - (xy)_B + b(xx)_B - b(xx)_T = 0$$

Hence,

$$b = \frac{(xy)_T - (xy)_B}{(xx)_T - (xx)_B} = \frac{(xy)_W}{(xx)_W}$$

In words, the least-square estimate of the parameter $\beta$ is given by the ratio of the *within-group* sum of product to the *within-group* sum of squares of $x$.

### 3. Test of significance of $b$

At this stage it is well to outline the routine analysis and tests before examining the situation in more detail. First of all, the sum of squares of $x$ and $y$ and the sum of products between and within groups are calculated as shown in Table 14.2A. Then the estimate of the regression coefficient is given by

$$b = \frac{(xy)_W}{(xx)_W} = \frac{45}{20} = 2.25$$

Then the within-group $(yy)_W = 138$ with 9 $df$ may be subdivided into two components, one due to regression and one residual, in exactly the same manner as in the preceding chapter. The geometrical meaning of this subdivision will be explained in a later section. Thus, with subscript $W$ understood,

Regression: $\quad (yy)' = \dfrac{45^2}{20} = 101.25 \qquad\qquad$ with 1 $df$

Residual: $\quad (yy)'' = 138.00 - 101.25 = 36.75 \qquad$ with 8 $df$

*Table 14.2*

*A.* The sum of squares and products, regression, and residual sum of squares, based on data of Table 14.1

| Source | df | (yy) | (xy) | (xx) | b | (yy)' | (yy)'' |
|--------|-----|------|------|------|------|--------|--------|
| Between, $B$ | 2 | 56 | 4 | 8 | ... | ... | 71.50† |
| Within, $W$ | 9 | 138 | 45 | 20 | 2.25 | 101.25 | 36.75 |
| Total, $T$ | 11 | 194 | 49 | 28 | 1.75 | 85.75 | 108.25 |

† Obtained by subtraction.

*B.* The analysis of variance based on the original and residual sum of squares

| Source | Original | | | Residual | | |
|--------|-----|------|------|-----|--------|------|
| | df | (yy) | msq | df | (yy)'' | msq |
| Between, $B$ | 2 | 56 | 28 | 2 | 71.50 | 35.75 |
| Within, $W$ | 9 | 138 | 15.3 | 8 | 36.75 | 4.59 |
| Total, $T$ | 11 | 194 | $F = 1.83$ | 10 | 108.25 | $F = 7.78$ |

as shown in the right-hand side of Table 14.2*A*. The first test to be made is the significance of this *b* by

$$F(1,8) = t^2 = \frac{101.25/1}{36.75/8} = 22$$

It is significant. If it were nonsignificant, no advantage would be gained by correcting the *y* means for variations in *x*. Then an analysis of variance of the *y* data alone, ignoring the existence of *x*, would be sufficient. Since the regression is significant in our example, we proceed further as follows.

## 4. Routine analysis of $(yy)''$

The total $(yy) = 194$ with 11 *df* may also be subdivided into regression and residual components by taking $b_T = {}^{49}\!/_{28} = 1.75$. Hence,

$$(yy)' = \frac{49^2}{28} = 85.75$$

and $(yy)'' = 194.00 - 85.75 = 108.25$, as shown in the bottom row of Table 14.2*A*. The rest of the analysis will be based on the residual sum of squares only. The residual $(yy)''$ for between groups is obtained by subtraction, *not* calculated independently from the first row of the table. Thus,

$$108.25 - 36.75 = 71.50$$

The reason for this procedure will be explained later. The final analysis is shown in Table 14.2*B*, which in practice may be combined with the preceding table. If we were to ignore the influence of $x$ on $y$, the group (treatment) effects would be insignificant, as shown in the left-hand half of the table. But when we base the analysis on the residual sum of squares, after correction for regression, it shows that the group effects are significant at the 5 per cent level. This completes the ordinary routine procedure of testing the significance of treatments in covariance analysis. Briefly, the ordinary analysis is a breakdown of the total sum of squares $(yy) = 194$, and the present analysis is a breakdown of the total residual sum of squares $(yy)'' = 108.25$. The arithmetic is simple enough. The remaining sections of this chapter deal with the meaning of the various quantities involved in the above analysis and thus provide us with understanding of the procedure.

## 5. Geometrical meaning

The various relationships of the quantities are most easily seen through diagrams and numerical examples. First plot the four points of each group separately and fit a straight line for each single group by the method of the preceding chapter. The corresponding pictures are shown in Figs. 14.1 to 14.3, and the arithmetic involved in calculating the individual slopes is given in the first three rows of Table 14.3. The slopes of these separate three groups are 2.0, 3.0, and 1.5, respectively.

Next, if we plot all of the 12 points in one diagram, ignoring the groups (that is, if we superimpose the first three diagrams on top of each other), the result is as shown in Fig. 14.6, for which the slope of the fitted line is calculated from the bottom row of Table 14.3. Thus, $b_T = {}^{49}\!/_{28} = 1.75$. Each group of four points has its own central point $(\bar{x}_i, \bar{y}_i)$. These central points are plotted in Fig. 14.5, showing a slope of 0.50 as calculated from that row in the table labeled "between." Finally, and this is the most important operation, if we superimpose the first three single-group diagrams on top of each other in such a way that the three central points coincide with each other, the resulting diagram is that of Fig. 14.4. Alternatively, if we plot the 12 points with coordinates $(x_{i\alpha} - \bar{x}_i, y_{i\alpha} - \bar{y}_i)$, we obtain the same scatter diagram, because then the central point for each group is $(0,0)$. Now we see that the least-square estimate $b = b_W = (xy)_W/(xx)_W = 2.25$ is the slope of the best

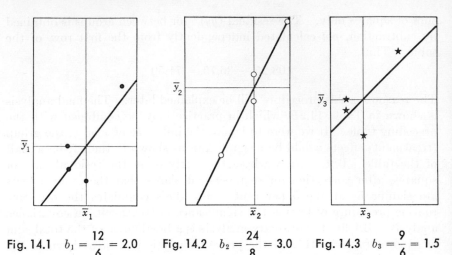

**Fig. 14.1** $b_1 = \dfrac{12}{6} = 2.0$   **Fig. 14.2** $b_2 = \dfrac{24}{8} = 3.0$   **Fig. 14.3** $b_3 = \dfrac{9}{6} = 1.5$

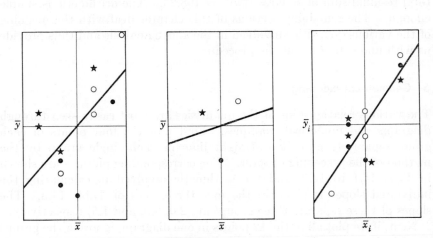

**Fig. 14.6** $b_T = \dfrac{49}{28} = 1.75$ **Fig. 14.5** $b_B = \dfrac{4}{8} = 0.50$ **Fig. 14.4** $b_W = \dfrac{45}{20} = 2.25$

fitted line with respect to the 12 points in Fig. 14.4.   It is convenient to pause here to make the important observation that the value

$$b = b_W = 2.25$$

is the weighted mean of the three single-group slopes ($b_1 = 2.0$, $b_2 = 3.0$, $b_3 = 1.5$) where the weights are the values of $(xx)_i$ of the corresponding group.   This is obvious from the way these $b$'s are calculated, and it also follows from the fact that $b$ has the same general property of a weighted mean. That is,

$$b_W = \frac{6b_1 + 8b_2 + 6b_3}{6 + 8 + 6} = \frac{45}{20} = 2.25$$

This completes the description of Figs. 14.1 to 14.6 and most of the entries of Table 14.3. Further subdivision of $(yy)$ is given in the next section.

*Table* 14.3   Detailed analysis of the data of Table 14.1

| Figure | Group | $df$ | Original $(yy)$ | $(xy)$ | $(xx)$ | $b$ | Regression $df$ | $(yy)'$ | Residual $df$ | $(yy)''$ | Components |
|---|---|---|---|---|---|---|---|---|---|---|---|
| 14.1 | (1) | 3 | 38 | 12 | 6 | 2.00 | 1 | 24.00 | 2 | 14.00 | |
| 14.2 | (2) | 3 | 71 | 24 | 8 | 3.00 | 1 | 72.00 | 2 | 2.00 | |
| 14.3 | (3) | 3 | 26 | 9 | 6 | 1.50 | 1 | 13.50 | 2 | 12.50 | |
| | | | | | | (sum) | 3 | 109.50 | 6 | 28.50 | $S_1$ |
| | | | | | | | | | 2 | 8.25 | $S_2$ |
| 14.4 | Within, $W$ | 9 | 138 | 45 | 20 | 2.25 | 1 | 101.25 | 8 | 36.75 | $S_1 + S_2$ |
| 14.5 | Between, $B$ | 2 | 56 | 4 | 8 | 0.50 | 1 | 2.00 | 1 | 54.00 | $S_3$ |
| | | | | | | (sum) | 2 | 103.25 | 9 | 90.75 | $S_1 + S_2 + S_3$ |
| | | | | | | | | | 1 | 17.50 | $S_4$ |
| 14.6 | Total, $T$ | 11 | 194 | 49 | 28 | 1.75 | 1 | 85.75 | 10 | 108.25 | $S_1 + S_2 + S_3 + S_4$ |

## 6. Components of within-group $(yy)''$

For each of the six diagrams in Figs. 14.1 to 14.6 (that is, for each of the six rows in the left-hand half of Table 14.3), we may calculate the sum of squares of $y$ due to regression and that due to residuals. Each of these six calculations is done in the manner developed in the preceding chapter for a single group, because the calculations refer to the six separate diagrams. The sum of the first three rows of Table 14.3:

|  | $df$ | $(yy)'$ | $df$ | $(yy)''$ |
|---|---|---|---|---|
| (Sum) | 3 | 109.50 | 6 | 28.50 $= S_1$ |

is to be distinguished from that calculated independently from the row labeled "within":

|  | $df$ | $(yy)'$ | $df$ | $(yy)''$ |
|---|---|---|---|---|
| $W$ | 1 | 101.25 | 8 | 36.75 $= S_1 + S_2$ |

The two procedures represent two different subdivisions of the same $(yy)_W = 138$. If the three individual slopes for each group are equal, then $b_1 = b_2 = b_3 = b_W$ and the values $(yy)'$ and $(yy)''$ of these two subdivisions are also equal. Hence, the difference

$$109.50 - 101.25 = 8.25 = S_2 \qquad \text{with 2 } df$$

is due to the variation of $b_1$, $b_2$, $b_3$ about their mean $b_W$. The algebraic proof of this statement will be given in Sec. 8.

It is clear that the residual $(yy)'' = 36.75$ with 8 $df$, as used in the routine analysis (Table 14.2), consists of two components: one is the total residual $(yy)''$ for the three single groups ($S_1 = 28.75$); another is due to the variation of the single $b$'s ($S_2 = 8.25$). The rationale of including this latter component as error is that we accept $b_W$ as the estimate of $\beta$ in our linear model, and the variation of the individual $b$'s from group to group is due to sampling rather than to treatment. However, if the individual $b$'s are drastically different from group to group, we must revise our linear model and the procedure of analysis accordingly. A discussion of this matter would be outside the scope of this chapter.

## 7. Components of between-group $(yy)''$

As in the preceding paragraph, we add the $(yy)'$ and $(yy)''$ for the within and between groups of Table 14.3, obtaining

| | $df$ | $(yy)'$ | $df$ | $(yy)''$ |
|---|---|---|---|---|
| (Sum) | 2 | 103.25 | 9 | $90.75 = S_1 + S_2 + S_3$ |

to be distinguished from that calculated independently from the total:

| | $df$ | $(yy)'$ | $df$ | $(yy)''$ |
|---|---|---|---|---|
| $T$ | 1 | 85.75 | 10 | $108.25 = S_1 + S_2 + S_3 + S_4$ |

These represent two different subdivisions of the same $(yy)_T = 194$. Note that $b_T = 1.75$ is the weighted mean of $b_W = 2.25$ and $b_B = 0.50$, where the weights are the corresponding values of $(xx)$. Then if the within-group and between-group slopes are the same, that is,

$$b_W = b_B = b_T$$

the values of $(yy)'$ and $(yy)''$ of these two subdivisions are equal. Hence the difference

$$103.25 - 85.75 = 17.50 = S_4 \qquad \text{with 1 } df$$

is due to the discrepancy between the slopes within and between groups. The algebraic proof of this statement will be given in the next section. If the treatments do not influence the linear relationship between $x$ and $y$, the slopes between and within groups should be the same $b$. Therefore the sum of squares $S_4 = 17.50$ is regarded as treatment effects. And we see that the treatment residual $(yy)'' = 71.50$ employed in the routine analysis (Table 14.2) also consists of two components: one is the residual of treatment means about their own regression line and the other is due to the difference of $b_W$ and $b_B$. The analysis of the total residual $(yy)''$ may be summarized as follows:

$$\left.\begin{array}{l} S_1 = 28.50 \\ S_2 = \phantom{0}8.25 \end{array}\right\} \quad \left.\begin{array}{r} S_1 + S_2 = 36.75 \end{array}\right\}$$
$$\left.\begin{array}{l} S_3 = 54.00 \\ S_4 = 17.50 \end{array}\right\} \quad \left.\begin{array}{r} S_3 + S_4 = 71.50 \end{array}\right\} \quad S_1 + S_2 + S_3 + S_4 = 108.25$$

The routine procedure for the analysis presented in Sec. 4 is a shortcut for obtaining the two major components $(S_1 + S_2)$ and $(S_3 + S_4)$ without further subdivision. The notations for the four components $(S_1, S_2, S_3, S_4)$ used here are the same as those adopted by M.G. Kendall ("The Advanced Theory of Statistics," vol. II, pp. 237–242, Charles Griffin & Co., Ltd., London, 1948). For more detailed tests, the reader may consult page 240 of this reference.

## 8. Algebra of $S_2$ and $S_4$

It is easy to convert the numerical values of Table 14.3 into algebraic expressions. For instance, the "sum" of the first $k$ lines of the $k$ groups and the "within" line under the "regression" column, are algebraically,

|         | $df$ | $(yy)'$         |
|---------|------|-----------------|
| (Sum)   | $k$  | $\Sigma b_i^2(xx)_i$ |
| $W$     | $1$  | $b_W^2(xx)_W$   |

where $(xx)_W = \Sigma(xx)_i$. Since $b_i$ is a number of weight $(xx)_i$ and $b_W$ is the weighted mean of the $b$'s, it follows that the difference, with $k - 1$ $df$, is

$$S_2 = \Sigma b_i^2(xx)_i - b_W^2(xx)_W = \Sigma(b_i - b_W)^2(xx)_i$$

which justifies the previous statement that $S_2$ is due to the variation of $b_i$.

Similarly, the "sum" of the within and between $(yy)'$ minus that for the total is

$$S_4 = b_W{}^2(xx)_W + b_B{}^2(xx)_B - b_T{}^2(xx)_T$$
$$= (b_W - b_T)^2(xx)_W + (b_B - b_T)^2(xx)_B$$

or

$$S_4 = (b_W - b_B)^2 \frac{(xx)_W(xx)_B}{(xx)_T}$$

as an alternate form, which justifies the statement in the preceding section that $S_4$ is due to the difference between $b_W$ and $b_B$.

## 9. Adjusted treatment means

It was shown in Sec. 2 that the estimate of the group (treatment) *effects* is

$$t_i = (\bar{y}_i - \bar{y}) - b(\bar{x}_i - \bar{x})$$

where $b = b_W = 2.25$ is the estimate of $\beta$ in the original linear model. Thus,

$$\bar{y} + t_i = \bar{y}_i - b(\bar{x}_i - \bar{x})$$

is the corrected or adjusted treatment *mean*. These values and their sum of squares in our example are as shown in the accompanying table.

| $\bar{y}_i$ | $\bar{x}_i$ | $b(\bar{x}_i - \bar{x})$ | $\bar{y} + t_i$ | $t_i$ | $t_i{}^2$ |
|---|---|---|---|---|---|
| 5 | 3 | 0 | 5.00 | −3.00 | 9.0000 |
| 10 | 4 | 2.25 | 7.75 | −0.25 | 0.0625 |
| 9 | 2 | −2.25 | 11.25 | 3.25 | 10.5625 |
| $\bar{y} = 8$ | $\bar{x} = 3$ | 0 | 24.00 | 0 | 19.6250 |
| $n = 4$ | | | | hence, $n\Sigma t_i{}^2 = 78.50$ | |

It is seen that the sum of squares of the adjusted treatment means, $n\Sigma t_i{}^2 = 78.50$, is neither equal to $S_3 = 54.00$ nor equal to $S_3 + S_4 = 71.50$, as might be thought intuitively, since the latter is being used in the routine analysis as the treatment residual $(yy)''$ obtained by subtraction. The value of $n\Sigma t_i{}^2$ is always larger than $S_3 + S_4$. The proof of this statement is given in Sec. 11. However, it may be observed from the setup of Table 14.3 that when the regression within and between the groups is the same, the component $S_4 = 0$ and hence $S_3$ would be equal to $n\Sigma t_i{}^2$. It is only in such a case that the treatment residual $(yy)''$

used in the routine analysis coincides with the sum of squares of the adjusted treatment means.

## 10. Points and distances

This and the next section may well be omitted by many readers who do not care for further details on the subject. The following demonstrations are not for practical calculation, but they do reveal the meaning of the components $S_1$, $S_2$, $S_3$, $S_4$ more vividly and show their relationship with the *ssq* of the adjusted treatment means more specifically. The reader has recognized that the so-called analysis of covariance is an analysis of variance based on the residual sum of squares of $y$. The four components of the total $(yy)''$ have been isolated previously (Table 14.3) in summarized form. However, they may be obtained directly and independently.

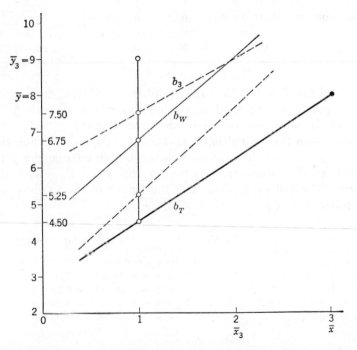

Fig. 14.7 The subdivision of the total vertical distance from a point to the total regression line into four segments by the other regression lines. The example point taken is $(x_{33}, y_{33}) = (1, 9)$. The sum of squares of these segments for all points are the components $S_1$, $S_2$, $S_3$, $S_4$ of $(yy)''$, as explained in the text.

*Table* 14.4   **Points and distances (based on data of Table 14.1)**

| | | Ordinate | | | | | Distance | | | |
|---|---|---|---|---|---|---|---|---|---|---|
| $P_0$ | $P_1$ | $P_2$ | $P_3$ | $P_4$ | $d_1$ | $d_2$ | $d_3$ | $d_4$ | | $d_T$ |
| 2 | 5 | 5.00 | 8.00 | 8.00 | $-3$ | 0 | $-3.0$ | 0 | | $-6.00$ |
| 3 | 3 | 2.75 | 5.75 | 6.25 | 0 | 0.25 | $-3.0$ | $-0.50$ | | $-3.25$ |
| 5 | 3 | 2.75 | 5.75 | 6.25 | 2 | 0.25 | $-3.0$ | $-0.50$ | | $-1.25$ |
| 10 | 9 | 9.50 | 12.50 | 11.50 | 1 | $-0.50$ | $-3.0$ | 1.00 | | $-1.50$ |
| 4 | 4 | 5.50 | 4.00 | 6.25 | 0 | $-1.50$ | 1.5 | $-2.25$ | | $-2.25$ |
| 9 | 10 | 10.00 | 8.50 | 9.75 | $-1$ | 0 | 1.5 | $-1.25$ | | $-0.75$ |
| 11 | 10 | 10.00 | 8.50 | 9.75 | 1 | 0 | 1.5 | $-1.25$ | | 1.25 |
| 16 | 16 | 14.50 | 13.00 | 13.25 | 0 | 1.50 | 1.5 | $-0.25$ | | 2.75 |
| 6 | 9 | 9.00 | 7.50 | 6.25 | $-3$ | 0 | 1.5 | 1.25 | | $-0.25$ |
| 8 | 7.5 | 6.75 | 5.25 | 4.50 | 0.5 | 0.75 | 1.5 | 0.75 | | 3.50 |
| 9 | 7.5 | 6.75 | 5.25 | 4.50 | 1.5 | 0.75 | 1.5 | 0.75 | | 4.50 |
| 13 | 12 | 13.50 | 12.00 | 9.75 | 1 | $-1.50$ | 1.5 | 2.25 | | 3.25 |
| 96 | 96 | 96.00 | 96.00 | 96.00 | 0 | 0 | 0 | 0 | | 0 |
| | | | | $\Sigma d^2$ | 28.50 | 8.25 | 54.00 | 17.50 | | 108.25 |
| | | | | | $S_1$ | $S_2$ | $S_3$ | $S_4$ | | $(yy)_{T''}$ |

The total $(yy)'' = 108.25$ is the *ssq* of the vertical distances from the observed points to the total regression line whose slope is $b_T = 1.75$. Each such distance may be subdivided into four segments by the other three regression lines with slopes $b_i$, $b_W$, $b_B$. To simplify notations, let $P_0$ be the $y$ value of the original points, etc., and $P_4$ be the $y$ value of the points on the total regression line, so that $P_0 - P_4 = d_T$ is the total distance. The full expressions and the length of the various segments are as follows:

| *Ordinate (y value) of point* | *Example point (Fig. 14.7)* |
|---|---|
| $P_0 = y_{i\alpha}$ | $= y_{33} = 9.00$ |
| $P_1 = \bar{y}_i + b_i(x_{i\alpha} - \bar{x}_i)$ | $= 9 + 1.50(-1) = 7.50$ |
| $P_2 = \bar{y}_i + b_W(x_{i\alpha} - \bar{x}_i)$ | $= 9 + 2.25(-1) = 6.75$ |
| $P_3 = \bar{y} + b_W(x_{i\alpha} - \bar{x}_i) + b_B(\bar{x}_i - \bar{x})$ | $= 8 + 2.25(-1) + 0.50(-1) = 5.25$ |
| $P_4 = \bar{y} + b_T(x_{i\alpha} - \bar{x})$ | $= 8 + 1.75(-2) = 4.50$ |

*Vertical distance*

$$P_0 - P_1 = d_1 = (y_{i\alpha} - \bar{y}_i) - b_i(x_{i\alpha} - \bar{x}_i) \qquad = 1.50$$
$$P_1 - P_2 = d_2 = (b_i - b_W)(x_{i\alpha} - x_i) \qquad = 0.75$$
$$P_2 - P_3 = d_3 = (\bar{y}_i - \bar{y}) - b_B(\bar{x}_i - \bar{x}) \qquad = 1.50$$
$$P_3 - P_4 = d_4 = b_W(x_{i\alpha} - \bar{x}_i) + b_B(\bar{x}_i - \bar{x}) - b_T(\bar{x}_{i\alpha} - \bar{x}) = 0.75$$
$$\overline{P_0 - P_4 = d_T = (y_{i\alpha} - \bar{y}) - b_T(x_{i\alpha} - \bar{x}) \qquad = 4.50}$$

The numerical values of these expressions are listed in Table 14.4. The total residual sum of squares of $y$ is then

$$(yy)'' = \Sigma d_T{}^2 = \Sigma(d_1 + d_2 + d_3 + d_4)^2$$
$$= \Sigma d_1{}^2 + \Sigma d_2{}^2 + \Sigma d_3{}^2 + \Sigma d_4{}^2$$
$$= S_1 + S_2 + S_3 + S_4$$

All product terms vanish on summation. These sums of squares are indicated at the bottom of Table 14.4.

## 11. Sum of squares of adjusted treatment means

The $ssq$ of the adjusted treatment means in algebraic form is

$$n\Sigma t_i{}^2 = n\Sigma(\bar{y}_i - \bar{y})^2 + b_W{}^2 n\Sigma(\bar{x}_i - \bar{x})^2 - 2b_W n\Sigma(\bar{x}_i - \bar{x})(\bar{y}_i - \bar{y})$$

or, briefly,

$$S_t = (yy)_B + b_W{}^2(xx)_B - 2b_W(xy)_B$$

and, in our example,

$$S_t = 56 + (2.25)^2 8 - 2(2.25)4 = 78.50$$

in agreement with what we have obtained previously by longhand calculation. The above expression for $S_t$ may be rewritten in two other forms. First, adding and subtracting the quantity $b_B{}^2(xx)_B$ and writing $(xy)_B$ as $b_B(xx)_B$ in the last term, we have

$$S_t = (yy)_B - b_B{}^2(xx)_B + b_B{}^2(xx)_B + b_W{}^2(xx)_B - 2b_W b_B(xx)_B$$
$$= S_3 + (b_W - b_B)^2(xx)_B,$$

and in our example, it is $54 + (2.25 - 0.50)^2 8 = 54 + 24.50 = 78.50$. This is probably the most concise form for $S_t$.

In order to show that $S_t$ is also greater than $S_3 + S_4$, we split the last term into two parts by multiplying it by $[(xx)_W + (xx)_B]/(xx)_T = 1$, so that

$$S_t = S_3 + \frac{(b_W - b_B)^2(xx)_B(xx)_W}{(xx)_T} + \frac{(b_W - b_B)^2(xx)_B{}^2}{(xx)_T}$$
$$= S_3 + S_4 + \frac{(b_W - b_B)^2(xx)_B{}^2}{(xx)_T}$$

This completes the proof. The reason that $S_t$ is larger than $S_3 + S_4$ is that the adjusted treatment means are correlated. For an appropriate $F$ test in the analysis of variance, the mean square $(S_3 + S_4)/(k - 1)$ for treatments should be used as outlined in the routine procedure.

## EXERCISES

**14.1**   Given the three groups of paired data in the first of the accompanying tables. Assuming that the regression of $y$ on $x$ is linear, perform an analysis of variance on the

| | I | | | | Group total | II | | | | Group total | III | | | | Group total | Grand total |
|---|---|---|---|---|---|---|---|---|---|---|---|---|---|---|---|---|
| $x$ | 3 | 2 | 2 | 5 | 12 | 2 | 4 | 4 | 6 | 16 | 2 | 1 | 1 | 4 | 8 | $X = 36$ |
| $y$ | 5 | 5 | 7 | 15 | 32 | 6 | 13 | 15 | 22 | 56 | 8 | 9 | 10 | 17 | 44 | $Y = 132$ |

residual *ssq* of $y$ after the elimination of the influence of $x$.   That is, the analysis of variance is to be made on $(yy)''$ rather than on $(yy)$ directly.   It is important to perform and tabulate the arithmetic systematically.   The very first thing to do is to calculate the three basic quantities (areas $A$, $B$, and $C$) for the $y$'s, for the $x$'s, and for the products $xy$.   In the second of the accompanying tables, some numbers have been given as partial check for your own calculation.   Tabulate your further calculations and analysis as indicated in the third of the accompanying tables.

| Area | $(yy)$ | $(xy)$ | $(xx)$ |
|---|---|---|---|
| $A$ | . . . . . | 473 | . . . |
| $B$ | . . . . . | . . . | 116 |
| $C$ | 1,452 | . . . | . . . |

| Source | df | $(yy)$ | $(xy)$ | $(xx)$ | $b$ | $(yy)'$ | $(yy)''$ | df | msq | F |
|---|---|---|---|---|---|---|---|---|---|---|
| Between | | . . . | 12 | . . . | . . . . | . . . . . . | † | | . . . | |
| Within | | 248 | . . . | . . . | . . . . | 211.25 | . . . . . . | 8 | . . . | |
| Total | | . . . | . . . | 28 | 2.75 | | 108.25 | | . . . | |

† Obtained by subtraction.

**14.2**   On comparing the $y$'s in Exercise 14.1 with those of Table 14.1, we observe that the former is the sum $x + y$ of the latter.   Suppose that $x$ is the initial weight of an animal before treatment and $y$ in Table 14.1 is the gain in weight after treatment. Then the $y$ in the preceding example is the final weight of the animal at the end of the experimental period.   The question that may be raised is which $y$ should be used in eliminating the influence of the initial difference in weight.

The result of the preceding example shows that we may use either the gain in weight or the final weight in the analysis of covariance (Li, 1957, p. 368).   Note that the right half of the table above (right of the heavy line) is identical with the results given in Table 14.2B.   Also note that the regression coefficients in the example above are greater than the previous $b$'s by unity.

In order to show these relationships, we shall refer the original $y$ (gain in weight in Table 14.1) as the old $y$ or plain $y$, and refer to the "$y$" $= x + y$ in Exercise 14.1 as the new $y$. It is sufficient to sketch the proof for a single group.

$$\text{New } (xy) = \Sigma x(x + y) - \frac{X(X + Y)}{n} = (xx) + (xy)$$

Hence,

$$\text{New } b = \frac{(xx) + (xy)}{(xx)} = 1 + \frac{(xy)}{(xx)} = 1 + \text{old } b$$

$$\text{New } (yy) = \Sigma(x + y)^2 - \frac{(X + Y)^2}{n} = (xx) + (yy) + 2(xy)$$

$$\text{New } (yy)'' = \text{new } (yy) - \frac{\text{new } (xy)^2}{(xx)}$$

$$- (xx) + (yy) + 2(xy) - \frac{[(xx) + (xy)]^2}{(xx)}$$

$$= (yy) - \frac{(xy)^2}{(xx)} = \text{old } (yy)''$$

Hence the residual *ssq* of $y$, that is, $(yy)'$, remains the same as before.

**14.3** It is desirable to have more practice with simple numbers on the procedure of the analysis as outlined in this chapter. The following is an example in which there are four groups of three pairs of observations each ($k = 4$ and $n = 3$, so that $N = 12$).

| | $x$ | $y$ | $x$ | $y$ | $x$ | $y$ | $x$ | $y$ | Total | |
|---|---|---|---|---|---|---|---|---|---|---|
| | 3 | 2 | 1 | 8 | 2 | 5 | 4 | 11 | | |
| | 2 | 4 | 1 | 9 | 5 | 10 | 6 | 16 | | |
| | 4 | 9 | 4 | 13 | 2 | 3 | 2 | 6 | $X$ | $Y$ |
| $X_i, Y_i$ | 9 | 15 | 6 | 30 | 9 | 18 | 12 | 33 | 36 | 96 |
| | | | | | | | | | $XY = 3456$ | |
| $X_iY_i$ | 135 | | 180 | | 162 | | 396 | | $\Sigma X_iY_i = 873$ | |
| $\Sigma xy$ | 50 | | 69 | | 66 | | 152 | | $\Sigma\Sigma xy = 337$ | |

After calculating the three basic quantities $A$, $B$, and $C$ for $y$, for $x$, and for $xy$, complete the accompanying table of analysis. After the elimination of the influence of $x$ on $y$, are the four groups significantly different by the $F$ test?

| Variation | $df$ | $(yy)$ | $(xy)$ | $(xx)$ | $b$ | $(yy)'$ | $(yy)''$ | $df$ | $msq$ |
|---|---|---|---|---|---|---|---|---|---|
| Between groups | 3 | 78 | 3 | ... | | | 88.43 | 3 | .... |
| Within groups | 8 | 116 | ... | 22 | 2.09 | 96.18 | ..... | 7 | 2.83 |
| Total | 11 | ... | 49 | 28 | 1.75 | ..... | 108.25 | 10 | |

**14.4** If you like, carry a more detailed analysis similar to Table 14.3 and identify the four independent components $(S_1, S_2, S_3, S_4)$ of the total residual $(yy)'' = 108.25$.

| Group | df | (yy) | (xy) | (xx) | b | df | (yy)' | df | (yy)'' |
|-------|----|------|------|------|----|----|-------|----|--------|
| (1) | 2 | 26 | 5 | 2 | 2.5 | 1 | 12.50 | 1 | 13.50 |
| (2) | 2 | ... | ... | ... | 1.5 | 1 | 13.50 | 1 | 0.50 |
| (3) | 2 | ... | ... | ... | 2.0 | 1 | 24.00 | 1 | 2.00 |
| (4) | 2 | 50 | 20 | 8 | 2.5 | 1 | 50.00 | 1 | 0 |
| | | | | | (sum) | 4 | 100.00 | 4 | $16.00 = S_1$ |
| | | | | | | | | 3 | $3.82 = S_2$ |
| Within | 8 | 116 | 46 | 22 | 2.09 | 1 | 96.18 | 7 | $19.82 = S_1 + S_2$ |
| Between | 3 | 78 | 3 | 6 | 0.50 | 1 | 1.50 | 2 | $76.50 = S_3$ |
| | | | | | (sum) | 2 | 97.68 | 9 | $96.32 = S_1 + S_2 + S_3$ |
| | | | | | | | | 1 | $11.93 = S_4$ |
| Total | 11 | 194 | 49 | 28 | 1.75 | 1 | 85.75 | 10 | $108.25 = \Sigma S$ |

It has been mentioned that the component $S_2$ is due to the variation of $b_i$ of the single groups. To test the significance of this variation (that is, testing $b_i = b_W$), we may use

$$F(3, 4) = \frac{S_2/(k - 1)}{S_1/k(n - 2)} = \frac{3.82/3}{16.00/4} = 0.32$$

Thus we conclude that the difference in within-group regression is insignificant and the groups may be taken as having a common regression of $b_W = {}^{23}\!/_{11} = 2.09$. This situation is usually assumed, so that in the routine analysis (Exercise 14.3) the estimate of error variance is

$$s^2 = \frac{S_1 + S_2}{N - k - 1} = \frac{19.82}{7} = 2.83$$

This may then be used to test the component $S_4$ due to the difference $b_W - b_B$.

part **2**

# Experimental designs

# 15

# Randomized experiments

In the fourteen short chapters of Part 1 are covered the underlying theoretical elements of the analysis of variance. In the next fifteen chapters, Part 2, we shall give a series of experimental designs that are of general application. Complicated designs for agricultural students are not included here. In the chapters of Part 2 the full details of the mathematical model and the algebraic expressions are not always explicitly given as in Part 1, but from the procedure of analysis it is clear that they are based on the same type of linear models, and the subdivision of sum of squares is based on the same type of algebraic identities. The main purpose of this chapter is to introduce some basic considerations and terminology in experimentation and statistical analysis.

## 1. Wisdom versus experimentation

I do not recall the source but I do remember the story that I read years ago. One day a king asks his wise men: *Does* a bottle weigh more when a fly gets into it?   One wise man thinks not, and gives his reasons; he is immediately supported by his group of wise men.   Another wise man thinks that the bottle will weigh more; his reasons are equally convincing and gain support from his group of wise men.   The arguments between the two groups are highly sophisticated, eloquent, philosophical, and scholarly.   But the king feels that no decision can be made with respect to his original question.   So, he suggests that the wise men get a bottle and a fly and resolve the problem.   This they are reluctant to do, I suppose because the ignorant fly will prove one or the other group wrong, and certainly no wise man wants to take that risk.

What they did find out I do not remember and hardly care.   The point is that the king is the scientist.   How can abstract arguments settle a point of fact?   Facts can be verified only by making actual observations.

The reader may say: "I can ask my physics professor.   He can tell me the correct answer without having to get a bottle and a fly."   Probably he can, I agree.   If he does know the answer from "theoretical" considerations, that is only because many similar events have been repeatedly observed previously.   He may not have to do a new experiment on this particular problem; that is only because many similar experiments have been done by previous scientists and the consistent results have already been summarized in the form of a general statement (a physical law) which includes the case of the bottle and the fly.   If nothing of a similar nature has ever been done before, your physics professor will be no wiser than the king's wise men.   A theory or a law is merely a summary of a body of consistent facts.   Nothing is a priori obliged to "obey" a law.   New experiments and new results lead to new laws.   In one word, theory follows facts, not facts follow theory.   Hence, experimentation is the *ultimate* source of new knowledge, new laws, and new applications.

## 2. Design and analysis

Design and analysis are inseparable.   When an experiment is carried out in a certain appropriate manner and the data are recorded accordingly, there may exist a corresponding appropriate method for analyzing the data.   I say "there may exist" because I admit that statistical science, like all other sciences, is continuously developing, and more comprehensive techniques are emerging everyday to cope with more complicated situations.   I sympathize with the experimenter's argument that his work should not be cut and trimmed to fit a certain statistical method, but that the statistician should develop techniques to suit his experiment.   Of course, the mathematical statisticians are doing just that—developing special techniques for special situations. But there must also be some limit to the experimenter's freedom in designing his experiments. There are such things as good or bad designs. Analyses can be very simple or very complicated according to their design. And the experimenter is obliged to consider the design of his experiment and its statistical consequences.

As an extreme example, let us imagine an experiment in which 300 Japanese children are fed fish and 300 Swedish children of the same age are fed beef for a certain period of time, say five years. The results are as follows: the mean height of the Japanese group is 62 inches and the mean height of the Swedish group is 67 inches. The experimenter wants to know the effect of diet on the youngsters' height. A statistician is consulted and promptly says: "The effect of diet and the racial difference are completely confounded and there is no way to separate them. From the data available I cannot say how important is the effect of diet or how pronounced is the racial difference." Turning to an authority in nutrition, the experimenter is told: "I believe most of the difference is due to the diet." An anthropologist responds with: "I believe most of the difference reflects the difference between the two races." What each believes may be right or may be wrong. There is no way to tell. The role of design is to make it possible to reach objective conclusions and to play down the opinions of "authorities." If I were the statistician in this case and the experimenter had said to me, "If you cannot tell us anything, what is a statistician for?", my answer would be, "It takes a statistician to insist that from your experiment no conclusion can be drawn."

In many medical experiments the diagnostic classification of patients often presents a difficult problem. Again, let us consider a hypothetical example. Five radiologists examine an X-ray picture of the chest of a patient; two of them say there is a lesion in the chest and the other three say there is none. They then expect the statistician to decide whether this patient has or has not a lesion. This situation is much like that involved in weighing mice with a defective balance or false scales. To ask a statistician to determine the right weight of mice from such data is to ask him to go beyond his capability, if not beyond his responsibility. Assuring accuracy in the weighing of a mouse or in the diagnosing of a patient is primarily the experimenter's job. To increase accuracy in determining the weight of a mouse, the only real improvement lies in using a good balance.

Those aspects of design which are within the responsibility of a statistician are discussed in the first few chapters of the textbooks listed at the end of this book. A most informative discussion is that of D. R. Cox (1958). The student must turn to such comprehensive books for detailed expositions. What I say in this introductory chapter is intended merely to direct the attention of students to some of the essential elements of the design of experiments.

## 3. Experimental unit

The first consideration in executing an experiment is a clear awareness of what constitutes the "experimental unit." An *experimental unit* is the object to which the treatment is applied and in which the variable under investigation is measured and analyzed. In an experiment involving egg-laying capacity in poultry, the unit is a hen. In testing the efficacy of an insect repellent, if the two sides of a cow are to be sprayed with two different repellents, each side of the cow is a unit, while the whole cow consists of a pair of units. In most agricultural field trials the experimental unit is a plot of land rather than an individual plant. However, it is not always clear what the unit should be, especially in human experiments. We have all heard of cases in which: "The operation was a success but the patient died" or "The tumor has regressed but the patient died." Even more difficult to evaluate are those experiments in which there are more tumors than one and some regress while others grow larger.

In most ordinary human experiments in which the treatment affects the whole individual, the individual is the unit. If the effect is local in nature, an individual may sometimes be used as a pair of units (right and left arms, etc.). Each unit will provide a measurement or measurements of the variable(s) under study. Sometimes, several treatments are given to the same individual at different times with intervening rest periods (Chap. 18). This individual is more than a unit; it may be viewed as a "block" (Chap. 16), which means a group of units. Throughout this book we assume that one unit gives one observation on the variable under study, except in Chap. 30, where several variables are measured on each unit.

## 4. Randomization

Randomization is a general term applicable to various processes. Children use it quite frequently in playing games where a random choice is needed. So they say:

> Eeny, meeny, myny, moe;
> Catch a tiger by the toe.
> If he hollers let him go;
> Eeny, meeny, myny, moe.

Since the number of words is fixed in this verse, this is not a true randomization, but that is what it is meant to be. The fact is that the children do not use this particular verse all the time but have at their disposal a collection of them with varying number of words. For instance:

> Aka baka soda cracker,
> Aka baka boo.
> In comes Uncle Sam,
> And out goes Y-O-U.

A random choice of these leads to randomization.

In experiments we adopt a similar procedure. The different treatments (drugs, diet, etc.) are to be assigned to experimental subjects at random. Even those of us who have never used the term know that we would not compare the height of Americans and Indians, at least not knowingly, by choosing especially tall Americans and especially short Indians. We are all aware that the choice should be "at random." Nor would we give one kind of food to heavy guinea pigs and another kind of food to thin guinea pigs in a nutrition experiment. Foods (or other treatments) must be assigned to animals at random. Randomization forms the basis of any valid statistical test. The role of randomization in executing an experiment is exceptionally well presented in R. A. Fisher's "The Design of Experiments" (1935 and 1960), a book that should be read by everyone who aspires to carry on experimentation.

To assist experimental scientists in assigning treatments at random to the units, many tables of random digits (commonly known as random numbers) have been prepared by statisticians. Table 15.1 gives a small sample of them for illustration. If, for instance, three treatments ($a$, $b$, $c$) are to be assigned at random to 12 animals with the restriction that each treatment should be applied to 4 animals, the assigning may be done in many different ways. We may line the 12 animals in a single file and assign them the corresponding numbers of the first row (or any other row) of Table 15.1 and then let the 4 animals with the smallest 4 numbers (13, 16, 19, 26) receive treatment $a$, and so on. Or we may give a random number (from the table) to each animal, and those ending in digits 1, 2, 3 then receive treatment $a$, those ending in 4, 5, 6 receive treatment $b$, and so on, ignoring the digit 0. When four animals are already in treatment $a$, we ignore digits 1, 2, 3 and read the rest. No fixed rules of randomization can be given, and the experimenter may devise his own randomization procedure. The numbers in Table 15.1 may be used horizontally as well as vertically. Although the table is printed in two-digit form, it may be combined in various ways. For instance, the first row may be taken as 12 numbers with 2 digits each,

*Table* 15.1   Random digits

| | | | | | | | | | | | |
|---|---|---|---|---|---|---|---|---|---|---|---|
| 26 | 19 | 13 | 27 | 95 | 69 | 73 | 54 | 97 | 86 | 65 | 16 |
| 90 | 29 | 81 | 74 | 34 | 98 | 16 | 95 | 03 | 65 | 94 | 61 |
| 08 | 53 | 56 | 52 | 49 | 15 | 44 | 16 | 86 | 30 | 72 | 18 |
| 13 | 20 | 59 | 95 | 66 | 97 | 84 | 72 | 73 | 16 | 83 | 42 |
| 81 | 27 | 36 | 83 | 55 | 12 | 43 | 83 | 05 | 88 | 57 | 87 |
| | | | | | | | | | | | |
| 35 | 78 | 84 | 93 | 94 | 14 | 10 | 56 | 11 | 50 | 64 | 12 |
| 61 | 88 | 74 | 60 | 22 | 09 | 94 | 69 | 36 | 04 | 99 | 84 |
| 12 | 31 | 46 | 97 | 62 | 75 | 56 | 98 | 94 | 05 | 25 | 97 |
| 16 | 79 | 76 | 11 | 42 | 25 | 48 | 69 | 34 | 95 | 67 | 54 |
| 53 | 13 | 27 | 84 | 57 | 65 | 23 | 27 | 64 | 73 | 11 | 55 |
| | | | | | | | | | | | |
| 58 | 87 | 03 | 56 | 33 | 46 | 00 | 23 | 79 | 76 | 36 | 62 |
| 48 | 18 | 85 | 65 | 10 | 25 | 54 | 18 | 45 | 24 | 66 | 79 |
| 79 | 12 | 53 | 67 | 21 | 77 | 12 | 01 | 16 | 30 | 72 | 76 |
| 41 | 35 | 26 | 91 | 94 | 70 | 72 | 10 | 46 | 68 | 93 | 19 |
| 37 | 63 | 00 | 43 | 89 | 06 | 97 | 92 | 67 | 83 | 52 | 14 |
| | | | | | | | | | | | |
| 28 | 26 | 66 | 25 | 62 | 99 | 76 | 45 | 73 | 32 | 96 | 07 |
| 73 | 11 | 20 | 67 | 70 | 19 | 65 | 80 | 69 | 01 | 80 | 47 |
| 66 | 24 | 48 | 12 | 13 | 94 | 93 | 47 | 54 | 64 | 03 | 40 |
| 21 | 97 | 59 | 73 | 66 | 29 | 74 | 39 | 30 | 89 | 05 | 10 |
| 62 | 34 | 99 | 79 | 54 | 37 | 02 | 69 | 91 | 83 | 70 | 68 |

or as 6 numbers with 4 digits each (2619, 1327, etc.), or as 24 single digits (2, 6, 1, 9, 1, 3, etc.).   The important point to remember is that every animal has the same chance to receive any of the treatments and there is no systematic rule to give a particular animal a particular treatment.   In a word, a table of random digits is a highly sophisticated and versatile version of "Eeny, meeny, myny, moe."

## 5. Repetition

Another essential feature of experimentation is *repetition* or, more technically, *replication*.   A single observation or a unique event seldom allows us to draw an inference.   (There are exceptions, of course.) Experimentation is essentially a process of learning, and we learn by repetition.   On repeating any observation, we find that the results of the second round are not identical with those obtained in the first. Indeed, they may be quite different.   Publius Syrus says: "When two do the same thing, it is not the same thing after all."   This variation has to be taken into account in the analyses of experiments.

There are actually many sources of variation.   Successive measurements on the same object will yield different readings owing to human variation in the act of taking observations, owing to instrument variation in indicating the exact measurement, or both. When measurements are taken not on the same object, but on a group of seemingly uniform objects, an additional source of variation is introduced owing to the fact that no two objects are really identical. A fundamental phenomenon in biology is variation; some of it is systematic and could be explained, and some is apparently random fluctuation. The nature of the random variation can be observed and described only by repeated observations under a given condition in which all the systematic variations are controlled.

The main purpose of experimentation is to separate the systematic and the random variations by repeated observations. The unexplained random variation is known as random error or, briefly, *error*, which is a technical term and does not mean a mistake or a goof. It is part of the innate biological characteristics.

From the brief discussion above, it is clear that repetition, randomization, and estimate of error are interlocked to become one comprehensive operation in experimentation, and we cannot have one without the other two. An example of the analysis of the simplest type of experiment will be given in the last two sections of the chapter.

## 6. Manipulation of data

A point which I often find difficult to get across to beginners is the futility of artificial manipulation of data. Such manipulation can only do harm to an appropriate interpretation. (Again, R. A. Fisher gives an excellent discussion of this point.) The amount of information contained in a body of data is fixed. No manipulation can create information that is not there. An experiment can answer only a limited number of questions. The investigator must spell out precisely what he wishes to learn from his experiment before he carries it out.

Eager but inexperienced research workers rightly feel that they must "find out something" from their experiments, but by "finding out something" they invariably mean establishing "significant results." If the results turn out to be nonsignificant, they often try various "corrections" (usually the exclusion of certain observations) which they hope will help them to achieve significance; for they feel that nonsignificance means failure. Nothing could be more wrong—or more harmful to experimentation—than this sort of a posteriori manipulation. Nonsignificant results tell the experimenter just as much as significant ones, and they are just as valuable to our accumulated knowledge. The quality of an experiment lies in its design and its accuracy of measurement, not in the significance or nonsignificance of the results. If you think that an important factor has been neglected in your present experiment, include it in your next experiment. The urge to achieve significant results may partially explain the frequent appearance of "effective" drugs which are rejected on further testing. Also, *ad hoc* explanation for particular experiments, though satisfying, perhaps, for the moment, seldom can be verified and thus contribute little to our understanding of the subject matter.

An experiment should be self-contained; that is, it should provide an independent piece of information about a certain question, and the conclusions should be based on the experimental data alone. Extraneous knowledge may help the investigator to understand the observed results, but it should not be injected into the statistical inferences. This does not mean one cannot use accumulated knowledge in planning further experiments. Experimentation is only one step in the continuous endeavor to learn.

So much for preliminary remarks. In the rest of the book all numbers are assumed to be bona fide measurements in some physical unit (grams, inches, etc.), and we shall not discuss whether the instruments employed are good or bad or how the reading is actually taken. This part of the problem cannot be discussed in general terms.

## 7. A randomized experiment

Suppose that there are five treatments (1, 2, 3, 4, 5) to be tested on 20 experimental units (mice, say). We shall further assume that these 20 units are uniform or homogeneous; they are all inbred mice from a certain strain and are of the same age, weight, color, and sex and were born at the same time. Apparently, we have no way to differentiate them at all and must take them as a homogeneous group.

We further decide that each of the five treatments is to be tested on four mice. That is, each treatment is to be repeated four times. The treatments are assigned to the mice according to some random scheme, or by using the random digits (Table 15.1) in some way. Then the five treatments naturally divide the animals into five treatment groups, and this is the only criterion for tabulating the data. The results of the experiment are shown in Table 15.2, which is usually known as a one-way classification table (that is, classification by treatments only).

The data throughout this book are hypothetical and given as whole numbers. There is no loss of generality in using whole numbers. The decimal point merely reflects the arbitrary unit employed by the investigator and has no other meaning. Suppose that the data in Table 15.2 are in units of centimeters. Had the meter been used as the unit of

*Table* 15.2   Data on a randomized experiment involving
five treatments, on each of which four observations have
been made

| Treatment | Observed values $y_{i\alpha}$ | | | | Total $Y_i$ | Mean $\bar{y}_i$ | Effect $t_i$ |
|---|---|---|---|---|---|---|---|
| (1) | 11 | 18 | 24 | 15 | 68 | 17 | −3 |
| (2) | 26 | 25 | 22 | 11 | 84 | 21 | +1 |
| (3) | 13 | 19 | 22 | 10 | 64 | 16 | −4 |
| (4) | 26 | 21 | 19 | 22 | 88 | 22 | +2 |
| (5) | 19 | 27 | 28 | 22 | 96 | 24 | +4 |
| Grand total and general mean | | | | | 400 $Y$ | 20 $\bar{y}$ | 0 |

measurement, the observed values would be 0.11, 0.18, 0.24, etc.   The
method of analysis and the conclusions remain the same.

The analysis of the data is based on the contents of Chaps. 5 to 8
and will not be repeated in detail here.   Briefly, each observed value is
regarded as consisting of three independent components (Chap. 6):

$$\begin{array}{ccccccc} \text{Observed} & = & \text{general} & + & \text{treatment} & + & \text{random} \\ \text{value} & & \text{constant} & & \text{effect} & & \text{error} \\ y_{i\alpha} & = & \bar{y} & + & t_i & + & e_{i\alpha} \end{array}$$

where $\bar{y}$ is the general mean of all observations, $t_i = \bar{y}_i - \bar{y}$ is the deviation of the group mean from the general mean, and $e_{i\alpha} = y_{i\alpha} - \bar{y}_i$ is the
deviation of the observed value from its own group mean.   The numeri-
cal values of these components for the data of Table 15.2 are given in
Table 15.3.

To perform an analysis of variance and test of the treatment effects,
we adopt the shortcut computational procedure of calculating the follow-

*Table* 15.3   Three components of observed values

| Treatment group | Observed value $y_{i\alpha}$ | Common element $\bar{y}$ | Treatment effect $t_i$ | Random errors $e_{i\alpha}$ | | | |
|---|---|---|---|---|---|---|---|
| (1) | $y_{1\alpha}$ | 20 | −3 | −6 | +1 | +7 | − 2 |
| (2) | $y_{2\alpha}$ | 20 | +1 | +5 | +4 | +1 | −10 |
| (3) | $y_{3\alpha}$ | 20 | −4 | −3 | +3 | +6 | − 6 |
| (4) | $y_{4\alpha}$ | 20 | +2 | +4 | −1 | −3 | 0 |
| (5) | $y_{5\alpha}$ | 20 | +4 | −5 | +3 | +4 | − 2 |

ing three basic quantities (from Table 15.2)

$$A = 11^2 + 18^2 + \cdots + 28^2 + 22^2 = 8{,}586 = \sum y_{i\alpha}^2$$

$$T = \tfrac{1}{4}(68^2 + 84^2 + 64^2 + 88^2 + 96^2) = 8{,}184 = \frac{1}{n} \sum Y_i^2$$

$$C = \tfrac{1}{20}(400^2) = 8{,}000 = \frac{1}{N} Y^2$$

It may easily be verified that the sum of squares between and within treatment groups are respectively

$$T - C = 184 = n \sum_i t_i^2 = 4(3^2 + 1^2 + 4^2 + 2^2 + 4^2)$$

$$A - T = 402 = \sum_{i\alpha} e_{i\alpha}^2 = 6^2 + 1^2 + 7^2 + \cdots + 3^2 + 4^2 + 2^2$$

and the total $ssq$ is $A - C$. In practice the individual values of $t_i$ and $e_{i\alpha}$ are never used for computation and we always calculate $A$, $T$, $C$ directly from the data table. The analysis of variance is given in Table 15.4, in which the references to the appropriate chapters are indicated. Thus, this section serves as a review of Chaps. 5 to 8, which lead to the analysis of variance and pave the way for subsequent chapters.

In this particular example, $F = 1.7$ approximately with 4 and 15 degrees of freedom for the numerator and denominator, respectively. The treatment differences are not much larger than the random variations within groups, and they could very well arise by chance. The tabulated value of $F$ with the same $df$ at the 5 per cent significance level is 3.06. Hence we conclude that the treatment differences are not statistically significant. Usually, the analysis stops here because of the very low value of $F$. Later chapters provide methods for further analysis of the treatment effects by further subdividing the treatment sum of squares, but that depends on the nature of the treatments and the predetermined hypothesis to be tested. For the time being we are merely testing the general null hypothesis $t_i = 0$ or $\sigma_t^2 = 0$. A nonsignificant

*Table* 15.4  **Analysis of variance (data from Table 15.2)**

| Source | $df$ | Chaps. 5 and 6<br>$ssq$ | Chap. 7<br>$msq$ | $E\{msq\}$ | Chap. 8<br>$F$ |
|---|---|---|---|---|---|
| Treatments | 4 | $T - C = 184$ | 46.0 | $\sigma_e^2 + n\sigma_t^2$ | 1.7 |
| Error | 15 | $A - T = 402$ | $s^2 = 26.8$ | $\sigma_e^2$ | |
| Total | 19 | $A - C = 586$ | | | $F_{.05} = 3.06$ |

result means that we have no reason to reject the null hypothesis. To put it the other way, we let the null hypothesis stand.

The estimate of the error variance is $s^2 = 26.80$ for single observations. The variance of a treatment mean based on $n = 4$ observations is (Chap. 4)

$$V(\bar{y}_i) = V(t_i) = \frac{s^2}{n} = \frac{26.80}{4} = 6.70$$

The variance of the difference between two treatment means is

$$V(\bar{y}_i - \bar{y}_j) = V(\bar{y}_i) + V(\bar{y}_j) = \frac{s^2}{n} + \frac{s^2}{n} = \frac{2s^2}{n} = 13.40$$

and the standard error is $s_d = \sqrt{2s^2/n} = \sqrt{13.40} = 3.66$. To compare the first two treatments, we may use either the $F$ or the $t$ statistics:

$$F = \frac{(21 - 17)^2}{13.40} = 1.194 \qquad t = \frac{21 - 17}{3.66} = 1.0927$$
$$\text{with 1, 15 } df \qquad\qquad \text{with 15 } df$$

Note that $(1.0927)^2 = 1.194$. The difference is nonsignificant. For multiple comparisons, see Chap. 31.

## 8. Unequal numbers

For a simple randomized experiment (one-way classification), the treatment groups need not have the same number of observations. When they have unequal numbers, the analysis is equally straightforward. As an example, the first two groups of Table 15.2 have been combined into one treatment group which then contains eight observations, while each of the remaining groups has four. The new data are presented in Table 15.5. The two basic quantities $A$ and $C$ remain the same as before:

$$A = 8{,}586 \qquad C = 8{,}000 \qquad A - C = 586$$

The basic quantity $T$ is calculated as follows:

$$T = \sum \frac{Y_i^2}{n_i} = \frac{152^2}{8} + \frac{64^2 + 88^2 + 96^2}{4} = 8{,}152$$

The analysis of variance is given in the lower portion of Table 15.5. Note the change in degrees of freedom. Since there are now four ($t = 4$) treatments, there are 3 $df$ for the treatment $ssq$ and $N - t = 20 - 4 = 16$ $df$ for error. The estimate of the error variance is $s^2 = 27.125$. The variance of the difference between treatment (1) mean and one of the others is ($i = 2, 3, 4$)

$$V(\bar{y}_1 - \bar{y}_i) = \frac{s^2}{n_1} + \frac{s^2}{n_i} = s^2\left(\frac{1}{n_1} + \frac{1}{n_i}\right) = s^2\left(\frac{1}{8} + \frac{1}{4}\right)$$

*Table 15.5*  A randomized experiment involving four
treatments with unequal number of observations

| Treatment | Observed values $y_{ia}$ | Size $n_i$ | Total $Y_i$ |
|:---:|:---:|:---:|:---:|
| (1) | 11  18  24  15<br>26  25  22  11 | 8 | 152 |
| (2) | 13  19  22  10 | 4 | 64 |
| (3) | 26  21  19  22 | 4 | 88 |
| (4) | 19  27  28  22 | 4 | 96 |
| Total | | $N = 20$ | $Y = 400$ |

Analysis of variance

| Source | $df$ | $ssq$ | $msq$ | $F$ |
|:---:|:---:|:---:|:---:|:---:|
| Treatments | 3 | $T - C = 152$ | 50.667 | 1.87 |
| Error | 16 | $A - T = 434$ | 27.125 | |
| Total | 19 | $A - C = 586$ | | $F_{.05} = 3.24$ |

For further analysis, however, various modifications will be required
(for example, Chap. 12, Sec. 7).

The most important requirement for conducting a simple randomized
experiment is a homogeneous group of experimental units.   It is not
that we do not know a homogeneous group is nonexistent, but sometimes
circumstances do not permit us to resort to further grouping of the units
(a subject to be discussed in the next chapter).   In many human experiments this is all an investigator can do.   Nevertheless, all the essential
ingredients of proper experimentation—randomization, replication, and
estimate of error—are present.

### EXERCISES

**15.1**   The data on four treatments, each replicated three times, are given in the
accompanying table.   Perform an analysis of variance and test the significance of
treatments.

Treatments

| | a | b | c | d |
|:---:|:---:|:---:|:---:|:---:|
| | 55 | 64 | 55 | 50 |
| | 47 | 55 | 49 | 41 |
| | 48 | 64 | 52 | 44 |
| Total | 150 | 183 | 156 | 135 |

*Partial Ans.:* (area) $A = 33,002$, $s^2 = {}^{152}\!/_8 = 19$, $F = 7.05$.

**15.2** If the last two groups of Table 15.5 are combined, the data are as given in the accompanying table. Perform an analysis of variance and test the significance of the treatments.

| Treatment | Observed values | Size | Total |
|:---:|:---:|:---:|:---:|
| (1) | 11  18  24  15  26  25  22  11 | 8 | 152 |
| (2) | 13  19  22  10 | 4 | 64 |
| (3) | 26  21  19  22  19  27  28  22 | 8 | 184 |
| | Total | 20 | 400 |

*Partial Check:* $T = 2,888 + 1,024 + 4,232 = 8,144$, $s^2 = 26$.

**15.3** Find the individual values of $e_{i\alpha}$ for the above data and verify that

$$\Sigma e_{i\alpha}^2 = 442 = A - T$$

**15.4** There are nine treatment groups, each consisting of four observations.

| (1) | (2) | (3) | (4) | (5) | (6) | (7) | (8) | (9) | |
|:---:|:---:|:---:|:---:|:---:|:---:|:---:|:---:|:---:|:---:|
| 12 | 13 | 26 | 8 | 11 | 28 | 19 | 17 | 24 | |
| 8 | 28 | 21 | 26 | 20 | 27 | 24 | 18 | 24 | |
| 11 | 13 | 33 | 18 | 22 | 28 | 10 | 14 | 23 | |
| 14 | 24 | 31 | 16 | 30 | 24 | 20 | 18 | 17 | |
| 45 | 78 | 111 | 68 | 83 | 107 | 73 | 67 | 88 | 720 |

Verify the following calculations:

$$(\text{area}) \; A = 12^2 + 8^2 + \cdots + 23^2 + 17^2 = 16,028.00$$
$$(\text{area}) \; B = \tfrac{1}{4}(45^2 + 78^2 + \cdots + 88^2) = 15,238.50$$
$$(\text{area}) \; C = \tfrac{1}{36}(720^2) = 14,400.00$$

| Source | df | ssq | msq |
|:---|:---:|:---:|:---:|
| Treatments (between groups) | 8 | $B - C = \quad 838.50$ | 104.81 |
| Error (within groups) | 27 | $A - B = \quad 789.50$ | 29.24 |
| Total | 35 | $A - C = 1,628.00$ | |

What is the value of $F$? Are the treatment differences significant?

# 16

# Randomized blocks

When the experimental units are not all alike, it should be intuitively clear that the variation in the units themselves may blur the true treatment effects. The setup known as the *randomized blocks*, to be described in this chapter, deals with a method for handling the heterogeneity of the experimental units, and it is the most fundamental design in all types of experimentation.

Historically, randomized blocks is the first valid design to estimate the experimental error and to test the significance of treatment effects in spite of the heterogeneity of the experimental units on which the observations are made. This is the design that revolutionized agricultural experiments from Rothamsted to Nanking more than three decades ago. It is no exaggeration to claim that it is still the backbone of the science of experimental design. However, no design can become popular and be accepted for general usage, no matter how sound it is statistically, if it is complicated in setting up and requires difficult procedures of analysis. The beauty of the randomized blocks is the happy combination of validity, simplicity, and flexibility.

## 1. Heterogeneity of units

When we say that the experimental units are not homogeneous, we always mean that they may not react or respond to the treatments in the same way or to the same extent because of their innate differences, and we do not necessarily mean that they do not look alike. Suppose that we wish to test the effect of several diets on the growth of mice. If females and males, regardless of their hair color, react to the diet differently, then we regard sex, not color, as a source of heterogeneity. Now, suppose that we wish to test the effects of heat treatments or insect repellents; it is conceivable that the hair color may make some difference and sex is irrelevant. If so, hair color rather than sex will be considered as a source of heterogeneity among the experimental units. These examples are hypothetical; the important point is that the term "heterogeneity" is one that is not absolute, but relative to the particular treatments under consideration for an experiment. Age, breed, and sex per se may or may not be sources of heterogeneity, depending on the nature of the experiment. However, in most biological experiments dealing with nutrition and disease, they happen to be relevant sources of heterogeneity and should be taken into consideration in designing the experiment.

## 2. Principle of blocking

We have seen in the preceding chapter the ideal situation in which there is no heterogeneity among the units and there is no need for any special design. Now, let us suppose that we are given a group of experimental animals to test the effects of several treatments (drugs, say). Each treatment is to be replicated the same number of times to facilitate comparison. But the units are not all alike in their possible response to the drugs. The heterogeneity has been symbolized by the various shapes of the units in Fig. 16.1.

The design of the randomized blocks consists of two steps. The first is to collect the like units together to form a homogeneous group; the group thus formed is called a block. This operation, known as blocking, is shown in the middle diagram of Fig. 16.1. Thus, we see that a block is simply a group of experimental units which are either homogeneous or at least more homogeneous than the original unclassified group.

The second step is to assign the various treatments at random to the animals within each block, as shown in the bottom diagram of Fig. 16.1. This is the chief difference between randomized blocks and the completely

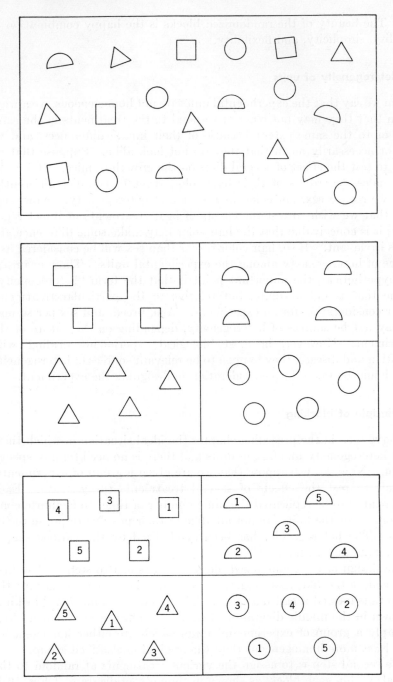

Fig. 16.1 For legend see opposite page.

randomized experiment of the preceding chapter. If the randomization of the latter is called complete, then we may say that the randomization of the former is restricted—restricted by blocks.

In the simple example depicted in Fig. 16.1, each block has five homogeneous units, and each unit therein receives a different treatment. In other words, the five treatments (1, 2, 3, 4, 5) are tried out four times, once in each block. In this simple circumstance, a block is also a replicate. A replication is a complete run for all the treatments to be tested in the experiment. Although the basic principle of blocking is as simple as that shown in Fig. 16.1, it is scarcely that simple in practice, and it is necessary for the investigator to have some prior knowledge of the animals he will use for experimentation.

Since the use of randomized blocks originates from agricultural experiments, we may mention here an example in that field to further illustrate the principle of blocking. The top of Fig. 16.2 shows the cross section of a strip of land to be used for comparison of crop yields. Let us assume that the heterogeneity of the soil is essentially along the direction from the western highland to the eastern lowland and that the strips in the north-south direction (parallel to the river) is more or less homogeneous in fertility and moisture content. The crop varieties (treatments) are to be replicated several times. How should this be done?

The novice may divide the given piece of land into a number of "blocks" along the east-west direction (perpendicular to the river) so that each block is just like any other block in its total fertility. In this case, there is no variation among the blocks but there is a great deal of variation within each block (middle diagram of Fig. 16.2). When the varieties of crops are planted within each block, some of them will be on the eastern fertile soil and others on the western dry soil, which will interfere with our evaluation of the true merit of the crop varieties. This is the wrong blocking for the purpose of experimentation. Incidentally, if this same piece of land is to be divided into several lots for residential

---

Fig. 16.1 Design of a randomized blocks experiment.

1. Given a group of 20 experimental units (mice, men, etc.) to test the effects of five treatments (chemicals, drugs, etc.). These units are not entirely alike; they are heterogeneous, as symbolized by the different shapes.

2. We sort out the experimental units according to their characteristics, so that all alike units form a group (a block). Then each block contains five homogeneous units.

3. The five treatments (denoted by 1, 2, 3, 4, 5) are assigned to the five units in each block at random. The obvious heterogeneity of the units is then controlled by the blocking process; the unseen, unknown, and uncontrolled heterogeneity of the units is randomized by the random assignment of treatment to units within each block.

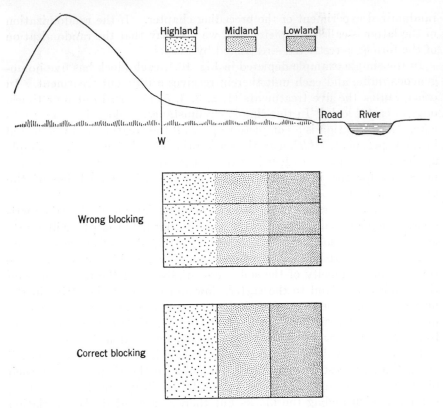

**Fig. 16.2** Top: Cross section of a piece of land. The section from W to E is to be used for experimentation. It is assumed that the fertility and moisture of the soil increases from high (W) to low (E). Middle: Wrong blocking of the land. This way, all blocks are alike, but the individual plots for treatments within the block are heterogeneous, which interferes with accurate evaluation of the treatment effects. Bottom: Correct blocking of the land. This way, the blocks are all different, but the individual plots within a block are homogeneous.

houses, this is precisely the way any realtor will divide it so that each lot is just like any other lot in having access to the road at the front and a hillside at the back.

The correct blocking for experimentation is shown at the bottom of Fig. 16.2, where the blocks are all different in fertility but the soil within each block is homogeneous. Then within each block, the varieties (say, 1, 2, 3, 4, 5) are to be planted, each occupying a "plot" at random. The situation then is analogous to that shown in Fig. 16.1.

The examples given in this section are all "idealized" situations. In practice, however, no strip of land is really homogeneous in fertility, nor are a number of animals identical in all respects. If every unit is more or less different from others, the principle is then to make the variation between blocks as large as possible and thus the variation within the blocks as small as possible. This point will be discussed in analytical terms in Sec. 5 in connection with the efficiency of the randomized blocks design.

### 3. Analysis of data

To be concrete, let us suppose that we have 4 treatments to be tested on a given lot of 36 animals. Each treatment may then be replicated 9 times. To accomplish this replication, the 36 animals are divided into 9 groups (blocks), each consisting of 4 very much alike animals. If the experiment is to be executed over a period of several weeks, then the different batches of animals received from time to time will form natural blocks. Then for each block of four animals, the four treatments are assigned at random—one to each animal. After the appropriate measurements of the variable under study are taken, the statistical analysis consists of three simple steps (Table 16.1):

1. Tabulation of data by blocks and treatments
2. Calculation of four basic quantities
3. Analysis of variance and test of significance

The entire procedure is based on the assumptions and techniques explained in Chap. 10 (Two-way classifications), which the reader may wish to review. In developing the general theory in that chapter, we merely speak of the row effects and the column effects, whatever they stand for. Now, for any concrete set of experimental data such as that exemplified in Table 16.1, we shall explicitly speak of the treatment effects and the block effects. Note that this is where the principle of blocking comes into play. Only when the blocks are different among themselves but homogeneous within can we legitimately speak of a block effect. This provides the correspondence between a physical setup and a mathematical model.

We shall not repeat the theoretical aspects of the analysis here except to reiterate that our basic assumption is

$$\begin{array}{c}\text{Each} \\ \text{observed} \\ \text{value}\end{array} = \begin{array}{c}\text{a general} \\ \text{constant}\end{array} + \begin{array}{c}\text{a block} \\ \text{effect}\end{array} + \begin{array}{c}\text{a treatment} \\ \text{effect}\end{array} + \begin{array}{c}\text{an} \\ \text{error}\end{array}$$

where the error is normally distributed with mean zero and variance $\sigma^2$. The details of practical computation are given in Table 16.1. The

*Table* 16.1   **Analysis of a randomized blocks experiment**

Tabulation of data ($Q_1$, $Q_2$, $Q_3$, $Q_4$ are treatments)

| Treatment | Blocks | | | | | | | | | Treatment total |
|---|---|---|---|---|---|---|---|---|---|---|
| | (1) | (2) | (3) | (4) | (5) | (6) | (7) | (8) | (9) | |
| $Q_1$ | 21 | 14 | 11 | 28 | 18 | 33 | 28 | 10 | 26 | 189 |
| $Q_2$ | 18 | 24 | 14 | 26 | 31 | 28 | 20 | 16 | 30 | 207 |
| $Q_3$ | 13 | 24 | 13 | 17 | 23 | 24 | 8 | 11 | 20 | 153 |
| $Q_4$ | 8 | 22 | 18 | 17 | 12 | 27 | 24 | 19 | 24 | 171 |
| Block total | 60 | 84 | 56 | 88 | 84 | 112 | 80 | 56 | 100 | 720 |

Calculation of basic quantities (areas in Chap. 10)

$$A = 21^2 + \cdots + 24^2 = 16{,}028 \qquad B = \tfrac{1}{4}(60^2 + \cdots + 100^2) = 15{,}168$$
$$T = \tfrac{1}{9}(189^2 + \cdots + 171^2) = 14{,}580 \qquad C = \tfrac{1}{36}(720)^2 = 14{,}400$$

Analysis of variance and the $F$ test

| Variation | df | ssq | | msq | F |
|---|---|---|---|---|---|
| Blocks | $9 - 1 = 8$ | $B - C$ | $= 768$ | 96.0 | |
| Treatments | $4 - 1 = 3$ | $T - C$ | $= 180$ | 60.0 | |
| | | | | | 2.12 |
| Error | $8 \times 3 = 24$ | $A - B - T + C =$ | 680 | 28.33 | |
| Total | $36 - 1 = 35$ | $A - C$ | $= 1628$ | | |

meaning and algebraic expressions for the various sums of squares have been given in detail in Chap. 10.   Here the symbols have been modified slightly: $B$ and $T$ for the quantities corresponding to blocks and treatments instead of $R$ and $K$ for rows and columns previously.   However, $R$ and $K$ will have to be used again in the next chapter in conjunction with $T$ for a more complicated design.

After the four basic quantities ($A$, $B$, $T$, $C$) have been calculated, the various sums of squares for total, blocks, and treatments are then obtained by taking differences $A - C$, $B - C$, and $T - C$, respectively. The error $ssq$ is obtained last by subtraction; thus,

$$1{,}628 - 768 - 180 = 680$$

Algebraically, it is equal to

$$(A - C) - (B - C) - (T - C) = A - B - T + C$$

For this particular experiment, $F = 60/28.33 = 2.12$ with 3 $df$ for the numerator and 24 $df$ for the denominator. The tabulated value of $F$ with corresponding number of degrees of freedom at the 5 per cent significance level is 3.01. Hence we conclude that the difference among the treatments shown in this experiment is not significant. This simply means that the difference observed in this experiment may have arisen from fluctuations caused by random sampling. The probability for such an event is greater than 5 per cent, so that we could not be reasonably sure to say that these treatments have any real difference.

It is advisable that the reader carry out and verify all the calculations shown in Table 16.1, especially the values of the basic quantities $A$ and $C$. These 36 numbers will be used again and again in later chapters to illustrate other designs or further analysis of the treatment effects.

## 4. Treatment means and their comparison

The treatment means are obtained by dividing the total by the number of blocks (also replications). The deviation of the treatment mean from the general mean is called the treatment effects. Thus, for the data of Table 16.1 we have:

|  | Treatment | | | | General mean |
|---|---|---|---|---|---|
|  | $Q_1$ | $Q_2$ | $Q_3$ | $Q_4$ | |
| Mean, $\bar{y}_i$ | 21.0 | 23.0 | 17.0 | 19.0 | 20.0 |
| Effect, $t_i$ | 1.0 | 3.0 | −3.0 | −1.0 | 0 |

The variance of the difference between any two treatment means (Chap. 12, Sec. 8) is

$$V(\bar{y}_i - \bar{y}_j) = \frac{2s^2}{r} = \frac{2 \times 28.33}{9} = 6.296$$

and the corresponding standard error is $\sqrt{6.296} = 2.509 = s_d$. The comparison of two treatment means is equivalent to the comparison of two treatment effects; for $\bar{y}_i - \bar{y}_j = t_i - t_j$. The value of $s^2 = 28.33$ is that given in the analysis of variance in Table 16.1. Thus, to compare the first two treatments $(\bar{y}_2 - \bar{y}_1) = 23 - 21 = 2.0$, we may use either the $F$ or the $t$ statistic:

$$F = \frac{2^2}{6.296} = 0.635 \qquad t = \frac{2}{2.509} = 0.797$$

where the $F$ has $(1, 24)$ $df$ and the $t$ has 24 $df$. The difference is nonsignificant.

Instead of comparing two treatments at a time, it is useful to calculate a standard magnitude for the difference between two means. In order to be significant, the value of $t$ with 24 $df$ must be at least as large as 2.064 at the 5 per cent significance level. Hence the difference between two means, $d = \bar{y}_i - \bar{y}_j$, must be such that

$$t = 2.064 < \frac{d}{2.509} = \frac{d}{s_d}$$

that is,

$$d > ts_d = 2.064 \times 2.509 = 5.18$$

Any difference smaller than 5.18 would not be significant. We have seen that the overall $F$ test (Table 16.1) gives nonsignificant results. Note, however, that $\bar{y}_2 - \bar{y}_3 = 23 - 17 = 6.0$. This is the only comparison out of the possible six pairs of comparisons that exceeds the standard magnitude 5.18. This is the difference between the largest and the smallest means. In this particular case, I would attach no significance to this difference. A more detailed discussion of the problem and a method of comparing the largest vs. the smallest mean (necessarily after examination of data) will be found in Chap. 31.

### 5. Efficiency of blocking

The advantage of grouping the experimental units into blocks over a completely randomized experiment may be further elucidated by examining and comparing the experimental errors in these two cases. Suppose that we are given $N = rt$ experimental units to test $t$ treatments, each with $r$ replications. These experimental units are given and fixed, whatever the design of the experiment. Each unit is associated with an error component $e$. Let the total sum of squares of these $N$ units be $S_T = \Sigma e^2$, which is a fixed quantity.

Under the linear model that treatment effect and random error are additive and with an orthogonal design in which the treatment effects may be isolated and eliminated from the estimation of error, we may simplify the situation by assuming that all treatment effects are zero, as would be the case if we have dummy treatments or uniformity trials. Then, if a completely randomized experiment were conducted, the error variance would simply be $E_T = S_T/(N - 1)$, that is, the total $ssq$ divided by total $df$.

Now, suppose that the units are grouped into $r$ blocks of size $t$. The total $ssq$, denoted above by $S_T$, will remain the same, but it is subdivided into two components: one is between blocks, depending on how the

*Table* 16.2 Comparison of error mean square with and without blocking

| Source | df | ssq | msq | Remark |
|--------|------|-----|-----|--------|
| Blocks | $r - 1$ | $S_H = (r - 1)H$ | $H$ | $F = H/E_W$ |
| Within | $r(t - 1)$ | $S_W = r(t - 1)E_W$ | $E_W$ | Error variance with blocking |
| Total | $rt - 1$ | $S_T = (N - 1)E_T$ | $E_T$ | Error variance without blocking |

blocking is done; and the other is within blocks. These two components of $S_T$ are designated as $(r - 1)H$ and $r(t - 1)E_W$, so that the corresponding mean squares are $H$ and $E_W$ (Table 16.2). The within-block *ssq* is, of course, smaller than the total $S_T$, but its degrees of freedom are also fewer. Hence, $E_W$ may be larger or smaller than $E_T$, depending on the actual blocking. The relationship between these two error variances is

$$E_T = \frac{S_T}{N - 1} = \frac{(r - 1)H + r(t - 1)E_W}{rt - 1}$$

And the ratio of $E_W$ to $E_T$ is

$$\frac{E_W}{E_T} = \frac{(rt - 1)E_W}{(r - 1)H + r(t - 1)E_W} = \frac{rt - 1}{(r - 1)F + r(t - 1)}$$

where $F = H/E_W$ is the ratio of the block mean square to that of error. Now it is clear from either of the above expressions that when $H = E_W$, the error variances with and without blocking will be equal, that is, $E_W = E_T$. In this case there is no gain or loss in blocking the experimental units. If $H < E_W$ because of poor blocking practice, the experimental error will be larger than that of a completely randomized block. Conversely, if $H > E_W$ as a result of blocking, the experimental error will be smaller than that without blocking, that is, $E_W < E_T$. This again emphasizes the importance of good blocking which makes $H$ as large as possible and thus makes $E_W$ as small as possible.

Translating this principle into practice, it means that we should put the units that are most nearly alike into the same block so that the blocking approaches the ideal situation depicted in Fig. 16.1.

### 6. The need for calculation

For a simple completely randomized experiment (one-way classification) the results are sometimes so obvious that an inspection of the data will give us a very good idea of the significance of differences among the treatments (see Exercise 8.4). With randomized-block experiments (two-way classification), mere inspection of the data will not be so helpful even when the data are of entirely different nature.

*Table* 16.3  Analysis of two sets of data with the same block and the same treatment effects

|  | Set I | | | | | | | Set II | | | | | |
|---|---|---|---|---|---|---|---|---|---|---|---|---|---|
|  | Treatments | | | | | | |  | Treatments | | | | |
|  | a | b | c | d | e | |  | a | b | c | d | e | |
| (i) | 17 | 19 | 18 | 20 | 21 | 95 | | 11 | 26 | 13 | 26 | 19 | 95 |
| (ii) | 21 | 22 | 16 | 23 | 28 | 110 | | 18 | 25 | 19 | 21 | 27 | 110 |
| (iii) | 19 | 25 | 17 | 25 | 29 | 115 | | 24 | 22 | 22 | 19 | 28 | 115 |
| (iv) | 11 | 18 | 13 | 20 | 18 | 80 | | 15 | 11 | 10 | 22 | 22 | 80 |
|  | 68 | 84 | 64 | 88 | 96 | 400 | | 68 | 84 | 64 | 88 | 96 | 400 |

$$A_1 = 8{,}388$$
$$R = 8{,}150 \qquad C = 8{,}000$$

$$A_2 = 8{,}586$$
$$K = 8{,}184$$

| Analysis of variance | | Set I | | | Set II | | |
|---|---|---|---|---|---|---|---|
| Source | df | ssq | msq | F | ssq | msq | F |
| Blocks (rows) | 3 | 150 | 50 | 11.1 | 150 | 50 | 2.38 |
| Treatments (columns) | 4 | 184 | 46 | 10.2 | 184 | 46 | 2.19 |
| Error | 12 | 54 | 4.5 | | 252 | 21 | |
| Total | 19 | 388 | | | 586 | | |

To illustrate the point, let us consider the two sets of data in Table 16.3. The block and treatment totals of the two experiments are identical. At first sight, one may be tempted to conclude that the experimental results are at least approximately the same.  Inspection of the single measurements shows that the value ranges from 11 to 29 in experiment I and from 10 to 28 in experiment II.  This may further convince the uncritical observer that the two sets of data are very much alike.  However, analysis (Table 16.3) shows that treatment effects in experiment I are significant at the 0.001 level and that those in experiment II are not significant even at 0.05 level.  The *ssq* and *msq* for blocks and treatments are, of course, the same for both experiments, but the error variance of experiment II is more than $4\frac{1}{2}$ times as large as that of experiment I. A difference of such magnitude could not have been guessed at by mere inspection of the data.  For more complicated experiments, we shall depend even more on calculation than on feeling derived from inspection.

### 7. Flexibility of the design

In an experiment with a randomized-block setup there is no formal restriction on the number of treatments to be tested nor is there any restriction on the number of blocks (replications) to be run.  It is this flexibility that renders the randomized blocks applicable almost to any situation. It even allows a slight change of techniques in treatments in separate blocks, although the technique must remain uniform within each block.

When there are different batches of animals arriving at, say, 2-week intervals, all the treatments should be tested on the same batch of animals. The different batches should be used as blocks, guarding against the possible heterogeneity between the batches.

Note that we say there is no formal (mathematical) restriction on the number of treatments to be tested, because in practice there is a physical limitation to the number of treatments. If the number is too large, there will be no group of animals or strip of land that is homogeneous enough to be regarded as a true block. Therefore, the "size" of a block is limited by Nature rather than by theory.

Much of the complicated design some of which will be described later in the book, arises from the fact that the number of treatments is so large that we cannot find whole blocks to accommodate them. For small experiments, however, the randomized blocks are to be highly recommended for general use.

## EXERCISES

**16.1** In agricultural field experiments the shape of blocks need not be regular (rectangles, squares) or all alike. Remember firmly that a block is determined by the homogeneity of the soil rather than by any predetermined shape. Figure 16.3

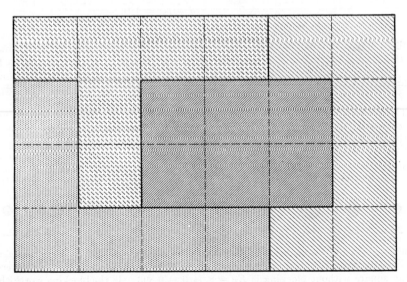

**Fig. 16.3** Area of experimental land is $4 \times 6$ square units to be divided into 4 blocks, each consisting of 6 units (plots).

shows the heterogeneity of a piece of experimental land. How would you divide it into four blocks?

**16.2** Six treatments ($a, b, c, d, e, f$) have been assigned at random to the six experimental units (animals, field plots, etc.) of each block and there have been four blocks or replications. The data are then collected and tabulated as shown in the accompanying table. Make an analysis of variance and test the significance of treatments. Also, calculate the standard magnitude for the difference between two treatment means in order to achieve significance at the 5 per cent level. (It is important to read Chap. 31 in this connection.)

| | | | Treatments | | | | Block |
|---|---|---|---|---|---|---|---|
| | $a$ | $b$ | $c$ | $d$ | $e$ | $f$ | total |
| I | 16 | 22 | 16 | 10 | 18 | 8 | 90 |
| II | 28 | 27 | 17 | 20 | 23 | 23 | 138 |
| III | 16 | 25 | 16 | 16 | 19 | 16 | 108 |
| IV | 28 | 30 | 19 | 18 | 24 | 25 | 144 |
| Treatment total | 88 | 104 | 68 | 64 | 84 | 72 | 480 |

*Partial Ans.:*

| Source | df | ssq | msq | F |
|---|---|---|---|---|
| Blocks | . . . | . . . | 108 | |
| Treatments | . . . | . . . | . . . | 7.00 |
| Error | 15 | 120 | $8 = s^2$ | |

$$s_d = \sqrt{\frac{2s^2}{r}} \qquad d = t s_d = 4.16$$

**16.3** Review thoroughly the content of Chap. 10 for the theoretical background of the analysis of variance. Find the block and treatment effects as well as the residual error. Calculate the various *ssq* components directly from these effects and see if they are in agreement with those obtained from the four basic quantities $A$, $B$, $T$, $C$.

The reader should get familiar with these 24 numbers, because they will be used as exercises for other designs in later chapters. Particularly should the reader verify (area) $A = 16^2 + \cdots + 25^2 = 10,324$.

**16.4** Given the following ten units of varying "size" (symbolizing their innate error response to the treatments) to test five treatments, each to be replicated twice, how would you divide the units into two groups (blocks) of five units each? How

| 2.3 | 3.2 | 4.2 | 2.6 | 3.6 | 4.5 | 2.8 | 3.9 | 4.8 | 2.1 |

would you make the *ssq* between the groups as large as possible? If your blocking is correctly done, the error mean square will be $E_W = 1.64/8 = 0.205$, while the error mean square without blocking is $E_T = 8.04/9 = 0.893$. Is the blocking worthwhile in this case?

**16.5** Two sets of experimental results are as given in the accompanying tables.

|  | Set I |  |  |  |  |  | Set II |  |  |  |
|---|---|---|---|---|---|---|---|---|---|---|
|  | Treatments |  |  |  |  |  | Treatments |  |  |  |
|  | *a* | *b* | *c* | *d* | *e* | *a* | *b* | *c* | *d* | *e* |
| B1 | 3.2 | 4.6 | 5.6 | 4.2 | 4.4 | 3.8 | 4.2 | 5.4 | 3.6 | 5.0 |
| B2 | 2.6 | 4.0 | 3.6 | 2.2 | 3.6 | 2.0 | 4.4 | 4.4 | 3.0 | 2.2 |
| B3 | 3.6 | 4.0 | 4.2 | 3.4 | 3.8 | 2.6 | 5.2 | 3.8 | 2.2 | 5.2 |
| B4 | 3.4 | 5.0 | 5.8 | 3.8 | 5.0 | 4.4 | 3.8 | 5.6 | 4.8 | 4.4 |

Make an analysis of variance and test the significance of treatment effects for each set of experiment. Which set gives significant results and which not?
HINT: Consult the results of Table 16.3 bottom.

**16.6** The randomized blocks design is so fundamental and so frequently used that the reader must be very familiar with the routine procedure of analysis. This is an additional exercise. Verify the four basic quantities $A = 10,324$, $B = 9,616$, $T = 10,032$, and $C = 9,600$. The reader must complete the rest of the calculations. We shall refer to this exercise again in Chap. 21 and elsewhere.

|  | Eight treatments |  |  |  |  |  |  |  | Block |
|---|---|---|---|---|---|---|---|---|---|
|  | (1) | (2) | (3) | (4) | (5) | (6) | (7) | (8) | total |
| I | 8 | 16 | 24 | 28 | 19 | 16 | 27 | 30 | 168 |
| II | 10 | 16 | 28 | 18 | 16 | 25 | 16 | 23 | 152 |
| III | 18 | 19 | 20 | 23 | 16 | 22 | 17 | 25 | 160 |
| Total | 36 | 51 | 72 | 69 | 51 | 63 | 60 | 78 | 480 |

**16.7** In Sec. 5 (Efficiency of blocking) a somewhat different system of notation has been employed in order to alert the reader to the fact that we are investigating the usefulness of blocking rather than analyzing a set of experimental data. It is good exercise for the reader to rewrite the results of that section in the ordinary system that has been followed throughout the book. Assuming no treatment effects, the total sum of squares due to the heterogeneity of the $N = rt$ units may be subdivided into two components: one between the blocks and one within the blocks with a corresponding subdivision of the degrees of freedom. Thus:

| Source | $df$ | $ssq$ | $msq$ = error variance |
|---|---|---|---|
| Between blocks | $f_H = r - 1$ | $ssq_H$ | $\dfrac{ssq_H}{f_H} = s_H{}^2$ |
| Within blocks | $f_W = r(t - 1)$ | $ssq_W$ | $\dfrac{ssq_W}{f_W} = s_W{}^2$ |
| Total | $f_T = rt - 1$ | $ssq_T$ | $\dfrac{ssq_T}{f_T} = s_T{}^2$ |

$$s_T{}^2 = \frac{ssq_T}{N - 1} = \frac{ssq_H + ssq_W}{rt - 1} = \frac{(r - 1)s_H{}^2 + r(t - 1)s_W{}^2}{rt - 1}$$

It is seen that when $s_H{}^2 = s_W{}^2$, we have $s_T{}^2 = (rt - 1)s_W{}^2/(rt - 1) = s_W{}^2$. In this case, the error variances with or without blocking are the same and there is no gain in blocking the units for experimentation. Indeed it would be expected to be so if the blocking were done at random.

**16.8** The purpose of blocking is to reduce the error variance by controlling the heterogeneity of the experimental units. In other words, our purpose is to make $s_W{}^2$ smaller, preferably much smaller, than $s_T{}^2$. Then the following inequality must hold:

$$\frac{ssq_W}{f_W} < \frac{ssq_T}{f_T} = \frac{ssq_H + ssq_W}{f_H + f_W}$$

This leads to the following inequalities:

$$ssq_W(f_H + f_W) < (ssq_H + ssq_W)f_W$$

$$ssq_W f_H < ssq_H f_W$$

That is,

$$s_W{}^2 = \frac{ssq_W}{f_W} < \frac{ssq_H}{f_H} = s_H{}^2$$

Hence, for efficient blocking, we must make the variation between blocks $s_H{}^2$ as large as possible and the variation within blocks $s_W{}^2$ as small as possible.

# 17

# Latin square

Continuing with the concept of blocking developed in the preceding chapter, we shall in this chapter discuss a more sophisticated method of blocking to control heterogeneity of the experimental units. If the ordinary blocking presented in the preceding chapter is called one-dimensional blocking, then the method to be described in this chapter may be called two-dimensional blocking.

## 1. Two-dimensional blocking

When the experimental units differ only in "shape" (Fig. 16.1), then the blocking is based on shape only. When the units differ both in "shape" and in "color" (shade) as shown at the top of Fig. 17.1, how are we to do the blocking? Strictly speaking, as exemplified by the 16 given units (imagine 16 animals) in the diagram, no two units are really alike and hence no ideal block of homogeneous units is possible. And yet this is

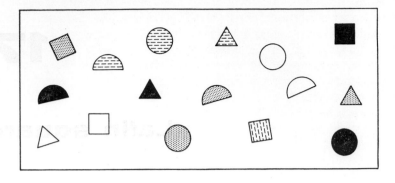

**Fig. 17.1** Two-dimensional blocking (or double blocking) of 16 experimental units which differ both in shape and in color.

probably a truer picture of the state of Nature than the idealized situation described in the preceding chapter. The present question is what can be done under the given circumstances.

The basic assumption we have to make here is that shape and color act independently and additively of each other (with respect to the treatments) and that there is no interaction between them. This point will be made more explicit later when we come to the topic of linear model. Under this basic assumption we may first do the blocking according to one characteristic and then according to the other. Thus, we may first collect the sixteen units according to their shape into four blocks (middle diagram in Fig. 17.1) just as we did in the preceding chapter. Having done that, we may then classify the animals according to color so that all animals of the same color belong to the same block. The result is shown in the bottom diagram of Fig. 17.1 (ignoring $a$ to $d$ for the time being), which exhibits the nature of a two-dimensional blocking. That is, the (vertical) columns are blocks based on shape and the (horizontal) rows are blocks based on color. Both columns and rows satisfy the requirement of a block—with respect to one characteristic at a time. In a case of double blocking like this, the simple single term "block" becomes inadequate and hence we shall speak of rows (that is, row blocks) and columns (that is, column blocks) instead.

### 2. Assignment of treatment

Now let us suppose that there are four treatments ($a$, $b$, $c$, $d$) to be tested on these sixteen animals. Obviously, each treatment should be replicated four times. According to the basic principle of randomized blocks, the four treatments should be assigned to the four units in a block at random. Since each row is a block, each row of four units should have $a$, $b$, $c$, $d$ in a random order. Similarly, each column, being a block, should also have $a$, $b$, $c$, $d$ in some random order. The "arrangement" (or assignment) of treatments $a$, $b$, $c$, $d$ shown in the bottom diagram of Fig. 17.1 satisfies such conditions. The reader should carefully check that $a$, $b$, $c$, $d$ appear in each row as well as in each column without duplication. In other words, no animals of the same shape receive the same treatment, nor do animals of the same color receive the same treatment. Such a double blocking of units and a corresponding doubly restricted random assignment of treatments is called a Latin-square design.

The name "Latin square" was originally given by R. A. Fisher [The Arrangement of Field Experiments, *J. Ministry Agric.*, **33**:503–513 (1926)]. The definition includes randomization of treatments. This point will be discussed more fully in Sec. 4. The reason for a Latin-square arrangement of treatments in an agricultural field experiment is particularly simple.   The fertility of any piece of land generally varies in all directions and not merely in one direction as depicted in Fig. 16.2. The two-dimensional blocking will enable us to control the soil heterogeneity in both directions; this design has been widely used in field experiments.   (In practice, the actual shape of the experimental land is, most of the time, a rectangle.)   It may also be well adapted to laboratory experiments based on a similar consideration of the heterogeneity of the animals or other units.

### 3. Orthogonality of design

To elucidate the property of orthogonality in experimental designs, let us return to randomized blocks for a moment.   Each treatment appears in each block once (or more generally, the same number of times), and conversely, each block contains each treatment once.   We say that the blocks and treatments are orthogonal to each other.   It is because of the orthogonality that the block effects ($b_i$) and treatment effects ($t_j$) may be estimated separately and simply (Chap. 10).

Now in a Latin-square arrangement, each row contains a unit from each column, and vice versa.   Hence rows and columns are orthogonal (perpendicular to each other).   Analogously to randomized blocks, each treatment appears in each row once and each row contains each of the treatments once.   Hence rows and treatments are orthogonal.   Similarly, columns and treatments are also orthogonal.   In short, a Latin square is a setup in which three factors (rows, columns, and treatments in this case) are mutually orthogonal.   The diagram of a Latin square (for example, Fig. 17.1 bottom) is simply a symbolic method of expressing the three mutually orthogonal factors.

In adopting a Latin-square design for experimentation, both the advantage and the disadvantage are quite obvious.   There is a more refined blocking of the units and hence more control of their heterogeneity, resulting in a smaller experimental error.   The inconvenience is that the number of treatments to be tested must equal the number of replications to be run.   For instance, if there are 4 treatments, they must be replicated 4 times in a 4 × 4 square.   The design is less flexible than the randomized blocks.   Designs that relax this restriction to a certain extent (at the sacrifice of orthogonality) will be discussed in Chap. 27.

### 4. A random Latin square

It is tempting, but we must resist delving into the many very interesting combinatorial properties of a Latin square.   Here we shall merely mention a few properties that are useful to the experimenter in his practical work. Whatever the method of construction employed, once a Latin square has been obtained, a number of others may readily be derived from it. An interchange of any two columns (or two rows) will result in another Latin square, and there are many possible *permutations* of the rows and columns. Similarly, an interchange of any two letters (for example, *a* into *b* and *b* into *a*) will result in another Latin square. The squares derived from the various permutations are, however, not necessarily all different.

Given a Latin square, it is always possible to rearrange the columns so that the first row reads *a, b, c, d, . . . .* Having accomplished this, we leave the first row fixed and rearrange the remaining rows so the first column also reads *a, b, c, d, . . . .* When a square is so arranged, we say that it is in its "standard" form (or "reduced" form). For example, 4 × 4 Latin squares, there are four standard forms, as shown here. It

$$
\begin{array}{cccc}
a & b & c & d \\
b & a & d & c \\
c & d & b & a \\
d & c & a & b
\end{array}
\qquad
\begin{array}{cccc}
a & b & c & d \\
b & c & d & a \\
c & d & a & b \\
d & a & b & c
\end{array}
\qquad
\begin{array}{cccc}
a & b & c & d \\
b & d & a & c \\
c & a & d & b \\
d & c & b & a
\end{array}
\qquad
\begin{array}{cccc}
a & b & c & d \\
b & a & d & c \\
c & d & a & b \\
d & c & b & a
\end{array}
$$

is these standard squares that are given in statistical tables. For each standard square of size $r \times r$ we may permute all the $r$ columns (yielding $r!$ different permutations) and the $r - 1$ rows, leaving the first row as still the first row without being involved in the permutation [yielding $(r - 1)!$ permutations] to ensure that all derived squares are different. Hence from each standard square we may obtain $r!(r - 1)!$ different squares. The number of standard forms for small Latin squares may be enumerated by various methods. When this has been done, the total number of different squares may be calculated, as shown in the accompanying table.

| Size of square | No. of standard forms | Value of $r!(r-1)!$ | Total no. of different squares |
|---|---|---|---|
| 3 × 3 | 1 | 12 | 12 |
| 4 × 4 | 4 | 144 | 576 |
| 5 × 5 | 56 | 2,880 | 161,280 |
| 6 × 6 | 9,408 | 86,400 | 812,851,200 |

Latin squares may be used for various purposes in experimental design other than two-dimensional blocking of units. We shall have occasion to use all the twelve 3 × 3 Latin squares later (Chap. 24) for a different purpose.

When using a Latin square for the purpose of assigning treatments to the doubly blocked units, we should use a random square obtained either by randomization of a standard form or by some other means. The important point is that we cannot use a fixed and the same square all the time.   The reason is analogous to that in the random assignment of treatments to the units in a block in the ordinary randomized blocks design.   Only when a random square is being adopted every time can a valid estimate of the experimental error be obtained.   The experimenter should have no difficulty in obtaining a random square.   Even for the size 6 × 6, there are more than 800 million possible squares to choose from.   For 7 × 7, there are billions.   To summarize, the combinatorial property of a Latin square is an arrangement of $r$ letters (or objects) into an $r \times r$ square, each replicated $r$ times, so that every row or column contains all the $r$ objects without duplication.   When used as the name of an experimental design, however, the term "Latin square" always implies randomization, that is, the selection of a random square for each experiment.

## 5. Analysis of data

We shall first give the arithmetic procedure of analysis for a particular case and then present the general situation in algebraic form.   Suppose that there are six treatments to be tested with a Latin-square experiment, each treatment being replicated six times.   In Table 17.1 the treatments are designated as (1) to (6) rather than $a$ to $f$.   The author personally finds that the numerals are easier to use than letters, especially when the square is large, but this is purely a matter of taste.   The actual arrangement of treatments shown in Table 17.1 is a random square. The six rows may well be the six batches of animals received from January to March.   The six columns may be the six days of a week (Monday to Saturday), assuming that each treatment requires a full day's work. Thus, for the first batch of animals, treatment (3) is tested on Monday, treatment (1) on Tuesday, treatment (4) on Wednesday, and so on.   In the next run (for the second batch of animals), treatment (2) will be used on Monday, etc.   The row and column totals are obtained directly in the usual way.   The treatment total is obtained by adding all the observed values of a particular treatment in the square, and this takes a small amount of searching.   Thus:

Treatment (1):    total = 14 + 24 + 20 + 16 + 8 + 26 = 108
Treatment (6):    total = 28 + 33 + 28 + 23 + 24 + 26 = 162

**Table 17.1   Data of a Latin-square experiment**

| (3) 21 | (1) 14 | (4) 11 | (6) 28 | (2) 18 | (5) 10 | Row total 102 |
|---|---|---|---|---|---|---|
| (2) 28 | (6) 33 | (3) 27 | (5) 18 | (1) 24 | (4) 14 | 144 |
| (5) 12 | (3) 31 | (6) 28 | (2) 30 | (4) 17 | (1) 20 | 138 |
| (4) 13 | (2) 24 | (5) 8 | (1) 16 | (3) 24 | (6) 23 | 108 |
| (1) 8 | (5) 11 | (2) 20 | (4) 22 | (6) 24 | (3) 17 | 102 |
| (6) 26 | (4) 13 | (1) 26 | (3) 24 | (5) 19 | (2) 18 | 126 |

| | | | | | | | |
|---|---|---|---|---|---|---|---|
| Column total | 108 | 126 | 120 | 138 | 126 | 102 | 720 |
| Treatment total | (1) 108 | (2) 138 | (3) 144 | (4) 90 | (5) 78 | (6) 162 | 720 |

These totals are listed at the bottom of Table 17.1. The grand total of the six treatments should, of course, be the same as the grand total of the six rows or of the six columns. This provides a simple check of the accuracy of addition.

The arithmetic procedure for analysis of variance and the $F$ test are given in Table 17.2. The two basic quantities $A$ and $C$ need no comment. The three similar quantities $R$, $K$, $T$ are based on the row totals, column totals, and treatment totals, respectively. From these basic quantities the various sum of squares in the analysis-of-variance table are obtained. The error *ssq* is in practice always obtained by subtraction. It is equal to

$$ssq_E = (A - C) - (R - C) - (K - C) - (T - C) = A - R - K - T + 2C$$

The error degree of freedom is also obtained by subtraction. If the square is $r \times r$, the total *df* is $r^2 - 1$ and the error *df* is

$$(r^2 - 1) - 3(r - 1) = (r - 1)(r - 2)$$

In our particular example the estimate of the error variance is

$$s^2 = {}^{284}\!/_{20} = 14.2$$

and the variance ratio is $F = 182.4/14.2 = 12.8$ approximately, with

*Table 17.2*   Analysis of a Latin-square experiment (data from Table 17.1)

Calculation of five basic quantities

$A = 21^2 + \cdots + 18^2 = 16{,}028$      $R = \frac{1}{6}(102^2 + 144^2 + \cdots) = 14{,}688$

$K = \frac{1}{6}(108^2 + 126^2 + \cdots) = 14{,}544$

$C = \frac{1}{36}(720)^2 = 14{,}400$      $T = \frac{1}{6}(108^2 + 138^2 + \cdots) = 15{,}312$

Analysis of variance, $r = 6$

| Source | df | ssq | msq | F |
|--------|-----|-----|-----|---|
| Rows | $r - 1 = 5$ | $R - C =$ 288 | (57.6) | |
| Columns | $r - 1 = 5$ | $K - C =$ 144 | (28.8) | |
| Treatments | $r - 1 = 5$ | $T - C =$ 912 | 182.4 | 12.8 |
| Error | $(r - 1)(r - 2) = 20$ | (subt) = 284 | 14.2 | |
| Total | $r^2 - 1 = 35$ | $A - C =$ 1,628 | | |

5 *df* in the numerator and 20 *df* in the denominator. The tabulated value of *F* with so many degrees of freedom is 2.71 at the 5 per cent significance level and 4.10 at the 1 per cent significance level. The difference between the treatments as observed in our experiments is highly significant.

The treatment means are calculated the usual way. The standard error between two treatment means is $s_d = \sqrt{2s^2/r}$, of the same form as in randomized blocks.

## 6. The linear model

The arithmetic procedure presented above makes it clear that the analysis for Latin square is similar in every respect to that for randomized blocks except that the *ssq* for blocks is replaced by one for rows and one for columns. This is due to the similarity between the two underlying models for these two designs. The linear model for a Latin-square experiment is that

$$\begin{array}{c} \text{Each} \\ \text{observed} \\ \text{value} \end{array} = \begin{array}{c} \text{general} \\ \text{constant} \end{array} + \begin{array}{c} \text{row} \\ \text{effect} \end{array} + \begin{array}{c} \text{column} \\ \text{effect} \end{array} + \begin{array}{c} \text{treatment} \\ \text{effect} \end{array} + \begin{array}{c} \text{random} \\ \text{error} \end{array}$$

$$y_{ij,k} = \mu + \rho_i + \kappa_j + \tau_k + \epsilon_{ij,k}$$

These effects all act additively, and there is no interaction between any two of the three factors (rows, columns, treatments). If we proceed the same way, using the method of least squares as we did in Chap. 10, it will be found that the estimate of the general constant is the general mean of the $r^2$ observed values. Let the total deviation be dev $= y_{ij,k} - \bar{y}$, where $y_{ij,k}$ is the observed value in the $i$th row and $j$th column that receives treatment $k$ and $\bar{y}$ is the general mean. Furthermore,

| | | |
|---|---|---|
| Effect of row $i$: | $\bar{y}_i - \bar{y} = r_i$ | $\Sigma r_i = 0$ |
| Effect of column $j$: | $\bar{y}_j - \bar{y} = c_j$ | $\Sigma c_j = 0$ |
| Effect of treatment: | $\bar{y}_k - \bar{y} = t_k$ | $\Sigma t_k = 0$ |

where $\bar{y}_i$, $\bar{y}_j$, $\bar{y}_k$ are the row, column, and treatment means, respectively. Hence the total deviation may be expressed in terms of four deviations:

$$\text{dev} = r_i + c_j + t_k + e_{ij,k}$$

where the error term is estimated by

$$e_{ij,k} = y_{ij,k} - \bar{y}_i - \bar{y}_j - \bar{y}_k + 2\bar{y}$$

Squaring both sides and summing over all the $r \times r$ values, we obtain the identity

$$\Sigma \text{dev}^2 = r\Sigma r_i{}^2 + r\Sigma c_j{}^2 + r\Sigma t_k{}^2 + \Sigma e_{ij,k}^2$$

all the product terms vanishing because of the orthogonality of the rows, columns, and treatments. This subdivision of the total *ssq* into four component *ssq*'s is that listed in the analysis of variance of Table 17.2. For instance, the *ssq* due to treatments is

$$r \sum t_k{}^2 = r \sum (\bar{y}_k - \bar{y})^2 = \frac{\Sigma Y_k{}^2}{r} - \frac{Y^2}{r^2} = T - C$$

where $Y_k$ is the total for treatment $k$ and $Y$ is the grand total. This is, of course, the same identity we have been using all along in preceding chapters.

## 7. Numerical demonstration of error *ssq*

For all practical purposes the error *ssq* $= \Sigma e^2$ is always obtained by subtracting the *ssq*'s for rows, columns, and treatments from the total *ssq*. However, at least once in a lifetime, we would like to calculate it directly to clarify its meaning and for peace of mind. The value of each single error term may be calculated from either the various means or the various deviations:

$$\begin{aligned}
e_{ij,k} &= y_{ij,k} - \bar{y}_i - \bar{y}_j - \bar{y}_k + 2\bar{y} \\
&= \text{dev} - r_i - c_j - t_k
\end{aligned}$$

*Table* 17.3  Values of error estimates $e_{ij,k}$ (data from Table 17.1)

| | | | | | | $r_i$ |
|---|---|---|---|---|---|---|
| (3) +2 | (1) −2 | (4) −1 | (6) 1 | (2) −3 | (5) +3 | −3 |
| (2) +3 | (6) +1 | (3) −1 | (5) −2 | (1) +1 | (4) −2 | +4 |
| (5) −2 | (3) +3 | (6) −2 | (2) +1 | (4) −2 | (1) 2 | +3 |
| (4) +2 | (2) +2 | (5) −3 | (1) −3 | (3) +1 | (6) +1 | −2 |
| (1) −5 | (5) 0 | (2) 0 | (4) +7 | (6) −1 | (3) −1 | −3 |
| (6) 0 | (4) −4 | (1) +7 | (3) −4 | (5) +4 | (2) −3 | +1 |
| $c_j$  −2 | +1 | 0 | +3 | +1 | −3 | 0 |
| (1) | (2) | (3) | (4) | (5) | (6) | |
| $t_k$  −2 | +3 | +4 | −5 | −7 | +7 | 0 |

These values are given in Table 17.3 for the numerical example under consideration. The general mean is $\bar{y} = {}^{720}\!/_{36} = 20$. The observed value (Table 17.1) in the upper left corner with treatment (3) is 21, and the total deviation is $21 - 20 = 1$. Thus, the error term is

$$e_{11,3} = 1 - (-3) - (-2) - (+4) = +2$$

As another example, the observed value in the fifth row and fourth column with treatment (4) is 22. Hence the error is estimated by

$$e_{54,4} = (22 - 20) - (-3) - (+3) - (-5) = +7$$

After all the 36 error terms have been calculated this way, it is simple enough to see that they add up to zero for each row as well as for each column. However, we should also observe that they add up to zero for each treatment. For instance, for treatment (4),

$$-1 - 2 - 2 + 2 + 7 - 4 = 0$$

Finally, squaring each of these 36 errors and adding them up, we find that

$$\text{Error } ssq = 2^2 + (-2)^2 + \cdots + (-3)^2 = 284$$

which is precisely the value obtained by subtraction in Table 17.2. This is the ultimate check of the accuracy of the arithmetic in our analysis. All the other components are much easier to verify. Thus, the treatment $ssq$ is

$$r\Sigma t_k{}^2 = 6(2^2 + 3^2 + 4^2 + 5^2 + 7^2 + 7^2) = 912$$

in agreement with that listed in Table 17.2.

## 8. Efficiency

The efficiency of the Latin-square design in comparison with the randomized blocks may be studied in the same way that we studied the efficiency of the latter in comparison with the completely randomized experiment. The treatment effects may be left out of consideration, since each treatment would combine with each experimental unit the same number of times in a long series of randomized experiments. In other words, in studying the efficiency of the design per se, we may assume that treatments have no effect at all and concentrate only on the estimate of the error variance due to different methods of blocking. The situation for randomized blocks and for Latin square is shown in Table 17.4, the left portion of which is the same as Table 16.2 except that $t$ has been replaced by $r$. The symbols also have the same meaning as before: $S_T$ is the total $ssq$ of the given $N = r^2$ experimental units, and $S_b$ and $S_1$ are the $ssq$'s for blocks and error, respectively ($S_b + S_1 = S_T$).

Now, if only the rows were used as blocks and there were no column blocking, the error variance would be $E_1 = S_1/r(r-1)$. When column blocking is used in addition to the existent rows, the subdivision of the sum of squares is like that shown in the right portion of Table 17.4. The error variance is $E_2 = S_2/(r-1)^2$. The entries for "columns" and

*Table 17.4* Efficiency of Latin square in comparison with randomized blocks

| Randomized blocks | | | | Latin square | | | |
|---|---|---|---|---|---|---|---|
| Source | $df$ | $ssq$ | $msq$ | Source | $df$ | $ssq$ | $msq$ |
| Blocks | $r-1$ | $S_b$ | $E_b$ | Rows | $r-1$ | $S_b$ | $E_b$ |
| Error | $r(r-1)$ | $S_1$ | $E_1$ | {Columns | $r-1$ | $S_c$ | $E_c$ |
| | | | | {Error | $(r-1)^2$ | $S_2$ | $E_2$ |
| Total | $r^2-1$ | $S_T$ | $E_T$ | Total | $r^2-1$ | $S_T$ | $E_T$ |

"error" for the Latin square is simply a subdivision of the entries for "error" for the randomized blocks, viz.,

$$df: \qquad r(r - 1) = (r - 1) + (r - 1)^2$$
$$ssq: \qquad \qquad S_1 = S_c + S_2$$

Hence

$$E_1 = \frac{S_1}{r(r - 1)} = \frac{(r - 1)E_c + (r - 1)^2 E_2}{r(r - 1)}$$

This is the relationship between $E_1$ (error variance with one-dimensional blocking) and $E_2$ (error variance with two-dimensional blocking).

The column mean square $E_c$ in the above formula may be replaced by the row mean square $E_b$ if columns are used as fixed blocks and we want to know the efficiency of the additional row blocking.

It is also clear from the above formula that $E_1 = E_2$ only when $E_c$ and $E_2$ are equal. If $E_c > E_2$, then $E_1$ is also larger than $E_2$, implying that the extra blocking is worthwhile. A small refinement of this statement is given in Exercise 17.4.

### 9. Graeco-Latin square

The concept of double blocking and Latin-square arrangement of treatment may be carried one step further. After the blocking of units according to color (rows) and shape (columns) as shown in Fig. 17.1, suppose that the same 16 units can be further blocked according to some other characteristics (to be symbolized by four Greek letters) so that no two units of the same color or of the same shape have the same Greek letter. In other words, each kind of the Greek characteristics appears among the various shapes once and among the various colors once. That is to say, the Greek characteristic is orthogonal to shape and to color simultaneously. We have seen that only an arrangement of the Latin-square type can satisfy these conditions. Hence, the Greek letters must form a Greek square by themselves. Then the treatments $a$, $b$, $c$, $d$ should not only be orthogonal to shape and color of the units but also orthogonal to the Greek characteristics at the same time. One such arrangement is shown in the upper portion of Table 17.5. The reader should check carefully that the Greek letters do form a Greek square by themselves and that the Latin letters also do so. Furthermore, for the four $\alpha$'s, there is one $a$, one $b$, one $c$, and one $d$. Similarly, for the four $\beta$'s, there is also one $a$, one $b$, one $c$, and one $d$, and so on. Conversely, for the four $a$'s, there is one $\alpha$, one $\beta$, one $\gamma$, and one $\delta$, and so on. Such an arrangement is called a Graeco-Latin square. That is, two Latin squares are superimposed upon each other so that they are orthogonal.

This kind of triple blocking has only an occasional application in practical experiments because of the manifold requirements. It is mentioned here partly because of its intrinsic interest and partly because of its applications in designing some other types of experiments which the reader may encounter in advanced texts. As a matter of fact, still another Latin square may be superimposed on the Graeco-Latin square orthogonally. Should the third square be denoted by Chinese characters, it would be called a Sino-Graeco-Latin square! In numerals, such a square is shown in the lower portion of Table 17.5. This is the limit for a 4 × 4 square, and it is impossible to superimpose another Latin

*Table 17.5* **A 4 × 4 Graeco-Latin square**

| $\alpha$ $a$ | $\beta$ $b$ | $\gamma$ $c$ | $\delta$ $d$ |
|---|---|---|---|
| $\gamma$ $b$ | $\delta$ $a$ | $\alpha$ $d$ | $\beta$ $c$ |
| $\delta$ $c$ | $\gamma$ $d$ | $\beta$ $a$ | $\alpha$ $b$ |
| $\beta$ $d$ | $\alpha$ $c$ | $\delta$ $b$ | $\gamma$ $a$ |

A complete set of 4 x 4 Latin squares

| 1 ① 1 | 2 ② 2 | 3 ③ 3 | 4 ④ 4 |
|---|---|---|---|
| 3 ② 4 | 4 ① 3 | 1 ④ 2 | 2 ③ 1 |
| 4 ③ 2 | 3 ④ 1 | 2 ① 4 | 1 ② 3 |
| 2 ④ 3 | 1 ③ 4 | 4 ② 1 | 3 ① 2 |

square orthogonally. Then it is called a complete set of orthogonal squares. In general, for an $r \times r$ square, there are $r^2 - 1 = (r - 1)(r + 1)$ degrees of freedom, and each orthogonal classification (blocking) has $r - 1$ degrees of freedom; therefore there are $r + 1$ possible orthogonal classifications. Hence, for a $4 \times 4$ square, the 15 *df* may be subdivided by 5 orthogonal classifications (rows, columns, and three Latin squares), each with 3 *df*. The complete set of orthogonal squares of various sizes have been listed in Fisher and Yates tables (1938–1963).

An important exception is that, for squares of $6 \times 6$, it is impossible to construct a set of orthogonal squares.

### EXERCISES

**17.1** Very large Latin squares are seldom used for practical experiments for two reasons. One is that the Latin square requires too many replications. Another is that it is difficult to find a large number of units that satisfy the physical conditions of blocking. Even in field experiments, squares larger than $12 \times 12$ are only occasionally used, because there is doubt if the artificial blocking can really eliminate the soil heterogeneity involved in a large area. On the other hand, very small squares are also impractical, unless they themselves are being replicated, because there are too few replications and too few degrees of freedom to estimate the experimental error. So the Latin square is most useful for experiments of intermediate size. The results of a $6 \times 6$ Latin square are given in the accompanying table. Make an analysis of variance to test the significance of treatments.

| (1) 1.1 | (6) 1.8 | (4) 0.8 | (3) 1.9 | (2) 1.9 | (5) 0.3 | 7.8 |
|---|---|---|---|---|---|---|
| (5) 1.3 | (4) 0.9 | (2) 2.3 | (1) 1.9 | (6) 2.8 | (3) 2.2 | 11.4 |
| (3) 1.9 | (2) 1.3 | (6) 2.1 | (5) 1.4 | (4) 0.8 | (1) 2.1 | 9.6 |
| (6) 2.3 | (5) 0.5 | (3) 1.6 | (2) 1.3 | (1) 0.9 | (4) 0.6 | 7.2 |
| (4) 1.7 | (3) 1.2 | (1) 0.3 | (6) 1.9 | (5) 0.6 | (2) 1.5 | 7.2 |
| (2) 2.5 | (1) 1.5 | (5) 0.7 | (4) 1.2 | (3) 2.6 | (6) 2.3 | 10.8 |
| 10.8 | 7.2 | 7.8 | 9.6 | 9.6 | 9.0 | 54.0 |

*Ans.:* $A = 97.28$; *ssq* same as those in Table 17.2 except for the decimals.

**17.2** The Latin square is so fundamental in experimental designs that a second numerical exercise is desirable. In the accompanying Latin square $L$ = litter, $P$ = pen, and $a$, $b$, $c$, $d$, $e$ are treatments. Make an analysis of variance and test the significance of treatments. Find the individual values of $e_{ij,k}$ and see that they add up to zero for each pen, for each litter, as well as for each treatment. Then verify $\Sigma e^2 = 48.00$.

|  | L1 | L2 | L3 | L4 | L5 | Pen total | Mean |
|---|---|---|---|---|---|---|---|
| P1 | $e$ 20 | $c$ 9 | $a$ 15 | $d$ 36 | $b$ 22 | 102 | 20.4 |
| P2 | $c$ 14 | $b$ 24 | $d$ 37 | $e$ 23 | $a$ 18 | 116 | 23.2 |
| P3 | $a$ 13 | $e$ 17 | $c$ 10 | $b$ 17 | $d$ 29 | 86 | 17.2 |
| P4 | $d$ 33 | $a$ 16 | $b$ 21 | $c$ 16 | $e$ 22 | 108 | 21.6 |
| P5 | $b$ 19 | $d$ 27 | $e$ 14 | $a$ 15 | $c$ 13 | 88 | 17.6 |
| Litter total | 99 | 93 | 97 | 107 | 104 | 500 |  |
| Mean | 19.8 | 18.6 | 19.4 | 21.4 | 20.8 |  | 20.0 |

| Treatments | $a$ | $b$ | $c$ | $d$ | $e$ |  |  |
|---|---|---|---|---|---|---|---|
| Total | 77 | 103 | 62 | 162 | 96 | 500 |  |
| Mean | 15.4 | 20.6 | 12.4 | 32.4 | 19.2 |  | 20.0 |

PARTIAL CHECK: $A = 11{,}374$, $s^2 = 4.00$, $F = 73.025$.

**17.3** Having practiced with numerical examples, the reader should get familiar with the general algebraic form of the breakdown of the sum of squares. In the accompanying table, $R_i$ = row total, $C_j$ = column total, $T_k$ = treatment total of a $r \times r$ Latin square, and $e_{ij,k} = y_{ij,k} - \bar{y}_i - \bar{y}_j - \bar{y}_k + 2\bar{y}$.

| Source | $df$ | $ssq$ | Practical form |
|---|---|---|---|
| Rows | $r - 1$ | $r\Sigma(\bar{y}_i - \bar{y})^2 \equiv \Sigma R_i^2/r - Y^2/r^2$ | |
| Columns | $r - 1$ | $r\Sigma(\bar{y}_j - \bar{y})^2 \equiv \Sigma C_j^2/r - Y^2/r^2$ | |
| Treatments | $r - 1$ | $r\Sigma(\bar{y}_k - \bar{y})^2 \equiv \Sigma T_k^2/r - Y^2/r^2$ | |
| Error | $(r-1)(r-2)$ | $\Sigma e_{ij,k}^2$ | by subtraction |
| Total | $r^2 - 1$ | $\Sigma(y_{ij,k} - \bar{y})^2 \equiv \Sigma y_{ij,k}^2 - Y^2/r^2$ | |

### 17.4   *Efficiency*

In Table 17.2 we have found that the error variance under double blocking is $E_2 = s^2 = 14.2$ and $E_c = 28.8$. If we had used only the rows as blocks without the additional column blocking, the error variance (of a long series of randomized blocks experiment) could be expected to be

$$E_1 = \frac{(r-1)E_c + (r-1)^2 E_2}{r(r-1)} = \frac{5(28.8) + 25(14.2)}{6 \times 5} = 16.63$$

which is larger than $E_2$. However, $E_1$ would have $df_1 = 25$, whereas $E_2$ has only $df_2 = 20$, and the smaller number of degrees of freedom of $E_2$ requires a larger value of $F$ to achieve the same significance level. Hence, $E_2$ should not merely be smaller than $E_1$ but should be smaller to a certain extent. One criterion is that

$$E_2 < \left(\frac{df_2 + 1}{df_2 + 3}\right)\left(\frac{df_1 + 3}{df_1 + 1}\right) E_1$$

In the example of $6 \times 6$ Latin square, the factor is

$$\left(\frac{20 + 1}{20 + 3}\right)\left(\frac{25 + 3}{25 + 1}\right) = \left(\frac{21}{23}\right)\left(\frac{28}{26}\right) = 0.983$$

In other words, $E_2$ should be about 2 per cent smaller than $E_1$. For a $5 \times 5$ Latin square, a similar calculation shows that $E_2$ should be about 3 per cent smaller than $E_1$. It is clear that this factor is important only for very small squares. Otherwise it is close to unity.

**17.5**   There are four treatments ($a$, $b$, $c$, $d$), each replicated four times but in different designs in three separate experiments. The treatment totals are all the same.

| Completely randomized | | | | Randomized blocks | | | | Latin square | | | |
|---|---|---|---|---|---|---|---|---|---|---|---|
| $a$ | $b$ | $c$ | $d$ | $a$ | $b$ | $c$ | $d$ | | | | |
| . | . | . | . | . | . | . | . | $c$ | $b$ | $a$ | $d$ |
| . | . | . | . | . | . | . | . | $a$ | $d$ | $c$ | $b$ |
| . | . | . | . | . | . | . | . | $d$ | $c$ | $b$ | $a$ |
| . | . | . | . | . | . | . | . | $b$ | $a$ | $d$ | $c$ |
| | | | | | | | | $a$ | $b$ | $c$ | $d$ |
| $Y_i$: 60 | 88 | 112 | 60 | 60 | 88 | 112 | 60 | 60 | 88 | 112 | 60 |

Calculate the sum of squares due to treatments for the three experiments.
*Ans.:* $ssq_T = T - C = 6{,}872 - 6{,}400 = 472$ for all three.

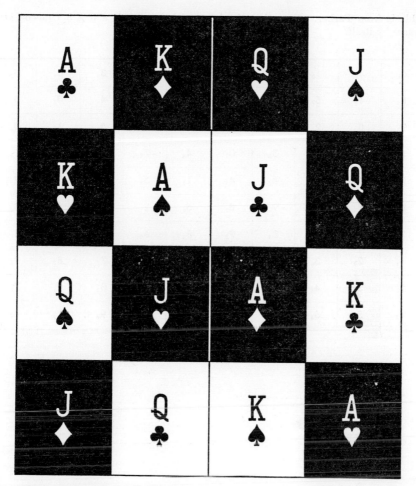

Fig. 17.2 A Graeco-Latin square of 16 cards. Permutation of rows and columns will yield other squares.

**17.6** A convenient way of getting acquainted with the idea of a Graeco-Latin square is as follows: From an ordinary deck of cards select the aces, kings, queens, and jacks of the four suits. First, arrange the A, K, Q, J into a 4 × 4 Latin square. Then adjust the suits so that C(club), D(diamond), H(heart), and S(spade) will also form a Latin square. The result is a Graeco-Latin square. One example is shown in Fig. 17.2.

It is really not necessary to use the A, K, Q, J. The reader may practice with 2, 3, 4, 5 of the four suits until he succeeds in obtaining a Graeco-Latin square.

**17.7** Graeco-Latin squares of dimension $4t + 2$ were thought generally impossible, and a 6 × 6 square is impossible. Recently, however, Bose and Shrikhande [*Proc. Natl. Acad. Sci. U.S.*, **45**:734–737 (1959)] and Parker [*ibid.*, 859–862] have constructed Graeco-Latin squares of size 22 × 22 and 10 × 10, respectively. The accompanying table is from Parker. Note that the lower right corner is a $3 \times 3$ Graeco-Latin square by itself.

| | | | | | | | | | |
|---|---|---|---|---|---|---|---|---|---|
| $0_0$ | $4_7$ | $1_8$ | $7_6$ | $2_9$ | $9_3$ | $8_5$ | $3_4$ | $6_1$ | $5_2$ |
| $8_6$ | $1_1$ | $5_7$ | $2_8$ | $7_0$ | $3_9$ | $9_4$ | $4_5$ | $0_2$ | $6_3$ |
| $9_5$ | $8_0$ | $2_2$ | $6_7$ | $3_8$ | $7_1$ | $4_9$ | $5_6$ | $1_3$ | $0_4$ |
| $5_9$ | $9_6$ | $8_1$ | $3_3$ | $0_7$ | $4_8$ | $7_2$ | $6_0$ | $2_4$ | $1_5$ |
| $7_3$ | $6_9$ | $9_0$ | $8_2$ | $4_4$ | $1_7$ | $5_8$ | $0_1$ | $3_5$ | $2_6$ |
| $6_8$ | $7_4$ | $0_9$ | $9_1$ | $8_3$ | $5_5$ | $2_7$ | $1_2$ | $4_6$ | $3_0$ |
| $3_7$ | $0_8$ | $7_5$ | $1_9$ | $9_2$ | $8_4$ | $6_6$ | $2_3$ | $5_0$ | $4_1$ |
| $1_4$ | $2_5$ | $3_6$ | $4_0$ | $5_1$ | $6_2$ | $0_3$ | $7_7$ | $8_8$ | $9_9$ |
| $2_1$ | $3_2$ | $4_3$ | $5_4$ | $6_5$ | $0_6$ | $1_0$ | $8_9$ | $9_7$ | $7_8$ |
| $4_2$ | $5_3$ | $6_4$ | $0_5$ | $1_6$ | $2_0$ | $3_1$ | $9_8$ | $7_9$ | $8_7$ |

# 18

# Change-over designs

It has been mentioned that many of the important experimental designs originated from considerations in agricultural field trials. The group of designs in this chapter, however, is especially suitable for experiments on large animals (cows, monkeys, man, etc.) with two, three, or four treatments. In most of our investigations on man, whether in the field of medicine or psychology, we study only a few treatments or tests at a time and on a comparatively small number of individuals. In such cases, the designs described in this chapter will be found most useful and accurate. There are numerous variations in the change-over design; only the simple and basic examples are given here.

## 1. Two treatments in change-over sequence

Suppose that twelve children are available for tests on two treatments $a$ and $b$. We could assign at random six of them to treatment $a$ and the other six to treatment $b$.   That would be a completely randomized experiment (Chap. 15) without the advantage of blocking (Chap. 16). Usually, however, the individuals differ very widely in their constitution and response to the treatments, and superficial blocking offers little help. Consequently, the experimental error is large and the comparison of treatments is not very sensitive.   It is therefore desirable to eliminate the heterogeneity between the individuals from the comparison of the treatments.   To do this, the individuals themselves may be used as blocks, provided that the two treatments are such that they can be applied to the same individual—at different times, of course.   There should be a *rest period* between the two treatments, so that the effect of one treatment does not carry over to the next treatment.   The intermediate rest period may be as short as a few hours or as long as a few weeks, depending entirely on the nature of the treatments.

In this chapter we assume that the rest period between two treatments is long enough for the treatment effects to wear off, and there is no carry-over effect from the first treatment period to the next.   If the residual effect of treatment is persistent or of a long-term nature (such as in a nutrition experiment), the simple analysis described in this chapter is not valid.   The estimation of and test for residual effects will be discussed in Chap. 29, because the analysis has certain similarity with the balanced incomplete block designs.

When each individual is employed as a block, the order of application of the two treatments on the individual should be at random with the restriction that there are equal number of individuals who receive treatment $a$ first and treatment $b$ next (the $ab$ sequence) and who receive treatment $b$ first and treatment $a$ next (the $ba$ sequence).   This is called the change-over design.   It is also widely known as the cross-over design. The sequence of treatments is either $ab$ or $ba$ for each individual.   The two individuals with two different treatment sequences are like the following:

|  | *Individual* (1) | *Individual* (2) |
|---|---|---|
| Period I | $a$ | $b$ |
| Period II | $b$ | $a$ |

This is simply a $2 \times 2$ Latin square.   Thus, the change-over design combines the features of small randomized blocks and small Latin squares.

The principle of change-over design is very analogous to that of tournament bridge, in which the north-south cards of one table are played as the east-west hands at another table, so that the *card effect* is eliminated from the contest.

## 2. An example

The top portion of Table 18.1 gives a numerical example of a two-period change-over design on twelve individuals, six of whom are of the *ab* sequence and six of the *ba* sequence. The sequences are assigned to the individuals at random, and we have left it that way in the table. If the reader likes to, the six *ab* and the six *ba* sequences may be grouped for ease of locating the treatments, but it is not necessary to do so.

Since individuals, periods, and treatments are all orthogonal to each other, the analysis is straightforward and simple. Regarding individuals as blocks, we first calculate the three basic quantities, *A*, *B*, *C*, in the usual way, as indicated in the middle portion of Table 18.1.

Each period total is the sum of the twelve observations in that period (row in the table). A treatment total is the sum of the twelve observations on a treatment. The total for treatment *a*, for instance, is the sum of the six *a*'s in the first period plus the sum of the six *a*'s in the second

*Table* 18.1  Data on twelve individuals, (1) to (12), each of whom received two treatments (*a* and *b*) in a change-over sequence design

| Period | (1) | (2) | (3) | (4) | (5) | (6) | (7) | (8) | (9) | (10) | (11) | (12) | Period total | Treat- ment total |
|---|---|---|---|---|---|---|---|---|---|---|---|---|---|---|
| I | *b* | *b* | *a* | *b* | *a* | *a* | *a* | *a* | *b* | *b* | *b* | *a* | | *a* |
| | 23 | 10 | 33 | 14 | 24 | 28 | 31 | 8 | 8 | 17 | 26 | 18 | 240 | 252 |
| II | *a* | *a* | *b* | *a* | *b* | *b* | *b* | *b* | *a* | *a* | *a* | *b* | | *b* |
| | 21 | 11 | 28 | 27 | 20 | 12 | 20 | 13 | 11 | 14 | 26 | 13 | 216 | 204 |
| Individual total | 44 | 21 | 61 | 41 | 44 | 40 | 51 | 21 | 19 | 31 | 52 | 31 | 456 | 456 |

### Calculations

$A = 23^2 + 10^2 + \cdots + 13^2 = 10{,}002$

$B = \frac{1}{2}(44^2 + 21^2 + \cdots + 31^2) - 9{,}672$

$C = (456)^2/24 = 8{,}664$

$P = \frac{1}{12}(240^2 + 216^2) = 8{,}688$

$T = \frac{1}{12}(252^2 + 204^2) = 8{,}760$

### Analysis of variance

| Source | df | ssq | msq | F |
|---|---|---|---|---|
| Individuals | 11 | $B - C = 1{,}008$ | | |
| Periods | 1 | $P - C = 24$ | | |
| Treatments | 1 | $T - C = 96$ | 96 | 4.57 |
| Error | 10 | (subt) 210 | 21 | |
| Total | 23 | $A - C = 1{,}338$ | | |

period.  The sum of squares due to periods and treatments may also be calculated the usual way, as indicated by the quantities $P$ and $T$ in the middle portion of the table.  However, when there are two equal groups, it is always simpler to make use of the identity (Chap. 3):

$$\frac{Y_1{}^2 + Y_2{}^2}{n} - \frac{(Y_1 + Y_2)^2}{2n} = \frac{(Y_1 - Y_2)^2}{2n}$$

Thus,

$$P - C = \frac{240^2 + 216^2}{12} - \frac{456^2}{24} = \frac{(240 - 216)^2}{24} = 24$$

$$T - C = \frac{252^2 + 204^2}{12} - \frac{456^2}{24} = \frac{(252 - 204)^2}{24} = 96$$

The analysis of variance is given at the bottom of Table 18.1.  The large variation between individuals in this experiment should be noted.  It makes the change-over design very much worthwhile.  The variance ratio test gives $F = 96/21 = 4.57$, just short of attaining the 5 per cent significance level ($F = 4.96$).

## 3. Possible further blockings

It may be argued that the 12 individuals in the preceding example may be grouped into 6 pairs, each pair forming a separate $2 \times 2$ Latin square with respect to treatments and periods, as indicated in the accompanying table.  This is a design of a series of (Latin) squares.  The comparison

|     | Square 1 | | Square 2 | | Square 3 | | Square 4 | | Square 5 | | Square 6 | |
| --- | --- | --- | --- | --- | --- | --- | --- | --- | --- | --- | --- | --- |
|     | (1) | (2) | (3) | (4) | (5) | (6) | (7) | (8) | (9) | (10) | (11) | (12) |
| I   | $b$ | $a$ | $a$ | $b$ | $a$ | $b$ | $b$ | $a$ | $b$ | $a$ | $a$ | $b$ |
| II  | $a$ | $b$ | $b$ | $a$ | $b$ | $a$ | $a$ | $b$ | $a$ | $b$ | $b$ | $a$ |

of this design with the preceding simple change-over design is given in Table 18.2.  The *ssq* between the squares has 5 *df*, and that between the individuals (columns) within the squares has 6 *df*.  These 11 *df* for the 12 column totals are the same 11 *df* between individuals of the simple change-over design.  The treatment *ssq* has 1 *df* in both cases.  The difference between the two designs therefore lies in the handling of period effects and error.  For the series of six squares, it is assumed that the period effect varies from pair to pair, and it is calculated from the period totals of each separate square, taking up to 6 *df* and leaving 5 *df*

*Table* 18.2  Comparison of a series of Latin squares with the simple change-over design

| Series of squares | | Simple change-over | |
|---|---|---|---|
| Source | df | df | Source |
| Between squares | 5⎫ | 11 | Between individuals |
| Individuals (within square) | 6⎭ | | |
| Treatments | 1 | 1 | Treatments |
| Periods (within square) | 6⎫ | ⎧ 1 | Periods (overall) |
| Error | 5⎭ | ⎩10 | Error |
| Total | 23 | 23 | Total |

for the error.  In the simple change-over design, it is assumed that the period effect is the same, or more or less the same, for all individuals; and it has 1 *df* based on the two overall period totals, leaving 10 *df* for the error.  Insofar as the period effect varies from individual to individual, this variation goes to the experimental error.  Briefly, in the series of squares, the degrees of freedom are split into 6 + 5 and in the simple change-over situation they are split into 1 + 10.

Generally speaking, we prefer that the error has at least 8 or 10 degrees of freedom in any experimental design.  This is one of the reasons that we use 12 individuals in our numerical example.  It is clear that the series of squares should be adopted only when there is really a biological basis for pairing the individuals *and* there is reason to believe that the period effect does vary from pair to pair.  Otherwise, the simple change-over design is to be preferred.

## 4. Method of difference

To pave the way for an even more sensitive design to be described in the next section, let us reanalyze the data of Table 18.1 in a different way. In Chap. 10, Sec. 10 it is shown that the paired data on two treatments may be analyzed either as an ordinary two-way classification or by Student's pairing method.  Likewise, with respect to the change-over data, we may take the difference between the two periods for each individual.  The procedure is shown in Table 18.3.  The difference *d* is the value in period I minus that in period II.  Note, however, that there are two kinds of *d*'s, according to whether the first period receives treatment *a* or treatment *b*.  To distinguish between them, we let

$$d_1 = Ia - IIb$$
$$d_2 = Ib - IIa$$

*Table* 18.3   Analysis of change-over data (Table 18.1) by the method of difference

| Individual | Period | | I − II | | $d^2$ |
|---|---|---|---|---|---|
| | I | II | $d_1$ | $d_2$ | |
| (1) | b 23 | a 21 | | 2 | 4 |
| (2) | b 10 | a 11 | | −1 | 1 |
| (3) | a 33 | b 28 | 5 | | 25 |
| (4) | b 14 | a 27 | | −13 | 169 |
| (5) | a 24 | b 20 | 4 | | 16 |
| (6) | a 28 | b 12 | 16 | | 256 |
| (7) | a 31 | b 20 | 11 | | 121 |
| (8) | a  8 | b 13 | −5 | | 25 |
| (9) | b  8 | a 11 | | −3 | 9 |
| (10) | b 17 | a 14 | | 3 | 9 |
| (11) | b 26 | a 26 | | 0 | 0 |
| (12) | a 18 | b 13 | 5 | | 25 |
| Total | 240 | 216 | 36 | −12 | 660 |
| Symbol | $P_1$ | $P_2$ | $D_1$ | $D_2$ | A |

Calculations: see text for three basic quantities, $A$, $B$, $C$.

Analysis of variance (of the $d$ values)

| Source | df | ssq | msq | F |
|---|---|---|---|---|
| Treatments | 1 | $B - C = 192$ | 192 | |
| Error | 10 | $A - B = 420$ | 42 | 4.57 |
| Total | 11 | $A - C = 612$ | | |

where the informal symbol I$a$ denotes the observed value of treatment $a$ in period I, etc.   In our example there are six ($r = 6$) $d_1$ and six $d_2$.   The rest of the analysis is based on these $d$ values.   We now observe that there are two groups of $d$'s, and the ordinary analysis of variance for one-way classification may be applied to test the significance of their difference.   To check the arithmetic, we may note that the grand total of the 12 $d$'s is

$$D_1 + D_2 = P_1 - P_2$$

that is,

$$36 - 12 = 240 - 216 = 24$$

Proceeding in the usual way we first find the three basic quantities:

$$A = \Sigma d^2 = 2^2 + (-1)^2 + 5^2 + \cdots = 660$$

$$B = \frac{1}{r}(D_1{}^2 + D_2{}^2) = \frac{1}{6}(36^2 + 12^2) = 240$$

$$C = \frac{(D_1 + D_2)^2}{2r} = \frac{24^2}{12} = 48$$

As noted previously, the *ssq* due to treatments may also be calculated directly:

$$B - C = \frac{(D_1 - D_2)^2}{2r} = \frac{48^2}{12} = 192$$

The analysis of variance and test of significance is given in the lower portion of Table 18.3. The reader should note that $F = {}^{192}\!/_{42} = 4.57$ obtained here is exactly the same as that obtained in Table 18.1 and has the same number of degrees of freedom. Comparing the analysis of variance in Tables 18.1 and 18.3, we make two important observations: First and most obvious is that the transform of the original observations into the *d* values has eliminated the variation between the periods as well as that between the individuals, so that the analysis of variance in Table 18.3 is merely the lower half (treatments and error) of that in Table 18.1. Second is that the *ssq* and *msq* for the *d* values are twice as large as those for the original observations, so that the value of $F$ remains the same. Since each $d = y_1 - y_2$, $V(d) = 2V(y)$. Our estimates are $s^2 = 21$ for the $y$'s and $S^2 = 42$ for the $d$'s. As far as test of significance is concerned, the two methods yield identical results. With this much background, we are ready to understand the next section.

## 5. Two treatments with double change-over sequence

Instead of having the simple *ab* and *ba* treatment sequences for two periods, the experiment may be further refined by having *aba* and *bab* treatment sequences on the individuals for three periods, still assuming that there is a rest period between the successive treatments and that there is no carry-over effect from one period to the next. Such a three-period design is called a double change-over, or double reversal, or switch-back in treatment sequences.

Table 18.4 gives an example of a double change-over experiment. The data for the first two periods are the same as those in Tables 18.1 and 18.3, to which a third period has been added. The analysis is the same as the method of differences outlined in the preceding section, except that now the difference takes the form $d = I - 2II + III$, that

*Table* 18.4  Data on twelve individuals in a double change-over design

| Individuals | Periods | | | I − 2II + III | | $d^2$ |
|---|---|---|---|---|---|---|
| | I | II | III | $d_1$ | $d_2$ | |
| (1) | $b$ 23 | $a$ 21 | $b$ 16 | | −3 | 9 |
| (2) | $b$ 10 | $a$ 11 | $b$ 17 | | 5 | 25 |
| (3) | $a$ 33 | $b$ 28 | $a$ 30 | 7 | | 49 |
| (4) | $b$ 14 | $a$ 27 | $b$ 19 | | −21 | 441 |
| (5) | $a$ 24 | $b$ 20 | $a$ 28 | 12 | | 144 |
| (6) | $a$ 28 | $b$ 12 | $a$ 24 | 28 | | 784 |
| (7) | $a$ 31 | $b$ 20 | $a$ 22 | 13 | | 169 |
| (8) | $a$  8 | $b$ 13 | $a$ 24 | 6 | | 36 |
| (9) | $b$  8 | $a$ 11 | $b$ 18 | | 4 | 16 |
| (10) | $b$ 17 | $a$ 14 | $b$ 24 | | 13 | 169 |
| (11) | $b$ 26 | $a$ 26 | $b$ 18 | | −8 | 64 |
| (12) | $a$ 18 | $b$ 13 | $a$ 24 | 16 | | 256 |
| Total | 240 | 216 | 264 | 82 | −10 | 2162 |
| Symbol | $P_1$ | $P_2$ | $P_3$ | $D_1$ | $D_2$ | $A$ |

Analysis of variance, $r = 6$

| Source | df | ssq | msq | F |
|---|---|---|---|---|
| Treatments | $1 = 1$ | 705.33 | 705.33 | 6.88 |
| Error | $2(r − 1) = 10$ | 1,024.67 | 102.47 | |
| Total | $2r − 1 = 11$ | 1,730.00 | | |

is, the sum of the observations of periods I and III minus twice that of period II.  Again, there are two kinds of $d$'s, according to whether the first period receives treatment $a$ or treatment $b$.  As before, we let

$$d_1 = \text{I}a − 2\text{II}b + \text{III}a$$
$$d_2 = \text{I}b − 2\text{II}a + \text{III}b$$

The rest of the analysis is based on these two groups of $d$'s by the usual method of one-way classification.

The first check of arithmetic is provided by the fact that the grand total of the $d$'s is

$$D_1 + D_2 = P_1 − 2P_2 + P_3$$
$$82 − 10 = 240 − 432 + 264 = 72$$

Then the three basic quantities are

$$A = \Sigma d^2 = (-3)^2 + 5^2 + \cdots = 2{,}162.00$$

$$B = \frac{1}{r}(D_1^2 + D_2^2) = \frac{1}{6}(82^2 + 10^2) = 1{,}137.33$$

$$C = \frac{(D_1 + D_2)^2}{2r} = \frac{72^2}{12} = 432.00$$

or, more directly,

$$B - C = \frac{(D_1 - D_2)^2}{2r} = \frac{92^2}{12} = 705.33$$

The analysis of variance is given at the bottom of Table 18.4, and $F$ is found to be 6.88, significant at the 5 per cent level.

As before, the transform of the original observations into the $d$ values has eliminated the variation between the periods as well as that between the individuals. If the significance of the difference between the two treatments is the only concern of the experimenter, this completes the analysis. This simple and ingenious method of analysis is due to Brandt [Tests of Significance in Reversal or Switch-back Trials, *Iowa Agr. Expt. Sta., Res. Bull.* 234 (1938)].

## 6. Model for the $d$ method

The reader may wonder why the contrast $D_1 - D_2$ gives the treatment effects. This may be seen most easily for the case of two treatments in a two-period change-over sequence. The assumption (linear model) is that each observed value is the sum of a general constant $u$, a treatment effect $a$ or $b$, a period effect $p_1$ or $p_2$, an individual effect $I$, and a random error $e$. Thus, the two observations on an individual in periods I and II for the $ab$ treatment sequence are, briefly,

$$y_1 = u + a + p_1 + I + e_1 \qquad y_2 = u + b + p_2 + I + e_2$$

so that

$$d_1 = y_1 - y_2 = (a - b) + (p_1 - p_2) + (e_1 - e_2)$$

and

$$D_1 = \Sigma d_1 = r(a - b) + r(p_1 - p_2)$$

where the error term vanishes on summation. Similarly, the sum of the $d_2$'s for the individuals receiving the $ba$ sequence is

$$D_2 = \Sigma d_2 = r(b - a) + r(p_1 - p_2)$$

Hence,

$$D_1 - D_2 = r(a - b) - r(b - a) = 2r(a - b)$$

where the period effect cancels out. It also follows that

$$\frac{D_1 - D_2}{2r} = a - b$$

is an estimate of the advantage of treatment $a$ over treatment $b$.

Since each $d = y_1 - y_2$,

$$V(d) = (1^2 + 1^2)V(y) = 2V(y)$$

In our example, the estimate of $V(d)$ is $S^2 = 42$ (Table 18.3) and the estimate of $V(y)$ is $s^2 = 21$ (Table 18.1). Now the variance of $D_1/r = \bar{d}_1$ is $S^2/r$, and the variance of the difference $(\bar{d}_1 - \bar{d}_2)$ is $2S^2/r$. Hence the variance of the quantity $(D_1 - D_2)/2r$ is

$$\left(\frac{1}{4}\right)\frac{2S^2}{r} = \frac{2 \times 2s^2}{4r} = \frac{s^2}{r}$$

The method for the switchback sequence ($aba$ and $bab$) is exactly the same, viz., the quantity $(D_1 - D_2)$ contains only the treatment effects, all others being canceled out. Since each $d$ is of the form

$$d_1 = y_1 - 2y_2 + y_3 = 2(a - b) + \cdots$$

we have

$$\frac{D_1 - D_2}{4r} = a - b$$

Also,

$$V(d) = (1^2 + 2^2 + 1^2)V(y) = 6V(y)$$

The estimate of $V(d)$ is $S^2 = 102.47$ in Table 18.4; then the variance on the basis of the original data is $s^2 = 102.47/6 = 17.08$. The variance of the quantity $(D_1 - D_2)/4r$ is therefore

$$\left(\frac{1}{16}\right)\frac{2S^2}{r} = \frac{2 \times 6s^2}{16r} = \frac{3s^2}{4r}$$

Our $F$ test in the analysis of variance is equivalent to testing the significance of $(a - b)$ against its standard error (see Exercise 18.4).

### 7. Three treatments in sequence

For two treatments ($a$ and $b$) the two different sequences form a 2 × 2 Latin square. Similarly, with three treatments ($a$, $b$, $c$) the three different sequences form a 3 × 3 Latin square, either of the form

|           | (1) | (2) | (3) |     | (4) | (5) | (6) |
|-----------|-----|-----|-----|-----|-----|-----|-----|
| Period I   | *a* | *b* | *c* |     | *a* | *b* | *c* |
| Period II  | *b* | *c* | *a* | or  | *c* | *a* | *b* |
| Period III | *c* | *a* | *b* |     | *b* | *c* | *a* |

The number of animals must be a multiple of 3, and there should be equal numbers of animals in each sequence.  If 6 or a multiple of 6 animals are available, we may use the 6 sequences of the 2 squares.  The treatment sequences are assigned to the animals at random.  There should be a rest period between the successive treatments, and it is assumed that the rest period is long enough to obliterate any carry-over effect from one period to the next.  Again, this design combines the features of randomized blocks and Latin square.  The analysis is straightforward (similar to Table 18.1), and it is indicated in Table 18.5.  No numerical example seems necessary.  As in the case of two treatments, further blocking of the animals into groups of three or six is possible when there is a biological reason for it.

*Table* 18.5  Change-over sequences for three treatments (*a*, *b*, *c*) in sets of six animals; treatment sequences are assigned to animals at random

| Period | Animals = columns = replicates | | | | | |
|--------|-----|-----|-----|-----|-----|-----|
|        | (1) | (2) | (3) | (4) | (5) | (6) |
| I      | *a* | *a* | *b* | *c* | *c* | *b* |
| II     | *c* | *b* | *c* | *b* | *a* | *a* |
| III    | *b* | *c* | *a* | *a* | *b* | *c* |

Analysis of variance

| Analysis   | Six animals *df* | Twelve animals *df* |
|------------|:----------------:|:-------------------:|
| Animals    | 5                | 11                  |
| Periods    | 2                | 2                   |
| Treatments | 2                | 2                   |
| Error      | 8                | 20                  |
| Total      | 17               | 35                  |

### 8. Three treatments in switchback sequence

Instead of each individual animal receiving all three treatments in some sequence as in the simple change-over design of the preceding section, we may adopt a three-period switchback or reversal design for two of the three treatments on each individual. There are six possible switch-back treatment sequences, viz.,

$$
\begin{array}{cc}
aba & bab \\
bcb & cbc \\
cac & aca
\end{array}
$$

The number of experimental animals required must thus be a multiple of 6, and there should be equal numbers of animals in all sequences. The reason for using the switchback sequence (instead of the simple sequence *abc*, etc.) is its high sensitivity, that is, smaller experimental error and thus greater chance of detecting treatment differences. The design and method of analysis, being a direct extension of the method of differences for two treatments, is due to Lucas [Switchback Trials for More Than Two Treatments, *J. Dairy Sci.*, **39**:146–154 (1956)].

Table 18.6 gives a numerical example based on twelve animals, there being two animals ($r = 2$) for each sequence. The treatment sequence is assigned to the animals at random, assuming no further blocking of the animals. To minimize the arithmetic, we use the same 36 numbers as in Table 18.4; only the treatment labels have been changed. The data on animals have been arranged according to sequence pattern for easy identification.

The first step in analysis is the calculation of the $d$ values, where each $d = \mathrm{I} - 2\mathrm{II} + \mathrm{III}$, written informally. The rest of the analysis is based on these twelve $d$ values. For instance, the total *ssq* is

$$
A - C = \sum d^2 - \frac{D^2}{12} = 2{,}162 - 432 = 1{,}730
$$

with eleven degrees of freedom. The calculation of the treatment *ssq* is new and requires careful arithmetic. For each treatment we calculate a quantity to be denoted by $Q$, which is the sum of the $d$'s for individuals receiving the treatment in periods I and III minus the sum of the $d$'s for individuals receiving the treatment in period II. For example, the $Q$ value for treatment $a$ is

$$
Q_a = \Sigma d(a - x - a) - \Sigma d(x - a - x)
$$

*Table* 18.6  Data on twelve individuals with switchback design for three treatments $(a, b, c)$

| Individual | Periods | | | I − 2II + III | |
|---|---|---|---|---|---|
| | I | II | III | $d$ | $d^2$ |
| (1) | $a$ 23 | $b$ 21 | $a$ 16 | −3 | 9 |
| (2) | $b$ 10 | $c$ 11 | $b$ 17 | 5 | 25 |
| (3) | $c$ 33 | $a$ 28 | $c$ 30 | 7 | 49 |
| (4) | $b$ 14 | $a$ 27 | $b$ 19 | −21 | 441 |
| (5) | $c$ 24 | $b$ 20 | $c$ 28 | 12 | 144 |
| (6) | $a$ 28 | $c$ 12 | $a$ 24 | 28 | 784 |
| (7) | $a$ 31 | $b$ 20 | $a$ 22 | 13 | 169 |
| (8) | $b$ 8 | $c$ 13 | $b$ 24 | 6 | 36 |
| (9) | $c$ 8 | $a$ 11 | $c$ 18 | 4 | 16 |
| (10) | $b$ 17 | $a$ 14 | $b$ 24 | 13 | 169 |
| (11) | $c$ 26 | $b$ 26 | $c$ 18 | − 8 | 64 |
| (12) | $a$ 18 | $c$ 13 | $a$ 24 | 16 | 256 |
| Total | 240 | 216 | 264 | 72 | 2,162 |
| Symbol | $P_1$ | $P_2$ | $P_3$ | $D$ | $A$ |
| | Correction term | | | $D^2/12 = C = 432$ | |

where the summation is taken over treatment $x = b, c$. To avoid arithmetic errors, it is best to collect the $d$ values according to the various sequence patterns, as in Table 18.7. Note that each $d$ value has been entered twice. Take, for instance, the $d = 7$ for individual (3), who receives the treatment sequence $cac$. It has been entered under $(cxc)$ to evaluate $Q_c$ and under $(xax)$ to evaluate $Q_a$. An arithmetic check is given by

$$Q_a + Q_b + Q_c = 0$$

The *ssq* for treatments is then

$$\frac{\Sigma Q^2}{2r \times t} = \frac{1}{4 \times 3} (51^2 + 11^2 + 40^2) = \frac{4{,}322}{12} = 360$$

where $r = 2$ is the number of replications of the sequences and $t = 3$ is the number of treatments. The error *ssq* is obtained by subtraction. The $F$ test is given in the lower part of Table 18.7; the difference between treatments is nonsignificant.

*Table* 18.7   Calculation of treatment effects and analysis of variance for three treatments in a switchback design (data of Table 18.6)

|            |      |       | \multicolumn{6}{c}{$d$ values of treatment sequences} |       |       |       |       |       |
|:----------:|:----:|:-----:|:---------:|:--------:|:---------:|:--------:|:---------:|:--------:|
| \multicolumn{3}{c}{Individuals} | $(axa)$ | $(xax)$ | $(bxb)$ | $(xbx)$ | $(cxc)$ | $(xcx)$ |
| (1)  | (2)  | (3)   | $-3$  | 7     | 5     | $-3$  | 7     | 5     |
| (4)  | (5)  | (6)   | 28    | $-21$ | $-21$ | 12    | 12    | 28    |
| (7)  | (8)  | (9)   | 13    | 4     | 6     | 13    | 4     | 6     |
| (10) | (11) | (12)  | 16    | 13    | 13    | $-8$  | $-8$  | 16    |
| \multicolumn{3}{c}{Total} | 54 | 3 | 3 | 14 | 15 | 55 |
| \multicolumn{3}{c}{$Q$} | \multicolumn{2}{c}{$Q_a = 51$} | \multicolumn{2}{c}{$Q_b = -11$} | \multicolumn{2}{c}{$Q_c = -40$} |

The analysis of variance

| Source      | $df$ | \multicolumn{2}{c}{$ssq$} | $msq$ | $F$  |
|:-----------:|:----:|:-----------:|:--------:|:-----:|:----:|
| Treatments  | 2    | $\Sigma Q^2/2rt =$ | 360 | 180   | 1.18 |
| Error       | 9    | (subt) | 1,370 | 152   |      |
| Total       | 11   | \multicolumn{2}{c}{$A - C = 1,730$} |       |      |

All of our calculations above are on the basis of the $d$ values. Some authors prefer to have them on the basis of the original observations (per individual per period). As explained in Sec. 6, they differ only by a factor of $1^2 + 2^2 + 1^2 = 6$. On a per period basis, the estimate of the error variance would be $1,370/(6 \times 9) = 25.37$, and all the other sums of squares should likewise be divided by 6. The $F$ value, however, remains the same.

The general mean of the 36 original observations is $\bar{y} = {}^{720}\!/_{36} = 20$. The mean for the $i$th treatment is $\bar{y} + Q_i/4rt$.

## 9. Possible further blocking of the animals

In the basic change-over designs, each individual plays the role of a block which receives a sequence of treatments. If the group of experimental animals can be further divided into subgroups by their biological characteristics or difference in experimental season, this can be incorporated into the analysis very easily. For purpose of illustration, let us assume that the first six individuals in Table 18.6 form one block containing all six treatment sequences and that the next six form another block. Then the only additional item we have to calculate is the *ssq* between these two blocks, each consisting of six individuals. The sum of the first six *d*'s is 28, and that of the next six is 44. Hence the *ssq* due to this blocking is

$$\frac{(28 - 44)^2}{2 \times 6} = \frac{256}{12} = 21.3$$

with one degree of freedom. The total and treatment *ssq* remain the same as in Table 18.7, but the error *ssq* is diminished by 21.3. The analysis-of-variance table is now as shown here. Incidentally, the above

| Source | df | ssq | msq |
|--------|----|-----|-----|
| Blocks | 1 | 21.3 | |
| Treatments | 2 | 360.2 | 180.1 |
| Error | 8 | 1,348.5 | 168.6 |
| Total | 11 | 1,730.0 | |

results provide us with an example in which the error *ssq* has been decreased because of the extra blocking while the error mean square has actually become larger (168 vs. 152), because the decrease in *ssq* is not sufficient to compensate for the corresponding decrease in degrees of freedom. This shows that extra blocking does not always increase the efficiency of an experiment, a point that we have discussed in the preceding two chapters

## 10. Other designs

The simple designs described in this chapter are of high sensitivity, and they should be used when they are applicable. This group of designs is especially useful for comparing a small number of treatments without carry-over effects. There are many variations to the basic design. One direction of generalization is to extend the number of periods; for example, a two-treatment four-period (I to IV) sequence may be of the form

I    II    III    IV

*a*    *b*    *a*    *b*

In such a case, the $d$ value should be

$$d = Ia - 3IIb + 3IIIa - IVb$$

The rest of the analysis is the same as those for two treatments.   There is, however, a practical limit to the number of periods each individual can be used for experimentation.

Another direction of generalization is to extend the number of treatments in the sequence design.   For four treatments (1, 2, 3, 4), the switchback sequences are as shown in the first of the accompanying tables (Lucas, 1956).   If the animals come in pairs, an alternative blocking scheme is as shown in the second of the accompanying tables.

| Period | Block | | | | Block | | | | Block | | | |
|--------|---|---|---|---|---|---|---|---|---|---|---|---|
| I   | 1 | 2 | 3 | 4 | 1 | 2 | 3 | 4 | 1 | 2 | 3 | 4 |
| II  | 2 | 3 | 4 | 1 | 3 | 4 | 1 | 2 | 4 | 1 | 2 | 3 |
| III | 1 | 2 | 3 | 4 | 1 | 2 | 3 | 4 | 1 | 2 | 3 | 4 |

| Period | Pair | | Pair | | Pair | | Pair | | Pair | | Pair | |
|--------|---|---|---|---|---|---|---|---|---|---|---|---|
| I   | 1 | 2 | 2 | 3 | 3 | 4 | 4 | 1 | 1 | 3 | 2 | 4 |
| II  | 2 | 1 | 3 | 2 | 4 | 3 | 1 | 4 | 3 | 1 | 4 | 2 |
| III | 1 | 2 | 2 | 3 | 3 | 4 | 4 | 1 | 1 | 3 | 2 | 4 |

If no blocking of the animals is necessary, we use all 12 sequences, which are to be assigned to the animals at random.   The method of analysis is the same as that for three treatments.

## EXERCISES

**18.1**   The scores of two tests ($a$ and $b$) on 18 school children are given in Table 18.8. The tests are four weeks apart (period I is September and period II is October).   Our purpose is to find out whether $a$ or $b$ yields a higher score.   Nine children receive the test sequence $ab$ and nine receive $ba$, the sequence being assigned to the children at random.   For ease of arithmetic, however, children receiving the same test sequence are tabulated together.

**a**   Do the analysis in terms of the original scores $y$.
**b**   Do the analysis in terms of the difference $d$ for each individual.
**c**   What is the conclusion about the difference between $a$ and $b$?

*Table* 18.8  Data on eighteen school children each of whom received two treatments ($a$ and $b$) in change-over sequence

| Individuals | Period I $a$ | Period II $b$ | Individual total | Difference (I-II) $d_1$ | |
|---|---|---|---|---|---|
| (1) | 21 | 22 | 43 | $-1$ | |
| (2) | 14 | 8 | 22 | 6 | |
| (3) | 11 | 17 | 28 | $-6$ | |
| (4) | 28 | 18 | 46 | 10 | |
| (5) | 18 | 12 | 30 | 6 | |
| (6) | 33 | 27 | 60 | 6 | |
| (7) | 28 | 24 | 52 | 4 | |
| (8) | 10 | 19 | 29 | $-9$ | |
| (9) | 26 | 24 | 50 | 2 | $D_1 = 18$ |
| | $b$ | $a$ | | $d_2$ | |
| (10) | 13 | 24 | 37 | $-11$ | |
| (11) | 24 | 18 | 42 | 6 | |
| (12) | 13 | 14 | 27 | $-1$ | |
| (13) | 17 | 26 | 43 | $-9$ | |
| (14) | 23 | 31 | 54 | $-8$ | |
| (15) | 24 | 20 | 44 | 4 | |
| (16) | 8 | 16 | 24 | $-8$ | |
| (17) | 11 | 28 | 39 | $-17$ | |
| (18) | 20 | 30 | 50 | $-10$ | $D_2 = -54$ |
| Total | $P_1 = 342$ | $P_2 = 378$ | $Y = 720$ | $D_1 + D_2 = -36$ | |
| Treatment | $(a)$ 396 | $(b)$ 324 | 720 | | |

If all the intermediate steps of calculation are done correctly, the various *ssq* are as shown in the accompanying table.

| Source | Based on $y$ | | Based on $d$ | | |
|---|---|---|---|---|---|
| | *df* | *ssq* | *df* | *ssq* | *F* |
| Individuals | 17 | 1,069 | | | |
| Periods | 1 | 36 | | | |
| Treatments | 1 | 144 | 1 | 288 | 6.08 |
| Error | 16 | 379 | 16 | 758 | |
| Total | 35 | 1,628 | 17 | 1,046 | |

**18.2**  The student must complete the preceding exercise before he does this one, so that a comparison of the results may be made.  Suppose that the 18 children in the experiment consist of 10 boys and 8 girls, and suppose that sex has been used as a basis for blocking of the individuals. The first 5 in each test sequence are boys **and** the remaining 4 are girls. For convenience of arithmetic, the data of Table 18.8 have been rearranged into the form of Table 18.9, in which the total scores (and $d$ values) are indicated for boys and girls separately.

|  | Period I | Period II | Total |
|---|---|---|---|
| Boys' total | 182 | 190 | 372 |
| Number | 10 | 10 | 20 |
| Girls' total | 160 | 188 | 348 |
| Number | 8 | 8 | 16 |
| Grand total | 342 | 378 | 720 |
| Number | 18 | 18 | 36 |

Let us analyze the original scores $(y)$ first. Consider the totals for the boys and girls. The $ssq$ for the four cell totals is calculated the usual way:

$$\frac{182^2}{10} + \frac{190^2}{10} + \frac{160^2}{8} + \frac{188^2}{8} - \frac{720^2}{36} = 140.4$$

with 3 $df$. It may be subdivided into three single components in two different ways:

| | | | | |
|---|---|---|---|---|
| Boys vs. girls | 1 | | Boys vs. girls | 1 |
| Periods in boys | 1 | | Periods (overall) | 1 |
| Periods in girls | 1 | | Interaction | 1 |

The first component (boys vs. girls) is the same for either system of subdivision. The $ssq$ is

$$\frac{372^2}{20} + \frac{348^2}{18} - \frac{720^2}{36} = 88.2$$

For the first system (left side), the $ssq$ for periods are calculated separately for the two sexes. Thus,

For boys: $\quad \dfrac{182^2}{10} + \dfrac{190^2}{10} - \dfrac{372^2}{20} = 3.2$

For girls: $\quad \dfrac{160^2}{8} + \dfrac{188^2}{8} - \dfrac{348^2}{16} = 49.0$

Note $88.2 + 3.2 + 49.0 = 140.4$ correctly. Under the second system (right side) of subdivision, the $ssq$ for overall period difference is

$$\frac{342^2}{18} + \frac{378^2}{18} - \frac{720^2}{36} = \frac{(342 - 378)^2}{36} = 36.0$$

The remaining $ssq$, $140.4 - 88.2 - 36.0 = 16.2$ is due to sex $\times$ period interaction. In the summary table we adopt the latter system of subdivision in order to compare with the analysis of the $d$ values.

The $ssq$ for individuals (within each sex) is calculated the usual way. Among

*Table* 18.9 Data on ten boys and eight girls, each of whom received two treatments (*a* and *b*) in change-over sequence

| Individuals | Period I | Period II | Sum $B_i$ | (I − II) $d_1$ | $d_2$ |
|---|---|---|---|---|---|
| | *a* | *b* | | | |
| (1) | 21 | 22 | 43 | −1 | |
| (2) | 14 | 8 | 22 | 6 | |
| (3) | 11 | 17 | 28 | −6 | |
| (4) | 28 | 18 | 46 | 10 | |
| (5) | 18 | 12 | 30 | 6 | |
| | *b* | *a* | | | |
| (10) | 13 | 24 | 37 | | −11 |
| (11) | 24 | 18 | 42 | | 6 |
| (12) | 13 | 14 | 27 | | −1 |
| (13) | 17 | 26 | 43 | | −9 |
| (14) | 23 | 31 | 54 | | −8 |
| Boys' total | 182 | 190 | 372 | 15 −23 | −8 |
| | *a* | *b* | | | |
| (6) | 33 | 27 | 60 | 6 | |
| (7) | 28 | 24 | 52 | 4 | |
| (8) | 10 | 19 | 29 | −9 | |
| (9) | 26 | 24 | 50 | 2 | |
| | *b* | *a* | | | |
| (15) | 24 | 20 | 44 | | 4 |
| (16) | 8 | 16 | 24 | | −8 |
| (17) | 11 | 28 | 39 | | −17 |
| (18) | 20 | 30 | 50 | | −10 |
| Girls' total | 160 | 188 | 348 | 3 −31 | −28 |
| Grand total | 342 | 378 | 720 | 18 −54 | −36 |

the boys, the *ssq* has 9 *df*, and among the girls, the *ssq* has 7 *df*. These plus the one for sex should be the total *ssq* for individuals in Exercise 18.1 (9 + 7 + 1 = 17 *df*). The treatment *ssq* remains the same as before. This completes the analysis of the *y* (see summary below).

The analysis of the *d* values follows the same procedure as that in the preceding exercise. For instance, the treatment *ssq* is $(D_1 - D_2)^2/18$. The only extra step is to calculate the *ssq* between the *d*'s of the two sexes. It is (from Table 18.9)

$$\frac{(-8)^2}{10} + \frac{(-28)^2}{8} - \frac{(-36)^2}{18} = 32.4$$

This component is due to the sex × period interaction. If all the intermediate steps of arithmetic are done correctly, the various *ssq* for the analysis of *y* and of *d* should be as shown in the accompanying table.   Compare these results with those obtained in Exercise 18.1.

| Source | On basis of $y$ | | On basis of $d$ | | $F$ |
|---|---|---|---|---|---|
| | df | ssq | df | ssq | |
| Sex | 1 | 88.2 | | | |
| Individuals | 16 | 980.8 | | | |
| Periods | 1 | 36.0 | | | |
| Sex $\times$ period | 1 | 16.2 | 1 | 32.4 | |
| Treatments | 1 | 144.0 | 1 | 288.0 | |
| Error | 15 | 362.8 | 15 | 725.6 | 5.95 |
| Total | 35 | 1,628.0 | 17 | 1,046.0 | |

$$88.2 + 980.8 = ?$$
$$16.2 + 362.8 = ?$$

What is the relationship between the *ssq* in terms of $d$ and those in terms of $y$?   In this particular experiment, is the extra blocking by sex worthwhile?   Does it make the error variance smaller than that without sex blocking?

**18.3**   In order to practice on the arithmetical procedure of analyzing a two-treatment experiment with a three-period double change-over sequence,

**a**   The reader may interchange the data of periods I and III in Table 18.4 and reanalyze the data.

**b**   Leaving the numbers unchanged, replace $a$ by $b$ and vice versa, so that the sequence *bab* becomes *aba*, and vice versa.   Reanalyze the data of Table 18.4.

**c**   Leaving the treatment labels as they are in Table 18.4, interchange the numbers of periods I and II and reanalyze the data.

*Ans:*   **a**   Everything remains the same.

   **b**   $d_1$ and $d_2$ are interchanged, so that $ab$ becomes $ba$.

   **c**   $D_1 = -26$, $D_2 = 26$, so that $D_1 + D_2 = 0$, and $D_1 - D_2 = -52$.

| Source | df | ssq | F | |
|---|---|---|---|---|
| Treatment | 1 | 225.3 | 1.25 | nonsignificant |
| Error | 10 | 1,798.7 | | |
| Total | 11 | 2,024.0 | | |

**18.4**   With the data of Table 18.4, calculate the numerical values of

$$\frac{D_1 - D_2}{4r} \qquad \text{and its standard error} \qquad \sqrt{\frac{S^2}{8r}}$$

where $r = 6$ and $S^2 = 102.467$.   Verify $t = 2.6238$, so that $t^2 = 6.88 = F$.

# 19

# Missing observations

It frequently happens in experimental work that one or more observations are missing from an otherwise complete set of data for various reasons beyond the investigator's control. If it happens in a completely randomized experiment (one-way classification, Chap. 15), we simply analyze the rest of the data the usual way. The missing observations merely change the size of the treatment groups and do not change the general procedure of analysis. However, if it is a randomized block design or Latin square, a missing observation destroys the orthogonality between the treatments and the blocks. Hence, strictly speaking, a long and complicated procedure for the analysis of unequal number of observations should be adopted. In most situations, especially when only one or two observations are missing, the analysis is not worth the labor. Instead, a much shorter and satisfactory method has been developed to remedy this situation. We begin with the simplest case.

## 1. One missing in randomized blocks

For concreteness let us assume that the observation for treatment $Q_2$ in block (7) of Table 16.1 (p. 180) is missing. Call the missing value $y$. Then Table 16.1 will look like the accompanying table. The dots in the

| Treatment | Blocks | | | | | | | | | Treatment total |
|---|---|---|---|---|---|---|---|---|---|---|
| | (1) | (2) | (3) | (4) | (5) | (6) | (7) | (8) | (9) | |
| $Q_1$ | · | · | · | · | · | · | · | · | · | · |
| $Q_2$ | · | · | · | · | · | · | $y$ | · | · | $187 + y$ |
| $Q_3$ | · | · | · | · | · | · | · | · | · | · |
| $Q_4$ | · | · | · | · | · | · | · | · | · | · |
| Block total | · | · | · | · | · | · | $60 + y$ | · | · | $700 + y$ |

table represent known numbers, and $y$ is the unknown. The shortcut method is to give $y$ some reasonable value so that the data table can be regarded formally complete and the conventional method of analysis may be employed. As a very rough approximation, we may substitute for $y$ the value of the general mean of the 35 observed numbers, or the block (7) mean, or the $Q_2$ treatment mean, etc. A highly refined method is the following.

The estimation of the parameters in our linear model is made by the method of least squares, which minimizes the *ssq* for the error term (Chap. 10). Following this principle, the missing value $y$ should also be given a numerical value that minimizes the *ssq* for error. Now the four basic quantities (areas) are, in terms of the unknown,

$$A = \cdots + y^2 + \cdots \qquad B = \cdots + \frac{(60 + y)^2}{4} + \cdots$$

$$C = \frac{(700 + y)^2}{36} \qquad T = \cdots + \frac{(187 + y)^2}{9} + \cdots$$

where the dots are known numbers. Then the error *ssq* is

$$ssq_E = A - B - T + C$$
$$= y^2 - \frac{(60 + y)^2}{4} - \frac{(187 + y)^2}{9} + \frac{(700 + y)^2}{36} + \cdots$$

We want the missing $y$ to take the value for which the $ssq_E$ is the smallest. Differentiating $ssq_E$ with respect to $y$ and setting the resulting derivative to zero, we obtain

$$y - \frac{60 + y}{4} - \frac{187 + y}{9} + \frac{700 + y}{36} = 0$$

Multiplying throughout by 36 and collecting terms,

$$y(36 - 4 - 9 + 1) = 4(187) + 9(60) - 700$$

and therefore

$$y = \frac{4(187) + 9(60) - 700}{(4 - 1)(9 - 1)} = \frac{588}{24} = 24.5$$

For practical calculation we may take $y = 24$ and proceed in the conventional way as if the data were complete. The missing $y$ acts as an extra parameter to be estimated. So the error $ssq$, losing one degree of freedom, should have 23 $df$ instead of the original 24 in Table 16.1. The total $df$ is 34 instead of 35.

The general formula for estimating one missing observation in a randomized block experiment is now obvious. Let $r = 9 =$ number of blocks, $R = 60 =$ incomplete total of the block in which the missing observation occurs, $t = 4 =$ number of treatments, $T = 187 =$ the incomplete treatment total, and $G = 700 =$ the incomplete grand total. Then the procedure of minimizing $ssq_E$ gives the equation

$$y(t - 1)(r - 1) = tT + rR - G$$

or

$$y = \frac{tT + rR - G}{(t - 1)(r - 1)}$$

In the next section we shall give only the first form, from which the second form is implied.

Using the numerical value of this expression for the missing observation, the rest of the calculation and analysis is the same as that for complete data, except that the total and error $df$ should each be diminished by 1. The error variance $s^2$ is then obtained the usual way. The variance between two treatment means—one with a missing observation and the other without—is larger than the usual $2s^2/r$, being

$$s^2 \left[ \frac{2}{r} + \frac{t}{r(r - 1)(t - 1)} \right]$$

The inflation is very slight when there are a moderate number of replications. It is common sense that the larger the experiment, the less the disturbance a missing observation causes.

## 2. Two missing in randomized blocks

When two observations are missing from an otherwise complete set of data in randomized blocks, suitable substitute values for the missing observations may be found by the same principle of minimizing the error sum of squares. There are two different situations. Case I: the two missing observations occur in the same block *or* in the same treatment. Case II: the two missing observations occur in different blocks *and* in different treatments. Let the two missing values be denoted by $y_1$ and $y_2$, wherever they occur, and let the corresponding incomplete treatment totals and block totals be $T_1$, $T_2$ and $R_1$, $R_2$. The situation for Case I—with both missing values in the same treatment—is as shown in the accompanying table. Proceeding exactly as before, we find that

|  |  | $r$ blocks |  |  |  | Total |
|---|---|---|---|---|---|---|
|  | . | . . | . | . | . | . |
| $t$ treatments | . | $y_1$ | . . | $y_2$ | . | $T_{12} + y_1 + y_2$ |
|  | . | . | . . | . | . | . |
| Total | . | $R_1 + y_1$ | . . | $R_2 + y_2$ | . | $G + y_1 + y_2$ |

the error *ssq*, being $A - B - T + C$ in general, is

$$ssq_E = y_1{}^2 + y_2{}^2 - \frac{(R_1 + y_1)^2 + (R_2 + y_2)^2}{t} - \frac{(T_{12} + y_1 + y_2)^2}{r}$$
$$+ \frac{(G + y_1 + y_2)^2}{rt} + \text{numbers}$$

Setting the partial derivatives of $ssq_E$ with respect to $y_1$ and $y_2$ to zero, we obtain two linear equations in two unknowns $y_1$ and $y_2$, viz:

$$y_1 - \frac{R_1 + y_1}{t} - \frac{T_{12} + y_1 + y_2}{r} + \frac{G + y_1 + y_2}{rt} = 0$$
$$y_2 - \frac{R_2 + y_2}{t} - \frac{T_{12} + y_1 + y_2}{r} + \frac{G + y_1 + y_2}{rt} = 0$$

which can be solved without difficulty. Leaving the algebraic details to the student as an exercise, we merely state the solutions:

$$y_1(t - 1)(r - 2) = tT_{12} + (r - 1)R_1 + R_2 - G$$
$$y_2(t - 1)(r - 2) = tT_{12} + R_1 + (r - 1)R_2 - G$$

If the two missing values occur in the same block but in different treatments, the same formulas hold except that $r$ and $t$ are interchanged and so are $R_1$, $R_2$ and $T_1$, $T_2$. A careful study of these formulas in connection with the illustration table will give the student an intuitive feeling of how the ball bounces.

Case II. If the two missing values are in different blocks as well as in different treatments, the situation will be something like that shown in the accompanying table. The procedure for finding the values of

| | $r$ blocks | | | | | Total |
|---|---|---|---|---|---|---|
| | $\cdot$ | $\cdot$ | $\cdot\quad\cdot$ | $\cdot$ | $\cdot$ | $\cdot$ |
| $t$ treatments | $\cdot$ | $y_1$ | $\cdot\quad\cdot$ | $\cdot$ | $\cdot$ | $T_1 + y_1$ |
| | | | | $y_2$ | $\cdot$ | $T_2 + y_2$ |
| | $\cdot$ | | $\cdot\quad\cdot$ | | | $\cdot$ |
| Total | $\cdot$ | $R_1 + y_1$ | $\cdot\quad\cdot$ | $R_2 + y_2$ | $\cdot$ | $G + y_1 + y_2$ |

$y_1$ and $y_2$ is exactly the same as before. To shorten the expressions, let us denote the known quantities by single letters. Thus

$$f = (t - 1)(r - 1)$$
$$W_1 = tT_1 + rR_1 - G \qquad W_2 = tT_2 + rR_2 - G$$

The solutions are then

$$(f^2 - 1)y_1 = fW_1 - W_2$$
$$(f^2 - 1)y_2 = fW_2 - W_1$$

Analogously to the case of one missing value, the two fundamental quantities here are of the type $W = tT + rR - G$. Once these two quantities are calculated, the numerical values of $y_1$ and $y_2$ are obtained immediately. The rest of the calculation and analysis of the data proceeds the usual way as if the data were complete. The $df$ for error and total $ssq$ are each to be diminished by 2, the number of missing observations.

## 3. Several missing in randomized blocks

When the number of missing observations is greater than two, the substitute values for the missing observations may still be found by minimizing the error sum of squares, but it is no longer practical to use explicit expressions, since they are very complicated. Suppose that there are three missing values, $y_1$, $y_2$, $y_3$. The usual method is by iterative calculation. First we assign some reasonable values, such as the observed treatment means, to two of the three missing values (say, $y_2$

and $y_3$), and then we calculate $y_1$ by the formula for one missing value. Adopting this value of $y_1$ and the assigned value of $y_3$, we can now calculate the value of $y_2$, again by the formula for one missing value. Finally, adopting the calculated values of $y_1$ and $y_2$, we now calculate the value of $y_3$ by the formula for one missing value. This completes one round of calculation. Then adopting the calculated values of $y_2$ and $y_3$, we can recalculate the value of $y_1$, the beginning of the second round, and so on, until the successive calculated values of the $y$'s stabilize. These are then the solutions.

Perhaps a faster procedure of iteration is to assign a reasonable value to one of the three missing values and calculate the other two by the formulas for two missing values. This would certainly facilitate the iteration when there are four or more missing values. Note that the iteration procedure also applies to the case of two missing values by repeated use of the formula for one missing value (that is, assign $y_1$ and calculate $y_2$, and then recalculate $y_1$, etc.).

Both as a numerical illustration of the iterative method and as a review of the formulas developed in the preceding sections, let us suppose that three values are missing from Table 16.1, for which $t = 4$ and $r = 9$. The data will then be as shown in the accompanying table. The dots

| | (1) | (2) | (3) | (4) | (5) | (6) | (7) | (8) | (9) | Total |
|---|---|---|---|---|---|---|---|---|---|---|
| $Q_1$ | . | $y_1$ | . | . | $y_2$ | . | . | . | . | $157 = T_{12}$ |
| $Q_2$ | . | . | . | . | . | . | $y_3$ | . | . | $187 = T_3$ |
| $Q_3$ | . | . | . | . | . | . | . | . | . | . |
| $Q_4$ | . | . | . | . | . | . | . | . | . | . |
| Total | . | $R_1 = 70$ | . | . | $R_2 = 66$ | . | $R_3 = 60$ | . | . | 668 |

denote the known values; they are the same as those in the original Table 16.1 and take no part in finding the substitute values of the missing observations. All the relevant numbers are shown above.

There are various ways to proceed with the iterative solution. Let us first assign a reasonable value to $y_3$ and proceed to solve for $y_1$ and $y_2$ by the formulas for two missing values in the same treatment. Now let $y_3 = 187/8 = 23$, accurate to the same unit of measurement as the original observation. The grand total is thus $G = 668 + 23 = 691$. Then

$$y_1(3 \times 7) = 4(157) + 8(70) + 66 - 691 = 563$$
$$y_2(3 \times 7) = 4(157) + 70 + 8(66) - 691 = 535$$

That is,

$$y_1 = 27 \qquad \text{and} \qquad y_2 = 25$$

Using these values for $y_1$ and $y_2$, which give $G = 668 + 27 + 25 = 720$, and regarding $y_3$ as the single missing observation, we obtain

$$y_3 = \frac{4(187) + 9(60) - 720}{3 \times 8} = \frac{568}{24} = 24$$

If we use this value of $y_3$, which gives $G = 668 + 24 = 692$, and solve for $y_2$ and $y_3$, we obtain the same values as before. Hence our final solutions are

$$y_1 = 27 \qquad y_2 = 25 \qquad y_3 = 24$$

An alternative procedure is to assign first a reasonable value for $y_1$ and solve for $y_2$ and $y_3$ by the formulas for two missing values in different blocks and in different treatments. Now let $y_1 = {}^{70}\!\!/_3 = 23$, yielding

$$T_2 = 157 + 23 = 180 \qquad G = 668 + 23 = 691$$

The two quantities of the type $tT + rR - G$ are

$$tT_2 + rR_2 - G = 4(180) + 9(66) - 691 = 623$$
$$tT_3 + rR_3 - G = 4(187) + 9(60) - 691 = 597$$

We also note that $f = (t - 1)(r - 1) = 3 \times 8 = 24$, and $f^2 - 1 = 575$. Thus,

$$575y_2 = 24(623) - 579 = 14{,}355$$
$$575y_3 = 24(597) - 623 = 13{,}705$$

yielding

$$y_2 = 25 \qquad y_3 = 24$$

Using these values for $y_2$ and $y_3$, which give new

$$T_1 = 157 + 25 = 182 \qquad G = 668 + 25 + 24 = 717$$

and regarding $y_1$ as a single missing value, we obtain

$$y_1 = \frac{4(182) + 9(70) - 717}{3 \times 8} = \frac{641}{24} = 27$$

Substituting this value of $y_1$ into the table and solving for $y_2$ and $y_3$, we obtain the same values as before. Hence the final solutions ($y_1 = 27$, $y_2 = 25$, $y_3 = 24$) are the same by either procedure.

## 4. Effective number of replicates

The analysis of variance may then be performed the usual way with the three substitute values inserted into the data table. The *df* for error and for total should each be diminished by 3 from the conventional values for a complete table. The error variance $s^2$ is then calculated.

Owing to the disturbance caused by the missing observations, the mean-square ($msq$) for treatments is always slightly larger than it would be otherwise. However, the $F$ test is seldom much in error unless a considerable number of missing observations occurred.

For a single missing observation we have given the formula for the variance of the difference between two treatment means. When there are several missing observations, the exact formula is very complicated and an approximation is needed. This is obtained by assigning each treatment an "effective" number of replications "$r$" relative to the other treatment with which it is to be compared. As an illustration, let us compare treatment $Q_1$ with $Q_2$ in the example above. The effective number of replications for $Q_1$ and $Q_2$ are calculated separately. When both treatments are present in a replication (block), the effective number is 1. When $Q_1$ is present but the other treatment is missing, the effective number for $Q_1$ is counted as $(t-2)/(t-1)$, where $t$ is the number of treatments in the experiment. When $Q_1$ is absent in a replication, the effective number is zero. The same method of counting applies to $Q_2$. Thus, the effective number of replications of these two treatments (when these two are compared) are as follows:

| Blocks | (1) | (2) | (3) | (4) | (5) | (6) | (7) | (8) | (9) | "$r$" |
|--------|-----|-----|-----|-----|-----|-----|-----|-----|-----|-------|
| $Q_1$ | 1 | 0 | 1 | 1 | 0 | 1 | $\tfrac{2}{3}$ | 1 | 1 | $6\tfrac{2}{3} = {}^{2}9\tfrac{2}{3}$ |
| $Q_2$ | 1 | $\tfrac{2}{3}$ | 1 | 1 | $\tfrac{2}{3}$ | 1 | 0 | 1 | 1 | $7\tfrac{1}{3} = {}^{2}2\tfrac{2}{3}$ |

Hence the variance of the difference between these two treatment means is

$$V(\bar{y}_1 - \bar{y}_2) = s^2\left(\frac{1}{\text{``}r_1\text{''}} + \frac{1}{\text{``}r_2\text{''}}\right) = s^2\left(\frac{3}{20} + \frac{3}{22}\right) = 0.286s^2$$

Note carefully that the effective number of replicates of a treatment varies according to the treatment with which it is to be compared. For instance, if we want to compare $Q_1$ with $Q_3$, which has no missing value, the calculation will be as follows:

| Blocks | (1) | (2) | (3) | (4) | (5) | (6) | (7) | (8) | (9) | "$r$" |
|--------|-----|-----|-----|-----|-----|-----|-----|-----|-----|-------|
| $Q_1$ | 1 | 0 | 1 | 1 | 0 | 1 | 1 | 1 | 1 | 7 |
| $Q_3$ | 1 | $\tfrac{2}{3}$ | 1 | 1 | $\tfrac{2}{3}$ | 1 | 1 | 1 | 1 | $8\tfrac{1}{3}$ |

The "$r$" value for $Q_1$ here is different from that in comparing with $Q_2$. After a little practice, the enumerative method of calculating the effective number of replications presented above may be dispensed with and a much shorter method used. Let $r_0$ be the actual number of replications of a treatment. Of these, suppose that there are $j$ observations missing in the other treatment. Then the effective number "$r$" of that treatment relative to the other is

$$"r" = r_0 - \frac{j}{t-1}$$

For example, in comparing $Q_1$ with $Q_2$, the actual number of replications for $Q_1$ is 7, of which one ($j = 1$) has no corresponding observation on $Q_2$. (This happened in block 7.) Hence

$$"r_1" \text{ for } Q_1 = 7 - \tfrac{1}{3} = 6\tfrac{2}{3}$$

Similarly,

$$"r_2" \text{ for } Q_2 = 8 - \tfrac{2}{3} = 7\tfrac{1}{3}$$

These "$r$" values are the same as those obtained by the enumerative method previously.

### 5. One missing in a Latin square

When a single observation is missing in a Latin-square experiment (Table 17.1), the principle and procedure of finding a substitute value is the same as in randomized blocks. Let $y$ be the missing value, and let $R$, $C$, and $T$ be the incomplete row, column, and treatment totals affected by the missing value. (The symbols $R$, $C$, $T$, here are not to be confused with the basic quantities called areas in Chap. 17. We use $R$, $C$, $T$ here merely because they are simpler to write than the more formal notations $Y_{i\cdot}$, $Y_{\cdot j}$, $Y_{\cdot\cdot k}$.) Again, let $G$ be the grand total of the $r^2 - 1$ actual observations in a $r \times r$ Latin square. The error sum of squares in terms of the unknown is then

$$ssq_E = y^2 - \frac{(R+y)^2}{r} - \frac{(C+y)^2}{r} - \frac{(T+y)^2}{r} + \frac{2(G+y)^2}{r^2} + \text{numbers}$$

To minimize the value of $ssq_E$, we equate $d(ssq_E)/dy$ to zero, obtaining the equation

$$y - \frac{(R+y) + (C+y) + (T+y)}{r} + \frac{2(G+y)}{r^2} = 0$$

$$y(r^2 - 3r + 2) = r(R + C + T) - 2G$$

$$y = \frac{r(R + C + T) - 2G}{(r-1)(r-2)}$$

Referring to Table 17.1 as a numerical example, let us assume that the observation in the fourth row and third column is missing. This obser-

vation received or would have received treatment (5) in this particular
6 × 6 experiment. Then the row, column, treatment, and grand totals
without this observation are respectively

$$R = 100 \qquad C = 112 \qquad T = 70 \qquad G = 712$$

and

$$y = \frac{6(100 + 112 + 70) - 2(712)}{5 \times 4} = \frac{268}{20} = 13$$

After this value is inserted for the missing observation, the rest of the
calculation and analysis is the same as that for a complete table. The
*df*'s for error and for total are to be diminished by 1 from the *df* for a
complete table. As in the case of randomized blocks, the variance of
the difference between two treatment means (involving one with the
missing value) is slightly larger than the usual $2s^2/r$, being

$$s^2 \left[ \frac{2}{r} + \frac{1}{(r - 1)(r - 2)} \right]$$

For $r = 6$, the inflation is only $\frac{1}{20}$ or 5 per cent.

## 6. Several missing in a Latin square

When two observations are missing in a Latin-square experiment, explicit
expressions for the substitute values are still obtainable with ease. The
two missing observations may occur in the same row, in the same column,
*or* in the same treatment. They may also occur in different rows, different columns, *and* different treatments. All the equations involve quantities of the type $r(R + C + T) - 2G$; they are given in the exercises.
In practice, however, it is not absolutely necessary to use these formulas.
As explained in preceding sections, when there are two or more missing
observations, the iterative procedure by repeated applications of the
formula for one missing will lead to correct solutions. In the analysis
of variance, the *df* for error should be diminished by the number of
missing values.

The variance of the difference between two treatment means involving
more than one missing value is very complicated. As before, an approximation is resorted to by calculating an "effective" number of replications
"*r*" for the treatments under comparison. The rule for counting replications for a treatment is as follows:

Count = 1 if the other treatment is present in the same row *and* column
Count = $\frac{2}{3}$ if the other treatment is present in the same row *or* column
Count = $\frac{1}{3}$ if the other treatment is absent in both row and column
Count = 0 if the treatment itself is missing

To illustrate the application of this approximation rule, let us consider the Latin square of Table 17.1, in which the three values indicated by circles have been assumed missing.

$$
\begin{array}{cccccc}
(3) & (1) & (4) & (6) & (2) & (5) \\
(2) & (6) & (3) & ⑤ & (1) & (4) \\
(5) & (3) & (6) & (2) & (4) & (1) \\
④ & (2) & ⑤ & (1) & (3) & (6) \\
(1) & (5) & (2) & (4) & (6) & (3) \\
(6) & (4) & (1) & (3) & (5) & (2)
\end{array}
$$

In comparing treatments (1) and (5), their effective numbers of replicates are counted as follows, proceeding from top row to bottom,

$$\text{``}r\text{''} \text{ for } (1) = 1 + \tfrac{2}{3} + 1 + \tfrac{1}{3} + 1 + \tfrac{2}{3} = 4\tfrac{2}{3} = \tfrac{14}{3}$$
$$\text{``}r\text{''} \text{ for } (5) = 1 + 0 + 1 + 0 + 1 + 1 = 4$$

The variance of the difference between these two treatment means is

$$
s^2 \left( \frac{1}{\text{``}r_1\text{''}} + \frac{1}{\text{``}r_5\text{''}} \right) = s^2 \left( \frac{3}{14} + \frac{1}{4} \right)
$$

Again we wish to emphasize that the effective number of replicates of a treatment varies according to the other treatment with which it is compared. Thus, in comparing treatments (4) and (5), the counts will be as follows, proceeding from top row to bottom:

$$\text{``}r\text{''} \text{ for } (4) = \tfrac{2}{3} + \tfrac{2}{3} + 1 + 0 + \tfrac{2}{3} + 1 = 4$$
$$\text{``}r\text{''} \text{ for } (5) = 1 + 0 + \tfrac{2}{3} + 0 + 1 + 1 = 3\tfrac{2}{3}$$

The effective number for treatment (5) relative to (4) is less than that relative to (1). Hence the variance of the difference between treatment (4) and (5) means is correspondingly larger.

When an entire replication (row or column) is missing in a Latin square, the analysis is slightly more complicated; it is given in Chap. 26.

## 7. Missing observations in change-over designs

It has been mentioned that change-over designs combine the features of randomized blocks and Latin square. So does the formula for a missing value ($y$). Let $t$ be the number of treatments ($t = 3$ in Table 18.5) and thus also the number of periods, and let $n$ be the number of animals ($n = 6$ or 12 in Table 18.5) where $n$ is always a multiple of $t$. If $R$, $P$, and $T$ are the incomplete totals of animals, periods, and treatments in which the missing observation is located, a similar procedure leads to the formula

$$y(t - 1)(n - 2) = nR + t(T + P) - 2G$$

The case of two treatments with change-over sequences (Table 18.1) deserves special attention. There are only two types of sequences and two periods, so that the number of animals may be written $n = 2r$, where $r$ is the number of replicates of treatment sequences. In Table 18.1, for example, $r = 6$ and $n = 12$. Substituting $t = 2$ and $n = 2r$, the formula for a missing value becomes

$$y(r - 1) = rR + T + P - G$$

As a numerical example, let us assume that the last observation in period I (treatment $a$, value 18) in Table 18.1 is missing, so that the incomplete totals are

$$R = 13 \qquad T = 234 \qquad P = 222 \qquad G = 438$$

Hence,

$$y = \frac{6(13) + 234 + 222 - 438}{6 - 1} = \frac{96}{5} = 19.2$$

It is instructive to consider an alternative method of finding the substitute value. Let us recall that the data of Table 18.1 may also be analyzed in terms of the $d$'s as was done in Table 18.3. Missing the last observation in period I implies missing the last $d_1 = 5$. The $d$'s are to be analyzed according to the method of one-way classification, for which missing values may simply be ignored and no substitute value is needed. This suggests that when there are missing values in a two-treatment change-over sequence experiment, the data may most conveniently be analyzed in terms of the $d$'s. However, in order to provide a comparison with the calculation of the original missing $y$, let us find the missing $d_1$ by the usual procedure, $D_1$ being the incomplete total. The error sum of squares for the $d$'s is

$$d_1{}^2 - \frac{1}{r}(D_1 + d_1)^2 + \text{numbers}$$

To minimize its value, we have the equation

$$d_1 - \frac{1}{r}(D_1 + d_1) = 0 \qquad \text{that is} \qquad d_1(r - 1) = D_1$$

In our numerical example (Table 18.3), the incomplete total is $D_1 = 31$, and

$$d_1 = \frac{D_1}{r - 1} = \frac{31}{5} = 6.2$$

The missing $d_1$ should simply be the mean of the five observed $d_1$'s. The substitution of this value does not, of course, contribute anything to the error *ssq*. In other words, the analysis is the same whether we insert the value $d_1 = 6.2$ or forget it altogether. This, incidentally, demonstrates in a most direct way why 1 *df* should be subtracted from the usual error *df*. Note that $d_1 = 6.2$ implies that the missing $y = 19.2$ as found previously, since the observation on treatment *b* in period II is 13. The conclusion is that the calculation of a missing $y$ is equivalent to the calculation of a missing $d$, and the latter is usually much simpler. For some other types of change-over designs, the missing value can also be found in terms of the $d$'s by applying the principle established in this chapter.

## 8. Discussion

We have used the word "missing" in a very general way; indeed, more as a technical term than in its colloquial sense. It covers a great variety of situations. The observation may have never been obtained at all, or perhaps it has been obtained and then lost. One important class of "missing data," however, consists of recorded values that are so obviously in gross error that they cannot be used in analysis. Discarding a value in obvious gross error is known as the "rejection" of a datum. In its place a substitute value may be inserted by the method described in this chapter.

While the author was in the eleventh grade, one of the laboratory assignments in physics was to determine empirically the ratio of the circumference to diameter by taking appropriate measurements on a wooden sphere. The determinations were to be replicated five (actually ten) times and the average was to be taken. The five determinations I got were something like the following:

$$3.18 \quad 3.12 \quad 3.09 \quad 8.17 \quad 3.21$$

While I was arguing about the 8.17 with my partner, the teacher came along and wanted to know what the trouble was. After a look at my data, he simply said: "Forget this [pointing at 8.17] and take the average of the other four." This was my first experience of rejecting an observation. The rejection was so fully justified in that case that no special explanation is needed. In much experimental work, no doubt, gross errors in taking observations also occur from time to time, but they are not necessarily as obvious as in my case of determining $\pi$. If it is a gross error, the experimental results will be made more accurate by rejecting it. On the other hand, if the investigator rejects a datum that does not fall into line with his thinking, he is simply ruining his own experiment and may miss an important discovery. It is difficult to make decisions in the twilight zone of whether a rejection of an observation is a blessing or cheating. The general advice is that a rejection should be made only with the utmost care.

The next question concerns the meaning of the substitute values calculated from the formulas for the missing observations. In the text I have tried to avoid a phrase such as "estimation of the missing value," lest the uncritical reader may gain the impression that although a certain value is missing, he can somehow get it back by a magic formula—very close to getting something for nothing! The truth is, of course, that the information on that treatment in that block is unknown and there is no way to get it back. The calculated $y$ is merely a reasonable number that enables us to complete the data table so that the standard method of analysis may be employed.

While what has been said above is generally true from the operational point of view, it may be shown that the $y$ calculated by applying the principle of minimizing the error *ssq* is actually the estimate of the average value of the missing observation.

Now, the reader may say: "Ah! It is an estimate of the missing observation after all. Then, why is it not getting something for nothing?" This paragraph is intended to reconcile the apparent paradox. The distinction can most easily be seen in the simple example of five observations on $\pi$ described above. Suppose that we want to preserve the superficial number of observations; it is fixed as five, and one observation is missing. Thus,

$$3.18 \qquad 3.12 \qquad 3.09 \qquad 3.21 \qquad y \qquad \text{total} = 12.60 + y$$

The sum of squares for the group of five numbers is

$$ssq = y^2 - \frac{(12.60 + y)^2}{5} + \text{numbers}$$

To minimize the value of *ssq*, we have the equation

$$5y - (12.60 + y) = 0 \qquad y = \frac{12.60}{4} = 3.15$$

Note that $y$ is simply the mean of the four known observations, and therefore it is an unbiased estimate of the true value of the parameter $\pi$ we are trying to measure. Inserting this value of $y$ to "complete" the data table, we have

$$3.18 \qquad 3.12 \qquad 3.09 \qquad 3.21 \qquad 3.15 \qquad \text{total} = 15.75$$

Superficially, there are now five numbers instead of four. But there is no change in our information on the true value of $\pi$. The estimate from the original four observations is the same as that given by the five observations, that is, $15.75/5 = 12.60/4 = 3.15$. There is no change in the sum of squares either.

$$3.18^2 + 3.12^2 + 3.09^2 + 3.21^2 - \frac{(12.60)^2}{4} = 0.0090$$

$$3.18^2 + 3.12^2 + 3.09^2 + 3.21^2 + 3.15^2 - \frac{(15.75)^2}{5} = 0.0090$$

The appropriate estimate of the variance in either case should be

$$s^2 = \frac{0.0090}{4 - 1} = \frac{0.0090}{5 - 2} = 0.0030$$

It is clear that the calculated $y$ is an estimate of the missing value, and yet it contributes nothing to our information or analysis. An understanding of this example shows that an estimation based on known observations of the same experiment implies no additional information; it is after all not a paradox but a necessary result.

## EXERCISES

**19.1** The set of data in the accompanying table is taken from Exercise 16.2 with two missing values.

| Block | Treatments | | | | | | Block total |
|---|---|---|---|---|---|---|---|
| | $a$ | $b$ | $c$ | $d$ | $e$ | $f$ | |
| I | 16 | 22 | 16 | $y_1$ | 18 | 8 | $80 + y_1$ |
| II | 28 | 27 | 17 | 20 | 23 | 23 | 138 |
| III | 16 | 25 | 16 | 16 | 19 | 16 | 108 |
| IV | 28 | $y_2$ | 19 | 18 | 24 | 25 | $114 + y_2$ |
| Treatment total | 88 | 74 $+ y_2$ | 68 | 54 $+ y_1$ | 84 | 72 | $440 + y_1 + y_2$ |

**a** First, give $y_1$ a reasonable value such as the block mean $80/5 = 16$ or the treatment mean $54/3 = 18$ and estimate the value of $y_2$ by the formula for one missing observation. Insert this calculated value of $y_2$ into the table and estimate the value of $y_1$ again by the formula for one missing value, and so on until the successive values of $y_1$ and $y_2$ remain the same.

**b**   Check your answer directly by using the formulas for two missing values in a randomized blocks experiment.   These two methods should yield the same estimates for $y_1$ and $y_2$.   Arithmetic should be carried to the same unit of measurement as the original data.   In this particular example, this means no decimal place.

**c**   Insert the estimates of $y_1$ and $y_2$ into the table and perform an analysis of variance, paying special attention to the degrees of freedom for error and total.

**d**   Calculate the standard error of the difference between the mean values of treatments *b* and *d*.

*Partial Ans.:*   **a** and **b**   $y_1 = 12$, $y_2 = 30$
   **c**   *df* for error $= (t - 1)(r - 1) - 2 = 15 - 2 = 13$
   **d**   "*r*" $= 2.80$ for each treatment

**19.2**   Two observations are missing in a $r \times r$ Latin-square experiment.

Case I.   The two missing values occur in the same treatment but in different rows and in different columns.   Let $y_1$ and $y_2$ be the two missing values, wherever they occur, and let their corresponding row, column, and treatment totals be

$$y_1: \quad R_1 + y_1 \qquad C_1 + y_1$$
$$y_2: \quad R_2 + y_2 \qquad C_2 + y_2 \qquad T_{12} + y_1 + y_2 \qquad G + y_1 + y_2$$

The procedure of minimizing the error sum of squares leads to the following two equations (after simplification)

$$(r - 1)(r - 2)y_1 - (r - 2)y_2 = r(R_1 + C_1 + T_{12}) - 2G = Q_1$$
$$-(r - 2)y_1 + (r - 1)(r - 2)y_2 = r(R_2 + C_2 + T_{12}) - 2G = Q_2$$

As in the text, we shall write the solutions in the form coefficient $\times y =$ known quantity, rather than in the form $y =$ known quantity $\div$ coefficient.   The solutions are then

$$r(r - 2)^2 y_1 = (r - 1)Q_1 + Q_2$$
$$r(r - 2)^2 y_2 = (r - 1)Q_2 + Q_1$$

The same set of formulas covers two other, similar cases.   If the two missing observations occur in the same row but in different columns and in different treatments, we only have to replace $T_{12}$ by separate $T_1$ and $T_2$ and use $R_{12}$ in place of $R_1$ and $R_2$ in calculating the quantities $Q_1$ and $Q_2$.

Case II.   If the two missing observations occur in different rows, different columns, and different treatments, we shall have $T_1$ and $T_2$ separately.   Since $(r - 1)(r - 2)$ occur frequently, the expressions may be shortened considerably by writing

$$f = (r - 1)(r - 2)$$

The equations after simplification are

$$fy_1 + 2y_2 = r(R_1 + C_1 + T_1) - 2G = Q_1$$
$$2y_1 + fy_2 = r(R_2 + C_2 + T_2) - 2G = Q_2$$

with solutions

$$(f^2 - 4)y_1 = fQ_1 - 2Q_2$$
$$(f^2 - 4)y_2 = fQ_2 - 2Q_1$$

These expressions are equivalent to those of Federer (1955, p. 165).

**19.3** Referring to Table 18.1 (or Table 18.3), let the observation on animal (12) in period II be missing, so that the data table is as follows.

| Period | Animals | | | | | | | | | | | | Period total | Treatment total |
|---|---|---|---|---|---|---|---|---|---|---|---|---|---|---|
| | (1) | (2) | (3) | (4) | (5) | (6) | (7) | (8) | (9) | (10) | (11) | (12) | | |
| I | $b$ | $b$ | $a$ | $b$ | $a$ | $a$ | $a$ | $a$ | $b$ | $b$ | $b$ | $a$ | | $a$ |
| | · | · | · | · | · | · | · | · | · | · | · | 18 | 240 | 252 |
| II | $a$ | $a$ | $b$ | $a$ | $b$ | $b$ | $b$ | $b$ | $a$ | $a$ | $a$ | $b$ | | $b$ |
| | · | · | · | · | · | · | · | · | · | · | · | $y$ | 203 | 191 |
| Animal total | · | · | · | · | · | · | · | · | · | · | · | 18 | 443 | 443 |

**a** Estimate the value of $y$ by an appropriate formula.

**b** Calculate the values $d = \text{I} - \text{II}$, and estimate the missing $d$.

**c** In the text we have assumed that the observation on animal (12) in period I is missing. Compare your answer for the present exercise with that given in the text. What do you notice?

**d** If both observations on animal (12) were missing, what would you propose to do?

*Partial Ans.:*  **a**  $y = {}^{59}\!/_{5} = 11.8$  **b**  $d_1 = 6.2$

# 20

# Concomitant observations

When the experimental units are heterogeneous in certain respects, the reader has been advised to use an appropriate blocking procedure so that the units within each block are homogeneous and those between blocks are heterogeneous. The variation from the latter source may then be eliminated by the analysis-of-variance method. Sometimes, however, especially in biological and medical research, the variation of experimental units is of such a nature that simple blocking of them is not feasible or is inadequate. In such cases, one method generally applicable to handling the heterogeneity is to measure and record for each unit the variate representing the heterogeneity and incorporate such information into the analysis of the final experimental results. The variate representing the heterogeneity of the experimental units (to be distinguished from the variable under study) is called a concomitant variate, here denoted by $x$.

## 1. Examples and assumptions

There are numerous occasions in biology when observations on a concomitant variable will' greatly increase the accuracy of the analysis and facilitate the interpretation and understanding of the experimental results. We shall mention only a few examples, chosen chiefly for their simplicity. In studying the effect on increase in body weight ($y$) under various diets (treatments), it would be ideal to have a group of units (mice, children, etc.) at a fixed age with identical *initial weights* before the commencement of treatments, but such a group is generally unobtainable. If you force a group of mice or children to have the same weight, they will no longer be the same age, and this will introduce an even more serious heterogeneity. A two-dimensional blocking may sometimes be resorted to, but in general it is difficult to obtain. Furthermore, with respect to a continuous variable like weight, blocking is inherently inadequate. The practical thing to do is to have a group of animals at a fixed age and record the initial weight $x$ of each animal. Then the variation in the $x$ values is certainly not due to treatments but represents the heterogeneity of the units with which the experiment begins. In the analysis of the increase in body weight, we shall try to adjust the experimental results $y$ on the basis of their $x$ values.

As another example of initial heterogeneity among the experimental units, we may consider drugs that supposedly reduce blood pressure of adults. Before the administration of the drugs (treatments), the blood pressure varies considerably from individual to individual. A rough blocking of the individuals according to their initial blood pressure is sometimes possible, albeit inconvenient in practice. As in the preceding example, we may record the blood pressure $x$ of each individual before the administration of the drugs and use this information to adjust the treatment effects $y$ in our analysis. Similar examples may be furnished by the readers themselves in their own respective fields.

Not all concomitant observations are taken before treatments. On certain occasions they are taken during the period of experimentation or even at the end of the experiment. The easiest-to-understand example is found again in the field of agriculture. Suppose that 30 plants (cotton, corn, etc.) had originally been planted in each plot (unit of land). At harvest time, however, some of the plots have only 20 or even 10 plants left, the rest of them being eaten up by crows as seeds or by wild animals as seedlings. The yield from such plots is naturally lower than that from 30 plants. The number of plants may be recorded as $x$ and later used to correct the yield $y$. We must note that here the assumption is that the number of plants left is *not* due to treatments but is due to an uncorrelated factor (crows, rabbits, etc.) which introduces heterogeneity to the experimental plots. In a word, $x$ represents the heterogeneity of units, not treatment effects.

There is no rule of thumb for correcting the $y$ values on the basis of the corresponding $x$ values; everything depends on the nature of the relationship between $x$ and $y$. If the relationship $y = f(x)$ is known a priori or is established on other grounds, the correction may be made accordingly. In most experimental situations, however, no a priori formulation of $y = f(x)$ is available and thus no straightforward correction of $y$ can be made. On the contrary, the relationship between $x$ and $y$ is to be determined empirically by the experimental data themselves. This results in circular reasoning, and we must break the circle by making an assumption somewhere. The most primitive assumption is that $y$ is linearly related to $x$; that is, the relationship is of the form

$$y = \bar{y} + b(x - \bar{x}) = (\bar{y} - b\bar{x}) + bx$$
$$= a + bx$$

where the values of $a$ and $b$ are to be determined by experimental data and the $y$ values are to be adjusted on that basis. This amounts to a compromise between a perfect a priori $y = f(x)$ on the one hand and a total neglect of the $x$ influence on the other. Here we assume the form of the relationship and determine the parameter empirically. In brief, we work in part by assumption and in part empirically.

The reader may, of course, raise the question of the validity of the linear assumption. Indeed, we may raise the question of the validity of the general linear model in the entire field of the analysis of variance. Within the scope of the so-called linear statistics, it is fair to say that the assumption $y = a + bx$ is valid to the same degree (whatever it is) as the assumption $y = u + b_i + t_j + e_{ij}$, and empirically the two are satisfactory to the same extent too.

In this chapter we shall deal essentially with the simple case of linear relationship. Toward the end of the chapter, some possible extensions are indicated.

## 2. Review of notations

This chapter should be regarded as a continuation of Chaps. 13 and 14 (Linear regression and Regression and analysis of variance) in its statistical aspects, just as Chap. 16 (Randomized blocks) is a continuation of Chap. 10 (Two-way classifications). The reader may well review those Chaps. 13 and 14 before engaging in this one. However, in order to provide some continuity, we shall review the essential notations (not the algebraic demonstrations) employed previously. In dealing with $x$ and $y$ simultaneously, a simplified notation is almost a necessity. To avoid the writing of deviations, their squares or products, and then the summation, we write in brevity for a group of $n$ pairs of $x$ and $y$,

$$(yy) = \Sigma(y - \bar{y})^2 \qquad (xy) = \Sigma(x - \bar{x})(y - \bar{y}) \qquad (xx) = \Sigma(x - \bar{x})^2$$

The regression coefficient for this group is then

$$b = \frac{\Sigma(x - \bar{x})(y - \bar{y})}{\Sigma(x - \bar{x})^2} = \frac{(xy)}{(xx)}$$

Let $y' = a + bx$ be the $y$ value on the straight line (to be distinguished from the actually observed $y$, which may be above, below, or on the line). The sum of squares of the observed $y$'s may be subdivided into two components: one is that of $y'$ on the straight line and one is that of $y$ from the straight line. The former is

$$\sum (y' - \bar{y})^2 = b^2 \sum (x - \bar{x})^2 = \frac{(xy)^2}{(xx)} = (yy)'$$

The latter is obtained by subtraction from the total

$$\sum (y - y')^2 = (yy) - \frac{(xy)^2}{(xx)} = (yy)''$$

The component $(yy)'$, denoted by a single prime, is entirely due to regression of $y$ on $x$ and has one degree of freedom. The second component, denoted by double prime, is called the residual (or reduced) sum of squares and has $n - 2$ degrees of freedom. This component is supposed to be free of the $x$ effects because it is based on the deviation $y - y'$. The algebraic expressions, the simplified notations, and the corresponding subdivision of degrees of freedom are summarized below:

$$\Sigma(y - \bar{y})^2 = \Sigma(y' - \bar{y})^2 + \Sigma(y - y')^2$$
$$(yy) = (yy)' + (yy)''$$
$$n - 1 = 1 + (n - 2)$$

This is the backbone formula in regression analysis. The application of linear regression in analysis of variance has been explained in some detail for the case of one-way classification in Chap. 14. Here we shall show how it may be applied to randomized blocks experiments and Latin squares. This involves, as before, repeated use of the basic formula reviewed above.

## 3. Preliminary calculations

In Chap. 16 we analyzed a $4 \times 9$ randomized blocks experiment, and we could have used that same example here. However, the calculations are so tedious for a beginner that we would rather use the $4 \times 6$ data

*Tabel 20.1* Data on $y$ and concomitant $x$ of a randomized blocks experiment

| | | | Treatments | | | | Block total | |
|---|---|---|---|---|---|---|---|---|
| | a | b | c | d | e | f | $Y_i$ | $X_i$ |
| Block I | 16 / 10 | 22 / 13 | 16 / 7 | 10 / 8 | 18 / 9 | 8 / 7 | 90 | 54 |
| Block II | 28 / 12 | 27 / 11 | 17 / 9 | 20 / 8 | 23 / 12 | 23 / 8 | 138 | 60 |
| Block III | 16 / 10 | 25 / 13 | 16 / 10 | 16 / 6 | 19 / 8 | 16 / 7 | 108 | 54 |
| Block IV | 28 / 16 | 30 / 15 | 19 / 10 | 18 / 6 | 24 / 11 | 25 / 14 | 144 | 72 |
| Treatment total $Y_j$ | 88 | 104 | 68 | 64 | 84 | 72 | 480 | |
| $X_j$ | 48 | 52 | 36 | 28 | 40 | 36 | | 240 |

Calculation of four basic quantities

| Quantity | For $y$ alone | For $x \times y$ | For $x$ alone |
|---|---|---|---|
| A | 10,324 | 5,077 | 2,582 |
| B | 9,924 | 4,890 | 2,436 |
| T | 9,880 | 4,956 | 2,496 |
| C | 9,600 | 4,800 | 2,400 |

Sum of squares and sum of products

| Source | df | Expression | $(yy)$ | $(xy)$ | $(xx)$ |
|---|---|---|---|---|---|
| Blocks | 3 | $B - C$ | 324 | 90 | 36 |
| Treatments | 5 | $T - C$ | 280 | 156 | 96 |
| Error | 15 | $A - B - T + C$ | 120 | 31 | 50 |
| Total | 23 | $A - C$ | 724 | 277 | 182 |

given in the exercises for Chap. 16. After the reader gets used to the calculations, he may analyze the larger experiment as an additional exercise. The data on $y$ have been reproduced in Table 20.1 with the corresponding concomitant variate $x$ values added. Our aim is to analyze the treatment effects after the influence of $x$ on $y$ has been eliminated by the regression method. Since the variation in $x$ represents either the initial heterogeneity of experimental units or heterogeneity caused by other uncorrelated factors during the experimentation, the elimination of the disturbing influence of $x$ by a statistical procedure, analogous to the elimination of block differences, enables us to obtain an unbiased estimate of treatment effects as if the influence of $x$ were nonexistent. For this reason, the procedure to be described in this chapter is also known as the "statistical control of heterogeneity" of experimental units in lieu of or in addition to the control by blocking.

The preliminary calculations are shown in the middle of Table 20.1. The four basic quantities ($A$, $B$, $T$, $C$) for the $y$'s alone and those for the $x$'s alone are calculated the usual way. The corresponding four basic quantities for the sum of products (spt) are calculated in the manner indicated in Chap. 13. To be quite specific, these are:

$$A = \Sigma xy = [(10 \times 16) + (13 \times 22) + \cdots + (14 \times 25)] = 5,077$$
$$B = \tfrac{1}{6}\Sigma X_i Y_i = \tfrac{1}{6}[(54 \times 90) + \cdots + (72 \times 144)] = 4,890$$
$$T = \tfrac{1}{4}\Sigma X_j Y_j = \tfrac{1}{4}[(48 \times 88) + \cdots + (36 \times 72)] = 4,956$$
$$C = \tfrac{1}{24}(XY) = \tfrac{1}{24}(240 \times 480) = 4800,$$

The subdivision of the total sum of products $A - C$ into various components is done the same way as that for the sum of squares. The complete tabulation of ssq and spt is shown at the bottom part of Table 20.1. This completes the preliminary step necessary for covariance analysis.

## 4. Analysis of residual ssq

The next step is to disregard the ssq and spt due to blocks and form a new table consisting of only the treatment and error components with a new "total" as shown in the left-hand portion of Table 20.2. The "total" with quotation marks means the total of the components for treatment and error only. The reader will now see the resemblance between Table 20.2 and Table 14.2 for data of one-way classification. Indeed, the procedure of eliminating the block components from consideration reduces the two-way classification data to a one-way situation. Had we adjusted the original data of Table 20.1 in order to eliminate the block (row) effects as we did in Table 10.4, we would have obtained the same results as shown here in Table 20.2. As an exercise, the reader should carry out the arithmetic to convince himself that this

*Table* 20.2  Calculation of residual ssq and test of significance

| Source | df | (yy) | (xy) | (xx) | Regression df | Regression (yy)' | Residual df | Residual (yy)'' | msq |
|--------|----|------|------|------|------|------|------|------|------|
| Treatments | 5 | 280 | 156 | 96 | ... | ... | 5 | 59.7 | 11.94 |
| Error | 15 | 120 | 31 | 50 | 1 | 19.2 | 14 | 100.8 | $7.20 = s^2$ |
| "Total" | 20 | 400 | 187 | 146 | 1 | 239.5 | 19 | 160.5 | $F = 1.66$ |

*Table* 20.2 (cont.)   Further analysis

| Source | df | (yy) | (xy) | (xx) | df | (yy)' | df | (yy)'' | msq |
|---|---|---|---|---|---|---|---|---|---|
|  |  |  |  |  | 1 |  | 1 | 33.2 | 33.2 |
| Treatments | 5 | 280 | 156 | 96 | 1 | 253.5 | 4 | 26.5 | 6.6 |
| Error | 15 | 120 | 31 | 50 | 1 | 19.2 | 14 | 100.8 | $7.2 = s^2$ |
| "Total" | 20 | 400 | 187 | 146 | 1 | 239.5 | 19 | 160.5 | $F = 4.61$ |

is so. For practical computation, however, the procedure outlined in Table 20.1 is the most convenient.

Having reduced the randomized blocks to a one-way classification situation, the rest of the analysis is exactly the same as that described in Chap. 14. That is, the method of subdividing the sum of squares into components $(yy)'$ and $(yy)''$ is then applied to the error $(yy) = 120$ and the "total" $(yy) = 400$ separately. Thus,

For error:    $(yy)' = \dfrac{(31)^2}{50} = 19.2,$    $(yy)'' = 120.0 - 19.2 = 100.8$

For "total":    $(yy)' = \dfrac{(187)^2}{146} = 239.5$    $(yy)'' = 400.0 - 239.5 = 160.5$

These values are entered into the right half of Table 20.2. Now the regression components $(yy)'$ are due to influence of $x$ on $y$ and should therefore be eliminated from further consideration. The residual components $(yy)''$ are free from the effect of $x$ and hence they form the new basis of the analysis of variance. Concentrating on the $(yy)''$ column alone, we see that the total is 160.5 and the error is 100.8, so that the sum of squares due to treatments is

$$160.5 - 100.8 = 59.7$$
with *df*                $$19 - 14 = \quad 5$$

Note that we do not calculate the regression component $(yy)'$ and residual $(yy)''$ for treatments, and the value 59.7 is obtained by subtraction.

At this stage it is convenient to introduce two terms to help describe the various kinds of sums of squares. The residual sum of squares $(yy)'' = (yy) - (xy)^2/(xx)$ is said to be "reduced," because it is always smaller than the original $(yy)$, disregarding the very special case in which $(xy) = 0$. Furthermore, their degrees of freedom are also reduced, usually by unity. Thus, 100.8 and 160.5, not only being smaller than 120 and 400, respectively, but also having fewer degrees of freedom, are all *reduced* sums of squares. On the other hand, the sum of squares 59.7, obtained by subtraction, is apparently of a different nature. It is not a simple $(yy)''$, and its *df* is not reduced either. Nevertheless, it is due to treatments. We shall call this kind of sum of squares "adjusted." It will be shown in a later section that the so-called adjusted *ssq* is always a mixture of two subcomponents. For our present purpose, it suffices to point out that an adjusted *ssq* is not necessarily smaller than the corresponding original *ssq*. In the present example, 59.7 with 5 *df* is smaller than 280 with 5 *df*. However, you will recall that in Table 14.2 the adjusted 71.50 with 2 *df* is larger than the original 56 with 2 *df*.

Accepting 59.7 as the adjusted sum of squares for treatments, we find the mean squares (*msq*) the usual way—dividing $(yy)''$ by its corresponding *df*. The final step is the variance-ratio test, $F = 11.94/7.20 = 1.66$. The treatment effects, after the elimination of the influence of $x$, are nonsignificant. In Exercise 16.2, without the information on $x$,

$$F = {}^{56}\!/_8 = 7$$

is significant. The interpretation is obvious. Most of the differences in the $y$'s are actually caused by (or may be explained by) the variations in the $x$'s rather than by the various treatments.

## 5. Significance of error regression

This section is mathematically the same as the corresponding section in Chap. 14, but it illustrates an entirely different situation. Concentrating on the error line of Table 20.2, we find that $b = {}^{31}\!/_{50} = 0.62$,

| Error | *df* | *ssq* | *msq* | *F* |
|---|---|---|---|---|
| Regression | 1 | 19.2 | 19.2 | 2.67 |
| Residual | 14 | 100.8 | 7.2 | |
| Total | 15 | 120.0 | | |

and its significance is judged by the accompanying $F$ test. The regression coefficient $b$ is nonsignificant. According to the usual advice, we may then forget the influence of $x$ and analyze the $y$ data only. Yet in this example we have seen that the results of analysis with $x$ and without $x$ are entirely different. This is due to the comparatively large value of $(xy)$ for treatments and the comparatively small value of $(xy)$ for error. Although the error variance has been reduced from $s^2 = 8.0$ without $x$ to $s^2 = 7.2$ with $x$, the reduction in the treatment mean square is much more pronounced and hence results in nonsignificant treatment effects. This means most of the differences in the various treatment groups may be accounted for by the corresponding differences in $x$

values. In such a case, the analysis of residual *ssq* is recommended regardless of the magnitude of the error regression, because it helps one to understand the variation of $y$.

To cite an analogy, we may recall that in an ordinary randomized block experiment, the variation among the blocks is eliminated from the estimation of treatment effects and the error variance, regardless of the significance of the block effects. In fact, most investigators do not even bother to test the block effect, because it merely represents the heterogeneity of the units and is of no inherent interest to the problem under investigation. By precisely the same philosophy, a routine procedure such as that shown in Table 20.2 may be recommended whenever concomitant observations are available, regardless of the significance of the error regression coefficient, because the variation in $x$ is thought of as a disturbance or heterogeneity that is not due to treatments under investigation.

## 6. Further analysis

A more detailed understanding of the relationship between $x$ and $y$ can be achieved only by plotting the various scatter diagrams and regression lines after the block differences have been eliminated. However, one outstanding component in the data may be obtained with little labor, and this is shown in Table 20.2$A$. Now we subdivide the treatment sum of squares $(yy) = 280$ into components $(yy)'$ and $(yy)''$ in the usual manner, viz.,

$$(yy)' = \frac{(156)^2}{96} = 253.5 \qquad (yy)'' = 280.0 - 253.5 = 26.5$$

In the $(yy)''$ column we obtain by subtraction

$$160.5 - 100.8 - 26.5 = 33.2$$

with 1 *df*. This amounts to subdividing the former 59.7 into 33.2 and 26.5. The value 33.2 is what we have called component $S_4$ in Chap. 14, and it is due to the difference between the treatment regression $156/96 = 1.625$ and the error regression $31/50 = 0.620$. It is obvious from the numerical table that if the treatment regression and the error regression are equal, the $S_4$ component will be zero, and the so-called adjusted and reduced sums of squares will be the same. In our example the difference between the two regression coefficients is tested by $F = 33.2/7.2 = 4.61$, barely reaching the 5 per cent significance level.

All the other algebraic details are the same as in Chap. 14. To summarize, the application of regression in analysis of variance (usually known as the analysis of covariance) in a randomized blocks experiment is first to reduce the data to a one-way situation. Hence, the detailed explanation for a one-way classification in Chap. 14 really forms the basis for all covariance analyses.

## 7. Adjusted treatment means

As mentioned previously (Chap. 14), the treatment means should be adjusted to free themselves from the influence of $x$. If $b = b_E$ is the error regression coefficient and $\bar{y}_j$ and $\bar{x}_j$ are the observed mean of treatment $j$ and its corresponding mean of $x$, the adjusted mean is then

$$\text{adj. } \bar{y}_j = \bar{y}_j - b(\bar{x}_j - \bar{x})$$

Thus,

$$\text{adj. } \bar{y}_1 = 22 - 0.62(12 - 10) = 20.76$$
$$\text{adj. } \bar{y}_6 = 18 - 0.62(9 - 10) = 18.62$$

The variance of an adjusted mean is greater than the usual $s^2/r$, being

$$V(\text{adj. } \bar{y}_j) = s^2 \left[ \frac{1}{r} + \frac{(\bar{x}_j - \bar{x})^2}{(xx)_E} \right]$$

where $s^2 = 7.2 =$ reduced error variance, $r = 4 =$ number of replications, and $(xx)_E = 50$ in Table 20.2 = sum of squares of $x$ for error. The variance of the difference between two adjusted means (say, treatments 1 and 2) is likewise greater than the usual $2s^2/r$, being

$$V(\text{adj. } d) = s^2 \left[ \frac{2}{r} + \frac{(\bar{x}_1 - \bar{x}_2)^2}{(xx)_E} \right]$$

It varies from pair to pair, depending on the $\bar{x}$'s. When there are six treatments, as in our example, there are fifteen possible pair comparisons, and it is convenient to calculate the average value of the variance of adjusted differences. Let $(xx)_T$ be the sum of squares of $x$ for treatments and $t$ be the number of treatments, so that $(xx)_T/(t-1)$ is the mean square of $x$ for treatments. Then the

$$\text{Average } V(\text{adj. } d) = \frac{2s^2}{r} \left[ 1 + \frac{(xx)_T/(t-1)}{(xx)_E} \right]$$
$$= \frac{2(7.2)}{4} \left( 1 + \frac{96\frac{2}{5}}{50} \right) = 4.98$$

and the average standard error of the difference between two adjusted treatment means is $s(\text{adj. } d) = \sqrt{4.98} = 2.23$.

## 8. Covariance in Latin square

Again, the Latin-square data should be reduced to a one-way classification by eliminating the row and column effects, and then the usual method of calculating the residual sum of squares $(yy)''$ may be used. The preliminary calculations yield a table like the accompanying one,

| Source | $df$ | $(yy)$ | $(xy)$ | $(xx)$ |
|--------|------|--------|--------|--------|
| Rows | $r - 1$ | * | * | * |
| Columns | $r - 1$ | * | * | * |
| Treatments | $r - 1$ | # | # | # |
| Error | $(r - 1)(r - 2)$ | # | # | # |
| Total | $r^2 - 1$ | * | * | * |

in outline form, where each * or # denotes a number (*ssq* or *spt*). The next step is to ignore the numbers for rows and columns and form a new table with a new "total" consisting of only the numbers for treatments and error. The calculation of residual $(yy)''$ is then based on this new table. In this second table, $f = (r - 1)(r - 2)$ is the original number

| Source | $df$ | $(yy)$ | $(xy)$ | $(xx)$ | $df$ | $(yy)''$ | $msq$ |
|--------|------|--------|--------|--------|------|----------|-------|
| Treatments | $r - 1$ | # | # | # | $r - 1$ | $(\text{subt})\#$ | $\#$ |
| Error | $(r - 1)(r - 2)$ | # | # | # | $f - 1$ | # | $s^2$ |
| "Total" | $(r - 1)^2$ | # | # | # | $r(r - 2)$ | # | |

of degrees of freedom for error and $(r - 1)^2 - 1 = r(r - 2)$. In practice it is unnecessary to write out the linear component $(yy)'$ with 1 $df$. Hence that part has been omitted from the second table. The rest of the analysis is then the same as that described in the preceding sections.

## 9. Multiple and curvilinear regressions

The assumption $y = a + bx$ may be extended in two ways. One is that there are two (or more) concomitant variates involved, say, $x$ and $z$. The simultaneous elimination of their influences on $y$ may be achieved by finding a multiple (but still linear) regression equation of $y$ on $x$ and $z$, viz.,

$$y = a + bx + cz$$

where $b$ and $c$ are called *partial* regression coefficients.

Another extension is to modify the linear relationship between $x$ and $y$ to a curvilinear relationship, the simplest of which is of the type

$$y = a + bx + cx^2$$

In spite of the difference in meaning, the arithmetic procedure of finding the regression equation in these two cases is the same, taking the $x^2$ of the latter as the variate $z$ of the former. Hence, in the following we shall refer to the multiple linear regression only.

First of all, we calculate the sum of squares $(yy)$, $(xx)$, $(zz)$ and then the sum of products $(xy)$, $(zy)$, $(xz)$. The (normal) equations for determining the values of $b$ and $c$ are

$$b(xx) + c(xz) = (xy)$$
$$b(xz) + c(zz) = (zy)$$

with solutions

$$b = \frac{(zz)(xy) - (xz)(zy)}{(xx)(zz) - (xz)^2}$$

$$c = \frac{(xx)(zy) - (xz)(xy)}{(xx)(zz) - (xz)^2}$$

After the values of $b$ and $c$ are calculated from the expressions above, the residual sum of squares of $y$ is obtained by the formula

$$(yy)'' = (yy) - b(xy) - c(zy)$$

which is a direct extension of $(yy)'' = (yy) - b(xy)$ for the case of one concomitant variate. The degrees of freedom for error residual $(yy)''$ and "total" residual $(yy)''$ should each be diminished by 2, because of the fitting of two constants ($b$ and $c$).

We shall not go into the details of multiple and curvilinear regressions in this elementary text. The half-page outline given above, however, serves two purposes. One is to show that the principle is the same whether one or two concomitant variates are involved. Another is to provide the reader with a "compass" so that when he consults more comprehensive texts dealing with such subjects, he is not lost in the midst of long and unwieldy arithmetic.

## EXERCISES

**20.1** From the data in Table 20.1 we obtain the block effects in the first of the accompanying tables. Using the block effects $b_i$ to adjust the $y$ values and $b_i'$ to adjust

| | Block total | | Block mean | | Block effects | |
| | $Y_i$ | $X_i$ | $\bar{y}_i$ | $\bar{x}_i$ | $b_i$ | $b_i'$ |
|---|---|---|---|---|---|---|
| I | 90 | 54 | 15 | 9 | $-5$ | $-1$ |
| II | 138 | 60 | 23 | 10 | 3 | 0 |
| III | 108 | 54 | 18 | 9 | $-2$ | $-1$ |
| IV | 144 | 72 | 24 | 12 | 4 | 2 |
| Total | 480 | 240 | | | 0 | 0 |
| Mean | | | 20 | 10 | | |

the $x$ values in Table 20.1, we obtain the data free of block effects, as in the second

|   | a | b | c | d | e | f | Total |
|---|---|---|---|---|---|---|---|
| I | 21 / 11 | 27 / 14 | 21 / 8 | 15 / 9 | 23 / 10 | 13 / 8 | 120 / 60 |
| II | 25 / 12 | 24 / 11 | 14 / 9 | 17 / 8 | 20 / 12 | 20 / 8 | 120 / 60 |
| III | 18 / 11 | 27 / 14 | 18 / 11 | 18 / 7 | 21 / 9 | 18 / 8 | 120 / 60 |
| IV | 24 / 14 | 26 / 13 | 15 / 8 | 14 / 4 | 20 / 9 | 21 / 12 | 120 / 60 |
| Total | 88 / 48 | 104 / 52 | 68 / 36 | 64 / 28 | 84 / 40 | 72 / 36 | 480 / 240 |

table. The treatment totals, however, remain the same. The sum of squares and sum of products between the blocks are now zero, and the data may be regarded as essentially a one-way situation consisting of six treatment groups. Calculations of the basic quantities and *ssq* and *spt* give us the results in the third and fourth of the accompanying tables. These calculations give us the results of Table 20.2 directly. The rest of the analysis is exactly the same as that of one-way classification.

| Group | For $y$ alone | For $x \times y$ | For $x$ alone |
|---|---|---|---|
| A | 10,000 | 4,987 | 2,546 |
| T | 9,880 | 4,956 | 2,496 |
| B = C | 9,600 | 4,800 | 2,400 |

| Source | df | value | $(yy)$ | $(xy)$ | $(xx)$ |
|---|---|---|---|---|---|
| Treatments | 5 | $T - C$ | 280 | 156 | 96 |
| Error | 15 | $A - T$ | 120 | 31 | 50 |
| Total | 20 | $A - C$ | 400 | 187 | 146 |

**20.2** Referring to the data in Table 20.1 once more, let us now assume that the six columns are blocks (replications) and the four rows represent four treatments. Perform an analysis of variance on the residual sum of squares after eliminating the effect of $x$ on $y$ by linear regression.

*Partial Ans.:*

| Source | df | $(yy)$ | $(xy)$ | $(xx)$ | df | $(yy)'$ | df | $(yy)''$ | msq | F |
|---|---|---|---|---|---|---|---|---|---|---|
| Treatments | 3 | · | · | · | . . . | . . . | 3 | 173.0 | 57.7 | 8.01 |
| Error | 15 | · | · | · | 1 | 19.2 | 14 | 100.8 | 7.2 | |
| Total | 18 | · | · | · | 1 | 170.2 | 17 | 273.8 | | |

**20.3** Let us continue to take the four rows of Table 20.1 as four treatments. Using the error regression coefficient, find the four adjusted treatment means and the average variance of the difference between two such adjusted means.

*Partial Ans.:*

$$\text{Average } V(\text{adj. } d) = \frac{2(7.2)}{6}\left(1 + \frac{36\frac{2}{3}}{50}\right) = 2.976$$

$$\text{Average } s(\text{adj. } d) = \sqrt{2.976} = 1.725$$

**20.4** The $4 \times 6$ data in the accompanying table are purely for your exercise.

Treatments

| | (1) | | (2) | | (3) | | (4) | | (5) | | (6) | |
|---|---|---|---|---|---|---|---|---|---|---|---|---|
| | $y$ | $x$ | $y$ | $x$ | $y$ | $x$ | $y$ | $x$ | $y$ | $x$ | $y$ | $x$ |
| I | 15 | 16 | 12 | 22 | 18 | 16 | 17 | 10 | 16 | 18 | 18 | 8 |
| II | 13 | 28 | 14 | 27 | 16 | 17 | 17 | 20 | 13 | 23 | 17 | 23 |
| III | 15 | 16 | 12 | 25 | 15 | 16 | 19 | 16 | 17 | 19 | 18 | 16 |
| IV | 9 | 28 | 10 | 30 | 15 | 19 | 19 | 18 | 14 | 24 | 11 | 25 |

Note carefully the negative regression coefficients. The four basic quantities for $x \times y$ are $A = 6,923$, $B = 7,110$, $T = 7,044$, and $C = 7,200$. The relevant portion for analysis of residual *ssq* is as shown in the second table.

| Source | *df* | $(yy)$ | $(xy)$ | $(xx)$ | *df* | $(yy)'$ | *df* | $(yy)''$ | *msq* |
|---|---|---|---|---|---|---|---|---|---|
| Treatments | 5 | 96 | −156 | 280 | ... | ... | 5 | 16.6 | 3.32 |
| Error | 15 | 50 | −31 | 120 | 1 | 8.0 | 14 | 42.0 | 3.00 |
| Total | 20 | 146 | −187 | 400 | 1 | 87.4 | 19 | 58.6 | |

**20.5** In the case of multiple regression with respect to two concomitant variates $x$ and $z$, although the amount of arithmetic is generally more than doubled, there is a very simple special situation and that is when $x$ and $z$ are uncorrelated. See that when $(xz) = 0$, the partial regression coefficients are simply

$$b = \frac{(xy)}{(xx)} \qquad c = \frac{(zy)}{(zz)}$$

That is, each of them is equal to the simple regression coefficient as if the other concomitant variate is nonexistent.

# 21

# Factors at two levels

In preceding chapters, except Chap. 12, we have been talking about a number of treatments being replicated and compared without specifying what these treatments are. Hence the preceding analysis is in general terms and applies to all situations. In this and the next few chapters we shall deal explicitly with certain particular kinds of treatments, namely, those involving more than one factor. Such experiments, usually called *factorial experiments*, are widely used in biological and agricultural research. It is hoped that some of the simpler types of factorial designs may be adopted for medical and human experiments.

## 1. Treatment combinations and notations

Previously the treatments have been denoted nonspecifically by $a, b, c, d,$ . . . or by (1), (2), (3), (4), . . . . When particular factors are involved, more specific notations are needed. We shall first give the longhand notation, and then, as the subject develops, introduce some convenient shorthand notations.

The simplest possible type of factorial experiment is one involving two factors, each at two levels. This has been mentioned in Example 3 of Chap. 12; and the reader may well review that whole chapter, because the so-called factorial analysis is nothing more than a systematic application of the method of making certain types of orthogonal contrasts among the treatments.

A *factor* is an ingredient or manipulation that goes into a treatment. The *level* is the dosage or amount of an ingredient employed in the treatment. The homely example of one or two tablets of aspirin and Bufferin serves to clarify the meaning of the *levels of factors*. Once this basic idea is clear, it may be extended to cover a variety of situations. Suppose that an experiment involves treatment temperatures at 25°C and 30°C as well as concentrations at 4 and 10 ppm. We may call temperature factor $a$—the lower temperature $a_0$ and the higher temperature $a_1$, and call concentration factor $b$—the lower concentration $b_0$ and the higher concentration $b_1$. With two factors, each at two levels, there are four possible treatment combinations, viz., $a_1b_1$, both factors at the higher level; $a_1b_0$ and $a_0b_1$, one factor at the higher level and the other at the lower level; and $a_0b_0$, both factors at the lower level. The lower level may, but not necessarily does, denote the absence of a factor. For instance, the two levels of concentration may be 0 and 10 ppm. The careful reader may have noticed that in Chap. 12 we used the notation $a_1$, $a_2$, and $b_1$, $b_2$, while in this chapter we switch to $a_0$, $a_1$ and $b_0$, $b_1$. This is not intended to confuse the reader; rather, it is intended to show him that it is irrelevant which system is used. In fact, both systems are in common use and the reader should be familiar with both.

## 2. Main effect and interaction

If we for convenience let $a_1b_1$ denote both the treatment combination and its measured value, etc., the four treatments and their corresponding values may be arranged in a $2 \times 2$ table. The difference between the

|        | $(b_1)$ | $(b_0)$ |  |
|--------|---------|---------|--------------------|
| $(a_1)$ | $a_1b_1$ | $a_1b_0$ | $a_1b_1 + a_1b_0$ |
| $(a_0)$ | $a_0b_1$ | $a_0b_0$ | $a_0b_1 + a_0b_0$ |
|        | $a_1b_1 + a_0b_1$ | $a_1b_0 + a_0b_0$ | Total |

two values in the same column having the same level of factor $b$ is presumably due to the difference in factor $a$. Thus,

$$a_1b_1 - a_0b_1 = \text{effect of factor } a \text{ at the fixed } b_1 \text{ level}$$
$$a_1b_0 - a_0b_0 = \text{effect of factor } a \text{ at the fixed } b_0 \text{ level}$$

The sum of these two differences (which is equal to the difference between the two row totals) is called the *main effect* of factor $a$ and is denoted by $A$.   It is the average effect of factor $a$ over the two levels of factor $b$. Thus,

$$A = \text{first row total} - \text{second row total}$$
$$= a_1b_1 + a_1b_0 - a_0b_1 - a_0b_0$$
$$= a_1(b_1 + b_0) - a_0(b_1 + b_0) \qquad \text{symbolically}$$

The last form is not an algebraic expression in the ordinary sense, but simply a convenient and formal way of representing the longer expression above it.   This is called a symbolic expression.   It is introduced at this early occasion for ease of understanding and acceptance, because there will be a lot more of symbolic writings later on.   The reason for resorting to this type of symbolic writing is that it is usually much easier to write and to remember than the long expressions.   In fact, if the "common factor" $(b_1 + b_0)$ is "taken out," the expression for the main effect of factor $a$ may be further abbreviated into

$$A = (a_1 - a_0)(b_1 + b_0)$$

meaning the sum of treatments with $a_1$ minus the sum of treatments with $a_0$.

If factors $a$ and $b$ have no influence on each other, then the effect of factor $a$ at $b_1$ level $(a_1b_1 - a_0b_1)$ and that at $b_0$ level $(a_1b_0 - a_0b_0)$ should be equal.   The *difference* between these two effects is called the *interaction* between factors $a$ and $b$; it is denoted by $AB$.   Thus,

$$AB = (a_1b_1 - a_0b_1) - (a_1b_0 - a_0b_0)$$
$$= a_1b_1 - a_0b_1 - a_1b_0 + a_0b_0$$
$$= (a_1 - a_0)(b_1 - b_0) \qquad \text{symbolically}$$

An interaction as defined above is always the difference between two differences.   In terms of the original $2 \times 2$ table, it is the difference between the sums of the two diagonally opposite corners.

In a similar way we can study the effects of factor $b$.   The difference between the two treatments in the same row of the $2 \times 2$ table is due to factor $b$ at a fixed level of factor $a$.   Thus,

$$a_1b_1 - a_1b_0 = \text{effect of factor } b \text{ at fixed } a_1 \text{ level}$$
$$a_0b_1 - a_0b_0 = \text{effect of factor } b \text{ at fixed } a_0 \text{ level}$$

Sum

$$B = a_1b_1 + a_0b_1 - a_1b_0 - a_0b_0 = \text{main effect of factor } b$$
$$= b_1(a_1 + a_0) - b_0(a_1 + a_0)$$
$$= (b_1 - b_0)(a_1 + a_0) \qquad \text{symbolically}$$

Difference

$$BA = (a_1b_1 - a_1b_0) - (a_0b_1 - a_0b_0) = \text{interaction between } b \text{ and } a$$
$$= (b_1 - b_0)(a_1 - a_0) \qquad \text{symbolically}$$

It is to be noted that $AB$ and $BA$ are identical. The interaction is the same whether we look at it from the viewpoint of factor $a$ or from that of factor $b$. The $2 \times 2$ combination table is symmetrical with respect to $a$ and $b$. Hence, from now on, we shall simply say the interaction between $a$ and $b$.

## 3. Orthogonal subdivision of treatment $ssq$

Consider the experiment with four treatments in six blocks (Exercise 16.2, blocks and treatments interchanged). The relevant portion under discussion is as shown in the accompanying table. You are supposed

| Blocks | | | | | | Treatment combination | Treatment total |
|---|---|---|---|---|---|---|---|
| (1) | (2) | (3) | (4) | (5) | (6) | | |
| . | . | . | . | . | . | $a_0b_0$ | 90 |
| . | . | . | . | . | . | $a_0b_1$ | 138 |
| . | . | . | . | . | . | $a_1b_0$ | 108 |
| . | . | . | . | . | . | $a_1b_1$ | 144 |
| . | . | . | . | . | . | Total | 480 |

to have previously calculated the total $ssq$ for treatments with 3 $df$, viz.,

$$\frac{90^2 + 138^2 + 108^2 + 144^2}{6} - \frac{480^2}{24} - 9{,}924 - 9{,}600 = 324$$

Our new assumption in this chapter is that the four treatments are the four combinations of two factors, each at two levels, as indicated above. According to our definitions the main effects of the factors and their interaction are

$$A = a_1b_1 + a_1b_0 - a_0b_1 - a_0b_0 = (a_1 - a_0)(b_1 + b_0) = 24$$
$$B = a_1b_1 - a_1b_0 + a_0b_1 - a_0b_0 = (a_1 + a_0)(b_1 - b_0) = 84$$
$$AB = a_1b_1 - a_1b_0 - a_0b_1 + a_0b_0 = (a_1 - a_0)(b_1 - b_0) = -12$$

These are three mutually orthogonal contrasts, and hence we may use the method developed in Chap. 12 to calculate the $ssq$ for each compo-

*Table* 21.1 Orthogonal subdivision of treatment sum of squares

|  | $a_1b_1$ 144 | $a_1b_0$ 108 | $a_0b_1$ 138 | $a_0b_0$ 90 | Effect $z$ | Divisor $D \times r$ | $ssq$ $z^2/Dr$ |
|---|---|---|---|---|---|---|---|
| $A$ | $+1$ | $+1$ | $-1$ | $-1$ | 24 | $4 \times 6$ | 24 |
| $B$ | $+1$ | $-1$ | $+1$ | $-1$ | 84 | $4 \times 6$ | 294 |
| $AB$ | $+1$ | $-1$ | $-1$ | $+1$ | $-12$ | $4 \times 6$ | 6 |
|  |  |  |  |  | Total treatment $ssq$ | | 324 |

*Table* 21.2 Analysis of variance of the entire experiment (Exercise 16.2)

| Variation | $df$ | $ssq$ | Subdivision | $df$ | $ssq$ | $F$ |
|---|---|---|---|---|---|---|
| Blocks | 5 | 280 | | | | |
| Treatments | 3 | 324 | $\begin{cases} A \\ B \\ AB \end{cases}$ | 1 1 1 | 24 294 6 | 3.00 36.75 |
| Error | 15 | 120 | $s^2 =$ | | 8 | |
| Total | 23 | 724 | | | | |

nent. The method has been used in Table 21.1, in which

$$D = 1^2 + 1^2 + 1^2 + 1^2 = 4 = \text{sum of squares of coefficients}$$

and $r = 6$, the number of replications.

The orthogonal subdivisions of the treatment $ssq$ may then be incorporated into the analysis of the entire experiment as shown in Table 21.2. In brief, the three single components take the place of the total treatment $ssq$ with three degrees of freedom. From this finer analysis we see that almost all the variation in treatments is due to the main effect of factor $b$. The effects of factor $a$ and interaction are quite negligible. This kind of analysis of treatment effects is known as the factorial analysis. The investigator, of course, does not have to perform this kind of analysis if he has some other comparisons of interest to make or some other special hypothesis to test. It all depends on the nature of the two factors, their relationship, and the particular purpose of the scientist in performing the experiment.

## 4. Treatment totals in tabular form

The subdivision of the total treatment $ssq$ may also be achieved through a method which perhaps has more intuitive appeal. The four treatment totals may be arranged into the form of a $2 \times 2$ table according to the

$$
\begin{array}{c|cc|c}
 & b_1 & b_0 & \\
\hline
a_1 & 144 & 108 & 252 \\
a_0 & 138 & 90 & 228 \\
\hline
 & 282 & 198 & 480
\end{array}
$$

levels of the factors. Then the ordinary method of calculating *ssq* may be directly applied to this table. The total *ssq* for the four numbers is calculated first, taking into account the fact that each number here is the sum of six original single numbers. It turns out to be 324. The *ssq* between the rows is that due to the main effect of factor $a$, and it is equal to

$$
\frac{252^2 + 228^2}{12} - \frac{480^2}{24} = 9,624 - 9,600 = 24
$$

or more simply,

$$
\frac{(252 - 228)^2}{24} = \frac{24^2}{24} = 24
$$

The sum of squares between columns is that due to the main effect of factor $b$, and it is equal to

$$
\frac{282^2 + 198^2}{12} - \frac{480^2}{24} = \frac{(282 - 198)^2}{24} = 294
$$

The remaining *ssq* for interaction may then be obtained by subtraction:

$$
324 - 24 - 294 = 6
$$

This is the most convenient practical computational procedure. The interaction component may, of course, be obtained independently by considering the sums of the diagonally opposite corners of the $2 \times 2$ table of treatment totals.

A comparison by tracing the numbers between the $2 \times 2$ tabular method and the formal method of laying out the orthogonal contrasts shows that they are in fact one and the same method — identical in every respect. This is precisely the point. We wish to bring out the equivalence of the two methods for this almost self-evident case in order to save lengthy explanations for more complicated cases. In all complete factorial designs, the breakdown of the treatment *ssq* may usually be achieved by simply arranging the treatment totals into a "factor $\times$ factor" two-way table without writing out explicitly the coefficients of the orthogonal contrasts. More about this point will be said later on when we deal with more than two factors.

## 5. Linear model for treatment effects

For the original observed values of a randomized blocks experiment, the linear model is that each number is the sum of a constant $u$, a block effect $b_i$, a treatment effect $t_j$, and a random error $e_{ij}$, viz.,

$$y_{ij} = u + b_i + t_j + e_{ij}$$

and the sum of squares due to treatments is $r\Sigma t_j^2$, where $r$ is the number of replications of the treatments. This we have demonstrated in quite some detail in preceding chapters. Now, in view of the factorial nature of the treatments, we have further subdivided the treatment $ssq = r\Sigma t_j^2$ into orthogonal components. What does this mean in terms of the underlying model?

The subscript $j$ of treatment effect $t_j$ usually takes the values 1, 2, 3, . . . up to the number of treatments. The four treatments in our present example are $a_0b_0$, $a_0b_1$, $a_1b_0$, $a_1b_1$, and the single subscript $j$ is not quite adequate to cover this case. However, for the sake of simplicity, we shall not introduce new and complicated symbols but simply let $j$ denote any one of the four treatment combinations. The treatment effects are found the usual way, as shown in the accompanying table.

| Treatment | Total $Y_j$ | Mean $\bar{y}_j$ | Effect $t_j$ |
|-----------|-------------|------------------|--------------|
| $a_0b_0$ | 90 | 15 | $-5$ |
| $a_0b_1$ | 138 | 23 | $+3$ |
| $a_1b_0$ | 108 | 18 | $-2$ |
| $a_1b_1$ | 144 | 24 | $+4$ |
| Total | $Y = 480$ | $\bar{y} = 20$ | 0 |

The subdivision of the treatment $ssq$ implies that $t_j$ itself is the sum of three further components. Indeed, each $t_j$ is the sum of an effect of factor $a$, an effect of factor $b$, and an interaction effect. Informally ($j = $ a combination of $l$ and $k$),

$$t_j = (a)_l + (b)_k + (ab)_{lk}$$

where $l$ and $k$ indicate the levels 0 or 1 of the factors. This is a linear model within the general linear model. The components of $t_j$ may be found exactly the same way as those of the original $y_{ij}$. We arrange the values of $t_j$ in a two-way table according to the factorial classifications, just as we arrange the values of $y_{ij}$ in a two-way table according

|       | $b_1$ | $b_0$ | Total | Mean |
|-------|-------|-------|-------|------|
| $a_1$ | $+4$  | $-2$  | $+2$  | $+1.0$ |
| $a_0$ | $+3$  | $-5$  | $-2$  | $-1.0$ |
| Total | $+7$  | $-7$  | $0$   |      |
| Mean  | $+3.5$ | $-3.5$ |      | $0$  |

to blocks and treatments, and the components are calculated from the marginal means. Since the general mean is zero, the marginal means give directly the effects of the corresponding factors. The effect of $a_1$ is 1.0 and that of $a_0$ is $-1.0$. Similarly, the effect of $b_1$ is 3.5 and that of $b_0$ is $-3.5$. The deviation of $t_j$ from the sum of the corresponding marginal effects is the interaction. The complete representation is as follows:

$$
\begin{array}{rrll}
t = & (a) & + (b) & + (ab) \\
a_1b_1: \quad +4 = & +1.0 & + 3.5 & - 0.5 \\
a_1b_0: \quad -2 = & +1.0 & - 3.5 & + 0.5 \\
a_0b_1: \quad +3 = & -1.0 & + 3.5 & + 0.5 \\
a_0b_0: \quad -5 = & -1.0 & - 3.5 & - 0.5
\end{array}
$$

If we calculate the sum of squares of the numbers in each column shown above, we shall find that

$$
\begin{array}{rcl}
\Sigma t_j^2 &=& \Sigma(a)^2 + \Sigma(b)^2 + \Sigma(ab)^2 \\
54 &=& 4 \;+\; 49 \;+ 1
\end{array}
$$

Multiplying by $r = 6$,

$$
324 = 24 \;+\; 294 \;+ 6
$$

These are precisely the three components of treatment *ssq* we have obtained by various methods previously. To summarize, the analysis of factorial treatments is based on a linear model for the treatment effects within the general linear model for the observed values.

## 6. Simplified notation and mechanical calculation

Anticipating more complicated cases, we wish to introduce a further simplified notation system and a highly routine method of calculating main effects and interactions at this early stage when the relationships may be so readily seen. The new system is as follows:

| Treatment combination: | $a_1b_1$ | $a_1b_0$ | $a_0b_1$ | $a_0b_0$ |      |
|------------------------|----------|----------|----------|----------|------|
| Simplified notation:   | $ab$     | $a$      | $b$      | $(1)$    |      |

The rule is that we write the letter if that factor is at the higher level and write nothing if that factor is at the lower level. Thus, for the combination $a_0b_1$, we simply write $b$, meaning that the factor $b$ is at the higher level and implying that the factor $a$ is at its lower level. When both factors are at the higher level, we write both letters. When both factors are at the lower level, we write the symbol (1), which, as the reader will see, plays the same role in symbolic expressions as the number 1 in algebraic expressions. The advantages of this new notation system are its extreme brevity and its amenability to symbolic writing and mechanical arrangement of the treatment combinations. Thus, the main effects and interaction may be written

$$A = (a - 1)(b + 1) = ab + a - b - 1$$
$$B = (a + 1)(b - 1) = ab - a + b - 1$$
$$AB = (a - 1)(b - 1) = ab - a - b + 1$$

The simplified notation also makes it clearer that the main effect of factor $a$ is the sum of treatments with letter $a$ minus that without. Similarly, the main effect of factor $b$ is the sum of treatments with letter $b$ minus that without. The interaction is the sum of treatments with even number (2 or 0) of letters minus that with an odd number of letters.

To obtain the numerical values of $A$, $B$, $AB$ from data, we arrange the four treatment totals in a column in the order (1), $a$, $b$, $ab$, as shown in the first column of Table 21.3. The order is very important. In the second column we put down the sum of pairs (two adjacent numbers)

**Table 21.3** Mechanical arithmetic procedure of calculating main effects and interaction for factors at two levels

| Treatment totals | | Pair sum and difference | Pair sum and difference | Effect |
|---|---|---|---|---|
| (1) | 90 | 198 | 480 | $Y$ (grand total) |
| $a$ | 108 | 282 | 24 | $A$ (main effect) |
| $b$ | 138 | 18 | 84 | $B$ (main effect) |
| $ab$ | 144 | 6 | $-12$ | $AB$ (interaction) |
| | 480 | | 576 | |

Algebraic basis of the mechanical method

| (1) | | $a + 1$ | $ab + a + b + 1$ | $Y$ |
|---|---|---|---|---|
| $a$ | | $ab + b$ | $ab + a - b - 1$ | $A$ |
| $b$ | | $a - 1$ | $ab - a + b - 1$ | $B$ |
| $ab$ | | $ab - b$ | $ab - a - b + 1$ | $AB$ |
| $Y$ | | | $4ab$ | |

of the preceding column; that is, $90 + 108 = 198$ and $138 + 144 = 282$. These fill up half of the second column. For the next half column, we put down the differences of pairs by subtracting the upper number from the lower number; that is, $108 - 90 = 18$, and $144 - 138 = 6$. This completes the second column. The third column is obtained in exactly the same way by operating on the numbers of the preceding column. In other words, the operation is repeated once more. Thus, the sums are $198 + 282 = 480$ and $18 + 6 = 24$. The differences are $282 - 198 = 84$ and $6 - 18 = -12$. This completes the third column.

Now we observe that the top number (480) in the third column is the grand total of the entire experiment. This fact furnishes a check to the accuracy of arithmetic operations. The next three numbers are precisely $A$, $B$, $AB$, in that order. Another simple arithmetic check is provided by the sum of the third column, 576 in our example. It should be equal to $4ab = 4 \times 144 = 576$ correctly. The algebraic basis for this mechanical operation is given in the lower portion of Table 21.3. The reader may check carefully each step of operation and see that the results in the third column are actually $Y$, $A$, $B$, $AB$ and that the total of the third column is $4ab$, all other quantities being canceled out.

Once the values of $A$, $B$, $AB$ are obtained, the calculation of their corresponding sum of squares is then the same as that shown in Table 21.1.

The reader may wonder why we should devote so much space on so simple a case. There are essentially two reasons. One is to provide a really thorough understanding of the basic principles involved in dealing with factorial analysis. Another is to save space. Everything can be completely exhibited for a $2 \times 2$ factorial situation by small tables and short operations. Once the basic idea is made clear, we shall merely extend the method to cases like $2 \times 2 \times 2$ or, generally, $2^n$ without lengthy explanations.

### 7. Three factors, each at two levels

This is the classical example of factorial designs. Nearly all the basic principles have been originated from this case. Historically the agriculturalist considers the three essential ingredients of fertilizers, namely, nitrogen (n), phosphate (p), and potash (k), each applied at two levels to the field. Even today many textbooks, especially those for agricultural students, still use the language of n-p-k, because it is so easy to comprehend. We shall use the symbols $a$, $b$, $c$ for the factors and employ the simplified notation from the beginning. For three factors, each at two levels, there are eight combinations:

| Combination: | $a_0b_0c_0$ | $a_1b_0c_0$ | $a_0b_1c_0$ | $a_1b_1c_0$ | $a_0b_0c_1$ | $a_1b_0c_1$ | $a_0b_1c_1$ | $a_1b_1c_1$ |
|---|---|---|---|---|---|---|---|---|
| Notation: | (1) | $a$ | $b$ | $ab$ | $c$ | $ac$ | $bc$ | $abc$ |

Should the reader have doubted the efficiency of the simplified notation system when he was dealing with the $2 \times 2$ factorial, that doubt would surely disappear when faced with the $2 \times 2 \times 2$ factorial. Without further laboring, we state that the main effects and interactions are, symbolically,

$$A = (a - 1)(b + 1)(c + 1)$$
$$B = (a + 1)(b - 1)(c + 1)$$
$$AB = (a - 1)(b - 1)(c + 1)$$
$$C = (a + 1)(b + 1)(c - 1)$$
$$AC = (a - 1)(b + 1)(c - 1)$$
$$BC = (a + 1)(b - 1)(c - 1)$$
$$ABC = (a - 1)(b - 1)(c - 1)$$

Expansion of these expressions will yield seven orthogonal contrasts among the eight treatments. These contrasts are listed in Table 21.4. Note that for each contrast there are four positive terms and four negative terms. Consider the main effect $A$, for instance. The four positive terms are those with letter $a$, and the four negative terms are those without it. As another example, let us consider the three-factor interaction $ABC$. The four positive terms are those with an odd number (1 or 3) of letters, and the four negative terms have an even number (0 or 2) of letters.

When the numerical values of $A$, $B$, etc. are obtained from such contrasts, the corresponding *ssq* may then be obtained the usual way (analogously to Table 21.1). Each *ssq* has a single degree of freedom. Thus the total treatment *ssq* with 7 *df* are subdivided into seven single components, each of which may be tested for significance (analogously to Table 21.2). These seven comparisons are of equal accuracy, and therefore the same $s^2$ is used for all the tests.

Table 21.4, giving explicit expressions for the various contrasts, is also useful for other purposes (Chap. 24) and may be used for numerical

*Table* 21.4   Seven orthogonal contrasts among $2 \times 2 \times 2 = 8$ treatments

|       | *abc* | *bc* | *ac* | *c* | *ab* | *b* | *a* | (1) |
|-------|-------|------|------|-----|------|-----|-----|-----|
| $A$   | +1    | −1   | +1   | −1  | +1   | −1  | +1  | −1  |
| $B$   | +1    | +1   | −1   | −1  | +1   | +1  | −1  | −1  |
| $AB$  | +1    | −1   | −1   | +1  | +1   | −1  | −1  | +1  |
| $C$   | +1    | +1   | +1   | +1  | −1   | −1  | −1  | −1  |
| $AC$  | +1    | −1   | +1   | −1  | −1   | +1  | −1  | +1  |
| $BC$  | +1    | +1   | −1   | −1  | −1   | −1  | +1  | +1  |
| $ABC$ | +1    | −1   | −1   | +1  | −1   | +1  | +1  | −1  |

calculation, provided you are careful with the positives and negatives. For practical computation, the following two procedures (Secs. 8 and 9) are in common use.

## 8. Numerical example

A convenient example is provided by the 3 × 8 data in Exercise 16.6. The eight treatments have not been specified earlier, but now they may be regarded as 2 × 2 × 2 factorial combinations. The relevant portion of the data is given in the accompanying table. The most straightfor-

|  | Treatments | | | | | | | | Block total |
|---|---|---|---|---|---|---|---|---|---|
|  | (1) | $a$ | $b$ | $ab$ | $c$ | $ac$ | $bc$ | $abc$ |  |
| I | . | . | . | . | . | . | . | . | . |
| II | . | . | . | . | . | . | . | . | . |
| III | . | . | . | . | . | . | . | . | . |
|  | 36 | 51 | 72 | 69 | 51 | 63 | 60 | 78 | 480 |

ward procedure is the mechanical method outlined in Table 21.5, which is a direct extension of Table 21.3. In using this method the order of

Table 21.5 Calculation of main effects and interactions and their corresponding sum of squares for a 2 × 2 × 2 factorial experiment

| Treatment total | | Pair sum and difference | | | | $z^2$ | Divisor $D \times r$ | $z^2/Dr$ |
|---|---|---|---|---|---|---|---|---|
|  |  | 1st | 2d | 3d |  |  |  |  |
| (1) | 36⎱ | 87⎱ | 228⎱ | 480 | $Y$ |  |  |  |
| $a$ | 51⎰ | 141⎰ | 252⎰ | 42 | $A$ | 1,764 | 8 × 3 | 73.50 |
| $b$ | 72⎱ | 114⎱ | 12⎱ | 78 | $B$ | 6,084 | 8 × 3 | 253.50 |
| $ab$ | 69⎰ | 138⎰ | 30⎰ | −12 | $AB$ | 144 | 8 × 3 | 6.00 |
| $c$ | 51⎱ | 15⎱ | 54⎱ | 24 | $C$ | 576 | 8 × 3 | 24.00 |
| $ac$ | 63⎰ | −3⎰ | 24⎰ | 18 | $AC$ | 324 | 8 × 3 | 13.50 |
| $bc$ | 60⎱ | 12⎱ | −18⎱ | −30 | $BC$ | 900 | 8 × 3 | 37.50 |
| $abc$ | 78⎰ | 18⎰ | 6⎰ | 24 | $ABC$ | 576 | 8 × 3 | 24.00 |
|  | 480 | | | 624 | |  |  | 432.00 |

Analysis of the whole experiment (Exercise 16.6)

| Source | df | ssq | msq |
|---|---|---|---|
| Blocks | 2 | 16 |  |
| Treatments | 7 | 432→ | Subdivided as above |
| Error | 14 | 276 | $s^2 = 19.7$ |
| Total | 23 | 724 |  |

writing the treatments is very important. Treatment (1) is always the first. Next, introduce one factor ($a$, $b$, or $c$; it is factor $a$ in the example) and then another, followed immediately by their interaction. This is the order of the first four treatments. The next four follow the same order, with the third factor added.

The four sums of pairs are entered in the upper half of the next column, and the four differences (lower number minus the upper) of pairs are entered in the lower half of the next column. This operation is to be repeated three times. The numbers in the "third" column under pair sum and difference are the values of the main effects and interactions as defined in Table 21.4, except the top number, 480, which is the grand total of the experiment. Another check of the arithmetic is that the sum of the "third" column is equal to $8\,abc = 8 \times 78 = 624$. The $ssq = z^2/Dr$ is then calculated for each of the contrasts. The ultimate check is that the sum of these seven $ssq$'s should add exactly up to the total treatment $ssq = 432$ with 7 $df$.

The analysis of the entire experiment without subdivision (Exercise 16.6) has also been shown in the lower portion of Table 21.5. The overall test for the eight treatments gives

$$F = \frac{432\%}{276\%_{14}} = \frac{61.7}{19.7} = 3.13 \qquad \text{with 7 and 14 } df$$

It is significant at the 5 per cent level. Now, with the subdivisions known, to test the significance of the main effect of factor $a$, we have $F = 73.5/19.7 = 3.73$ with 1 and 14 $df$. It is nonsignificant at the 5 per cent level (which requires $F = 4.60$). Hence, the only significant component is that of the main effect of factor $b$, for which

$$F = \frac{253.5}{19.7} = 12.9$$

This finding is much more specific than that from the overall test.

### 9. Three-dimensional classification

The $2 \times 2 \times 2 = 8$ treatment totals may also be regarded as a three-way classification of the eight numbers (Table 21.6). Three-way classification has not been dealt with per se as an overall experimental design in our text, partly because experimental data of that nature in biological and medical fields are not too common and partly because it is merely a natural extension of the two-way classification. The analysis of the $2 \times 2 \times 2$ treatments, however, gives us an opportunity to show how it is to be handled. Although it is a special case, the procedure employed is the same for a general three-way classification.

One point that often confuses the beginner may well be clarified here. The (3 blocks) × (8 treatments) experiment under discussion is a two-way classification for the experiment as a whole. The eight treatments themselves, $2 \times 2 \times 2$, form a three-dimensional classification. The factorial analysis is an analysis of variance applied to the treatments alone. The structure of the experiment as a whole and the structure of the treatments are two different things and have no necessary relationship. The same eight ($2 \times 2 \times 2$) treatments may be tested in a completely randomized experiment (one-way classification) or in an $8 \times 8$ Latin square or any other design. The analysis of single treatment effects is within the general analysis of the entire experiment, just as the linear model for treatment effects is within the general linear model of the experiment as a whole (as shown for the $2 \times 2$ case).

In Table 21.6 the eight treatment totals are actually arranged into a three-dimensional figure showing the $a$ dimension, the $b$ dimension, and the vertical $c$ dimension. In practice this is clearly unnecessary, because the upper layer ($c_1$ level) and the lower layer ($c_0$ level) can be written side by side as two tables. The sum of squares of these eight treatment totals has been calculated previously in the general analysis of the treatment effects. It is equal to

$$\tfrac{1}{3}(78^2 + \cdot \cdot \cdot) - \frac{480^2}{24} = 10{,}032 - 9{,}600 = 432$$

denoted by $ssq_T = ssq_{a \times b \times c}$, where the subscript $a \times b \times c$ indicates the factors included in this sum of squares.

Next, three two-way tables are constructed from the original three-dimensional table. For instance, the $a \times b$ table is obtained by telescoping (or collapsing or condensing) the $c$ dimension. In terms of arithmetic, this means taking the sum of the corresponding cells of the two layers. The other two tables are constructed in the same way by eliminating the $b$ dimension or the $a$ dimension. Then, for each of the three tables, we calculate the $ssq$ of the four numbers in the body of the table. For instance, for the $b \times c$ table, the sum of squares is

$$ssq_{b \times c} = \frac{138^2 + 141^2 + 114^2 + 87^2}{6} - \frac{480^2}{24} = 315$$

where the subscript $b \times c$ indicates the factors included in this $ssq$. The number $138 = 78 + 60$, etc., is the sum of six single observations. The other two sums of squares are calculated the same way. These sums of squares are overlapping; by that we mean, for instance, both $ssq_{a \times b}$ and $ssq_{a \times c}$ contain the sum of squares due to factor $a$.

*Table 21.6* Treatment totals of $2 \times 2 \times 2$ factorial as a three-way classification and the three corresponding two-way tables and sums of squares

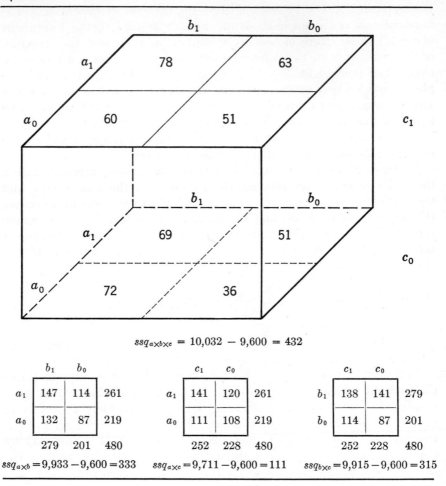

$$ssq_{a \times b \times c} = 10{,}032 - 9{,}600 = 432$$

|       | $b_1$ | $b_0$ |     |
|-------|-------|-------|-----|
| $a_1$ | 147   | 114   | 261 |
| $a_0$ | 132   | 87    | 219 |
|       | 279   | 201   | 480 |

$ssq_{a \times b} = 9{,}933 - 9{,}600 = 333$

|       | $c_1$ | $c_0$ |     |
|-------|-------|-------|-----|
| $a_1$ | 141   | 120   | 261 |
| $a_0$ | 111   | 108   | 219 |
|       | 252   | 228   | 480 |

$ssq_{a \times c} = 9{,}711 - 9{,}600 = 111$

|       | $c_1$ | $c_0$ |     |
|-------|-------|-------|-----|
| $b_1$ | 138   | 141   | 279 |
| $b_0$ | 114   | 87    | 201 |
|       | 252   | 228   | 480 |

$ssq_{b \times c} = 9{,}915 - 9{,}600 = 315$

The final step is to calculate the sum of squares due to the main effect of each single factor from the appropriate marginal totals.   Thus,

$$A: \quad ssq_a = \frac{(261 - 219)^2}{24} = 73.50$$

$$B: \quad ssq_b = \frac{(279 - 201)^2}{24} = 253.50$$

$$C: \quad ssq_c = \frac{(252 - 228)^2}{24} = 24.00$$

The sums of squares for interactions are then obtained by subtraction.

$$AB: \quad ssq_{a\times b} - ssq_a - ssq_b = 333 - 73.50 - 253.50 = 6.00$$
$$AC: \quad ssq_{a\times c} - ssq_a - ssq_c = 111 - 73.50 - 24.00 = 13.50$$
$$BC: \quad ssq_{b\times c} - ssq_b - ssq_c = 315 - 253.50 - 24.00 = 37.50$$
$$ABC: \quad ssq_{a\times b\times c} - \text{all others} = 432 - 73.50 - \cdots = 24.00$$

Comparing these results with those in Table 21.5, we see that they are identical. While some investigators prefer the routine method of "pair sum and difference" (Table 21.5) because it may be easily programmed for a high-speed calculator, others prefer the successive reducing method (Table 21.6) because the picture for each pair of factors may be separately and explicitly examined.

Analogously to the case of $2 \times 2$, the way we subdivide the total treatment $ssq$ into seven orthogonal components implies the linear model that each treatment effect $(t)$ is the sum of seven parts: main effects and interactions. Indeed, it may be written, omitting the subscripts for levels of factors,

$$t = (A) + (B) + (AB) + (C) + (AC) + (BC) + (ABC)$$

This completes the analysis of $2 \times 2 \times 2$ factorial treatments. When there are four factors, each at two levels, the sixteen treatment combinations may be analyzed the same way.

## 10. Advantages of factorial experiments

In the physical sciences the experiments usually allow only one factor to vary and keep all others constant. The one-factor experiment has indeed become more or less the standard pattern of scientific investigation, because it has been highly successful. In the biological world, unfortunately, the various factors interact with each other in a complicated and unexpected manner, and the one-factor experiment is frequently not only inadequate in scope but actually misleading in inference. Suppose that factor $b$ increases blood pressure in the absence of factor $a$ but decreases it or has no effect when factor $a$ is present. This interaction may be immediately revealed by appropriate factorial experiments, while any conclusion with respect to one factor alone is either misleading or useless.

Possibly to the surprise of students, a factorial experiment is usually more economical from both the physical and the statistical point of view. Consider once more the $3 \times 8$ experiment, where $8 = 2 \times 2 \times 2$ factorial. The main effect of factor $a$, for instance, is based on a contrast between 12 observations with $a$ and 12 observations without $a$. Every one of the original 24 observations plays a role in evaluating this effect. The same is true in determining the effect of factor $b$ and all the others. Hence, each observation of a factorial experiment plays a multiple role and is used several times for different purposes. If we had conducted three separate one-factor experiments for factors $a$, $b$, $c$, the observations in the experiment for factor $a$, say, would play no role at all in determining the effects of factors $b$ or $c$. They are useful only in evaluating one factor (besides estimating error). Furthermore, even after all three separate experiments are done, the information on interaction is still missing. To summarize, a factorial experiment broadens the basis for valid inference in a very economical way.

There are, however, limitations to factorial experiments. Too many factors in one project would make the experiment too large; one remedy to this situation will be dealt with in Chap. 24. A possibly more serious limitation is the validity of the linear model for treatment effects. The various factors may not even approximately act additively in a biological system. And there is always the difficulty in interpretating the higher interactions (for example, the practical meaning of $ABCD$). The investigator must use his judgment and knowledge of the factors under consideration in designing a factorial experiment.

### EXERCISES

**21.1** The following are the results of a $2 \times 2$ factorial treatment in five randomized blocks. Subdivide the treatment $ssq$ into single components and test the significance of each. Use all three different methods to achieve the subdivision of the total treatment sum of squares, viz., (**a**) directly from the set of orthogonal contrasts,

|  |  | Blocks | | | | | Treatment total |
|---|---|---|---|---|---|---|---|
|  |  | (1) | (2) | (3) | (4) | (5) |  |
| Treatment combinations | $a_1b_1$ | 17 | 19 | 18 | 20 | 21 | 95 |
|  | $a_1b_0$ | 21 | 22 | 16 | 23 | 28 | 110 |
|  | $a_0b_1$ | 19 | 25 | 17 | 25 | 29 | 115 |
|  | $a_0b_0$ | 11 | 18 | 13 | 20 | 18 | 80 |
| Block total |  | 68 | 84 | 64 | 88 | 96 | 400 |

(**b**) by the method of pair sum and difference, and (**c**) through a $2 \times 2$ table of the treatment totals. Check your answers with the accompanying table. The average effects ($A$ and $B$) are not significant, but the interaction ($AB$) is highly significant. What does this mean in this example? What does it tell you? Examine the four treatment totals (or means) closely. Either factor $a$ or factor $b$ alone will yield a higher response; but when they are both present (at the higher level) or both absent, the response is low. What practical example like this in your own field can you think of?

| Source | df | ssq | Sub-division | df | ssq = msq | F |
|---|---|---|---|---|---|---|
| Blocks | 4 | 184 | | | | |
| Treatments | 3 | 150 | $A$ | 1 | 5 | 1.11 |
| | | | $B$ | 1 | 20 | 4.44 |
| | | | $AB$ | 1 | 125 | 27.78 |
| Error | 12 | 54 | | | $s^2 = 4.50$ | |
| Total | 19 | 388 | | | | |

**21.2**  In our definitions of main effects and interactions, we did not bother to put everything on a per observation basis, because the arithmetic procedure of calculating the final sum of squares (the divisor $D \times r$) will automatically take care of the matter of units. Thus, we simply write

$$0 = ab + a + b + (1)$$
$$A = ab + a - b - (1)$$
$$B = ab - a + b - (1)$$
$$AB = ab - a - b + (1)$$

The last three expressions have been given in the text where $ab$, etc., denote treatment totals. Now we add the first expression so that $ab$, etc., denote *deviations* from mean treatment totals, without affecting values of $A$, $B$, $AB$.

If we regard these as four equations, we may also express the quantities $ab$, $a$, $b$, $(1)$ in terms of $A$, $B$, $AB$. This is easily done by noting that the sum of these four expressions is $4ab$. By changing the signs of two of the four expressions, we obtain in a similar way the values of $4a$, $4b$, $4(1)$. Thus, using the numerical values of $A$, $B$, $AB$ given in Table 21.1, we have

$$4ab = +A + B + AB = +24 + 84 - 12 = +96$$
$$4a = +A - B - AB = +24 - 84 + 12 = -48$$
$$4b = -A + B - AB = -24 + 84 + 12 = +72$$
$$4(1) = -A - B + AB = -24 - 84 - 12 = -120$$

Since each treatment total is the sum of $r = 6$ observations, division of the quantities above by $4 \times 6 = 24$ will put them on a per observation basis. The results are

$$+1.0 + 3.5 - 0.5 = 4 = t_{ab}$$
$$+1.0 - 3.5 + 0.5 = -2 = t_a$$
$$-1.0 + 3.5 + 0.5 = 3 = t_b$$
$$-1.0 - 3.5 - 0.5 = -5 = t_1$$

Note that these are precisely the components of treatment effects we have obtained in the text through the $2 \times 2$ table of treatment effects. The general conclusion is that the treatment effect components are directly given by the main effects and interactions defined by the various contrasts. In practice, only the sums of squares of these components are obtained.

### 21.3    The routine reverse computation

In Table 21.3 is represented a routine method of calculating the main effects and interactions from treatment totals. It is seen from Exercise 21.2 that treatment totals may be obtained from main effects and interactions in a very similar way. Indeed, the same "pair sum and difference" method may be adopted by writing the interactions and main effects in the reverse order as shown in the accompanying table.

| Effects (Table 21.3) | | Pair sum and difference | Pair sum and difference | Treatment totals | |
|---|---|---|---|---|---|
| $AB$ | $-12$ | 72  | 576 | 144 | $ab$ |
| $B$  | 84   | 504 | 552 | 138 | $b$ |
| $A$  | 24   | 96  | 432 | 108 | $a$ |
| $Y$  | 480  | 456 | 360 | 90  | $(1)$ |

The treatment total is one-fourth of the value obtained by the second round of taking pair sums and differences; thus, $\frac{1}{4}(576) = 144$, $\frac{1}{4}(552) = 138$, etc.

The reverse computation has little practical application for orthogonal experiments. For certain confounded experiments (exercises for Chap. 24), however, it may be used to reconstruct unbiased treatment totals (or means) from corrected interactions.

**21.4**    Calculate the main effects and interactions and their corresponding sums of squares by the method of "pair sum and difference" (Table 21.5) but with a different ordering of the treatment totals. Are your results identical with those shown in Table 21.5?

| Treatment total | | Pair sum and difference | | | Effect $z$ |
|---|---|---|---|---|---|
| | | 1st | 2d | 3d | |
| $(1)$ | 36 | | | | $Y$ |
| $b$   | 72 | | | | $B$ |
| $c$   | 51 | | | | $C$ |
| $bc$  | 60 | | | | $BC$ |
| $a$   | 51 | | | | $A$ |
| $ab$  | 69 | | | | $AB$ |
| $ac$  | 63 | | | | $AC$ |
| $abc$ | 78 | | | | $ABC$ |

**21.5** The eight treatments of $2 \times 2 \times 2$ factorial combinations are assigned to 24 experimental units at random without blocking, each treatment combination being replicated 3 times. The results are shown in the first of the accompanying tables.

| (1) | a | b | ab | c | ac | bc | abc |
|-----|-----|-----|-----|-----|-----|-----|-----|
| 10 | 16 | 27 | 23 | 16 | 16 | 25 | 28 |
| 18 | 19 | 17 | 18 | 25 | 16 | 23 | 24 |
| 8 | 16 | 16 | 28 | 22 | 19 | 30 | 20 |
| 36 | 51 | 60 | 69 | 63 | 51 | 78 | 72 |

|  | df | ssq | Single components | |
|---|---|---|---|---|
|  |  |  | $A$ | 1.50 |
|  |  |  | $B$ | 253.50 |
|  |  |  | $AB$ | 0 |
| Treatments | 7 | 432 | $C$ | 96.00 |
|  |  |  | $AC$ | 73.50 |
|  |  |  | $BC$ | 1.50 |
|  |  |  | $ABC$ | 6.00 |
| Error | 16 | 292 | | $s^2 = 18.25$ |
| Total | 23 | 724 | | |

Use all three different methods to subdivide the total treatment sum of squares into seven single components and test the significance of each. The three methods are **a** orthogonal contrasts directly (Table 21.4), **b** pair sum and difference (Table 21.5), and **c** two-way tables for all pairs of factors (Table 21.6). Check your results with the second table.

**21.6** The treatment effects of Exercise 21.4 are found as in the accompanying table. It has been mentioned in the text that each of these effects is the sum of seven components. These components may be found either through tables of treatment effects

| Treatment | (1) | a | b | ab | c | ac | bc | abc |
|-----------|-----|-----|-----|-----|-----|-----|-----|-----|
| Total | 36 | 51 | 72 | 69 | 51 | 63 | 60 | 78 |
| Mean | 12 | 17 | 24 | 23 | 17 | 21 | 20 | 26 |
| Effect | −8 | −3 | +4 | +3 | −3 | +1 | 0 | +6 |

(three-dimensional and two-dimensional) or by using the values of $A$, $B$, $AB$, etc. listed in Table 21.5 in a manner illustrated in Exercise 21.2.    The results are:

$$
\begin{array}{ccccccccc}
 & (A) & (B) & (AB) & (C) & (AC) & (BC) & (ABC) \\
abc\colon & +6 = +1.75 & + 3.25 & - 0.50 & + 1.00 & + 0.75 & - 1.25 & + 1.00 \\
bc\colon & 0 = -1.75 & + 3.25 & + 0.50 & + 1.00 & - 0.75 & - 1.25 & - 1.00 \\
ac\colon & +1 = +1.75 & - 3.25 & + 0.50 & + 1.00 & + 0.75 & + 1.25 & - 1.00 \\
c\colon & -3 = -1.75 & - 3.25 & - 0.50 & + 1.00 & - 0.75 & + 1.25 & + 1.00 \\
ab\colon & +3 = +1.75 & + 3.25 & - 0.50 & - 1.00 & - 0.75 & + 1.25 & - 1.00 \\
b\colon & +4 = -1.75 & + 3.25 & + 0.50 & - 1.00 & + 0.75 & + 1.25 & + 1.00 \\
a\colon & -3 = +1.75 & - 3.25 & + 0.50 & - 1.00 & - 0.75 & - 1.25 & + 1.00 \\
(1)\colon & -8 = -1.75 & - 3.25 & - 0.50 & - 1.00 & + 0.75 & - 1.25 & - 1.00 \\
\end{array}
$$

Verify that the sums of squares of these seven columns of numbers are precisely those listed in the last column of Table 21.5, after being multiplied by $r = 3$.

Note that the components of treatment effects on a per observation basis are numerically $\frac{1}{24}$ of $A = 42$, $B = 78$, etc. listed in Table 21.5.    In finding the sums of squares for these components, we may either multiply $(1.75)^2$, $(3.25)^2$, . . . by 24 or divide $(42)^2$, $(78)^2$, . . . by 24, which is what we did in Table 21.5.    (This is analogous to the relationship $n\bar{y}^2 = Y^2/n$.)    In practice there is no need to calculate these components for each $t_j$.

**21.7**    Expand the symbolic expression for the four-factor interaction

$$ABCD = (a - 1)(b - 1)(c - 1)(d - 1)$$

Separate the eight positive terms from the eight negative terms and list them in two different rows.    Do you notice any general rule as to which terms are positive and which are negative?    Why?    Can you formulate the rule directly from the symbolic product form above?    (Save your results for Chap. 24.)

**21.8**    Another notation system for the treatment combinations is writing out the levels (subscripts) of the factors only:

| Treatments: | $a_1b_1$ | $a_1b_0$ | $a_0b_1$ | $a_0b_0$ |
|---|---|---|---|---|
| Notation: | 11 | 10 | 01 | 00 |

The first digit always refers to factor $a$ and the second to factor $b$.    For three factors, the treatment $a_1b_0c_1$ will be 101.    Write out the eight combinations in this notation.

# 22

# Factors at three levels

Having familiarized ourselves with the basic notions of factorial treatment combinations in the preceding chapter, we shall now extend the method to factors at three levels. Here we shall need somewhat less verbal explanation, because the analysis is quite analogous to that at two levels.

## 1. One factor at three levels

The three levels of factor $a$ may be denoted by $a_0$, $a_1$, $a_2$, with equal intervals. In dealing with factors at two levels we have not mentioned the matter of equality of interval because there is only one interval in that case. Levels with unequal intervals can also be analyzed (Chap. 12), but we shall limit ourselves to equal intervals in this book. Figure 22.1 shows the situation for one factor at three levels, $a_0$, etc., also denoting the response at that level. We observe the following:

Increment from 0 to 1: $\qquad\qquad\ a_1 - a_0$
Increment from 1 to 2: $\qquad\qquad\ a_2 - a_1$
Sum of increments: $\qquad\qquad\quad\ (a_2 - a_1) + (a_1 - a_0) = a_2 - a_0$
Difference between increments: $\ (a_2 - a_1) - (a_1 - a_0) = a_2 - 2a_1 + a_0$

The first contrast, $a_2 - a_0$, measures the linear effect of the factor. The second contrast, $a_2 - 2a_1 + a_0$, measures the deviation from linearity, because if the three points in Fig. 22.1 are collinear, this quantity equals zero. The method of subdividing the *ssq* with two degrees of freedom

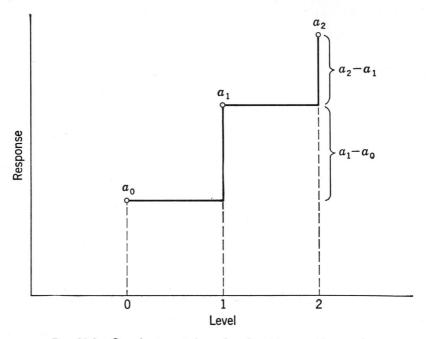

**Fig. 22.1** One factor at three levels with equal intervals.

*Table* 22.1 Subdivision of *ssq* of treatments at three levels with equal intervals

|           | Treatment total | | | Effect | Divisor | *ssq* |
|           | $a_2$ 300 | $a_1$ 228 | $a_0$ 192 | $z$ | $D \times r$ | $z^2/Dr$ |
|-----------|------|------|------|--------|--------------|----------|
| Linear    | 1    | 0    | $-1$ | 108    | $2 \times 12$ | 486      |
| Deviation | 1    | $-2$ | 1    | 36     | $6 \times 12$ | 18       |
|           |      |      |      |        | Total        | 504      |

into single components has been given in Table 12.2.   To provide some continuity, let us consider the three treatment totals 300, 228, 192, totaling 720, each treatment being replicated 12 times.   The total *ssq* for the three treatments (with 2 *df*) is

$$\frac{300^2 + 228^2 + 192^2}{12} - \frac{720^2}{36} = 504$$

The subdivision is given in Table 22.1.

## 2. Two factors, each at three levels

When there are two factors, each at three levels, there will be nine treatment combinations, an example of which is given in Table 22.2.   The *ssq* between blocks (with 3 *df*), the *ssq* between treatments (with 8 *df*), and the *ssq* for error (with 24 *df*) may be calculated the usual way.   In fact, these preliminary results may be found in Table 16.1, with blocks and treatments interchanged and rearranged.

*Table* 22.2   Data on 3 × 3 factorial treatment combinations in four randomized blocks

|     | $a_2$ | | | $a_1$ | | | $a_0$ | | | Total |
| --- | --- | --- | --- | --- | --- | --- | --- | --- | --- | --- |
|     | $b_2$ | $b_1$ | $b_0$ | $b_2$ | $b_1$ | $b_0$ | $b_2$ | $b_1$ | $b_0$ | |
| I | 26 | 33 | 28 | 14 | 18 | 21 | 28 | 10 | 11 | 189 |
| II | 30 | 28 | 26 | 24 | 31 | 18 | 20 | 16 | 14 | 207 |
| III | 20 | 24 | 17 | 24 | 23 | 13 | 8 | 11 | 13 | 153 |
| IV | 24 | 27 | 17 | 22 | 12 | 8 | 24 | 19 | 18 | 171 |
| Total | 100 | 112 | 88 | 84 | 84 | 60 | 80 | 56 | 56 | 720 |

The first step toward subdividing the treatment *ssq* into components is to form the factorial two-way table of treatment totals.   The sum of

|     | $b_2$ | $b_1$ | $b_0$ | |
| --- | --- | --- | --- | --- |
| $a_2$ | 100 | 112 | 88 | 300 |
| $a_1$ | 84 | 84 | 60 | 288 |
| $a_0$ | 80 | 56 | 56 | 192 |
| | 264 | 252 | 204 | 720 |

squares for the nine numbers in the body of the table is, of course, the total treatment *ssq* with 8 *df*:

$$\frac{100^2 + \cdots + 56^2}{4} - \frac{720^2}{36} = 768$$

This *ssq* may be subdivided into three major components in the same manner as for any two-way classification table. The *ssq* for the rows ($a_2$, $a_1$, $a_0$) is 504, with 2 *df* as obtained in the preceding section. This, we say, is due to the main effects of factor *a*. Similarly, the *ssq* for the columns ($b_2$, $b_1$, $b_0$) is

$$\frac{264^2 + 252^2 + 204^2}{12} - \frac{720^2}{36} = 168$$

with 2 *df* which, we say, is due to the main effects of factor *b*. The remaining portion, $768 - 504 - 168 = 96$ with 4 *df* is due to inter-actions between factors *a* and *b*. The results obtained so far have been summarized in Table 22.3. This completes the preliminary subdivision of treatment effects. For practical purposes, the analysis up to this stage is frequently sufficient for the investigator to draw conclusions as to the effects of these factors. However, we shall present the further subdivision into eight single components in the next section for completeness.

*Table* 22.3   Analysis of variance of a 3 × 3 factorial experiment (data of Table 22.2)

| Source | df | ssq | Subdivision | df | ssq | msq | F |
|---|---|---|---|---|---|---|---|
| Blocks | 3 | 180 | | | | | |
| Treatments | 8 | 768 | A | 2 | 504 | 252 | 8.89 |
| | | | B | 2 | 168 | 84 | 2.96 |
| | | | AB | 4 | 96 | 24 | |
| Error | 24 | 680 | | | | 28.3 | |
| Total | 35 | 1,628 | | | | | |

### 3. The eight contrasts

For factors at two levels, the symbolic expressions and expansions for the various contrasts have been greatly simplified by replacing $a_1$ and $a_0$ by *a* and 1, etc. Unfortunately, there is no equally simple notation for factors at three levels. We may, however, adopt a half-simplified system, writing $a_2$, $a_1$, $a_0$ as $a_2$, *a*, 1. Even this half-simplified notation will save us a lot of writing labor.

Now the sum of squares for the three row totals (300, 228, 192) may be subdivided into two single components, as was done in Sec. 1. These are totals over all levels of factors *b*. Similarly, the *ssq* for the three column totals (264, 252, 204) over all levels of factor *a* may also be subdivided into two single components by the same technique. These four contrasts may be written symbolically

$$A_1 = (a_2 - 1)(b_2 + b + 1)$$
$$A_2 = (a_2 - 2a + 1)(b_2 + b + 1)$$
$$B_1 = (a_2 + a + 1)(b_2 - 1)$$
$$B_2 = (a_2 + a + 1)(b_2 - 2b + 1)$$

where factors like $(a_2 + a + 1)$ and $(b_2 + b + 1)$ simply mean marginal totals. The remaining four expressions for different types of interactions are

$$A_1B_1 = (a_2 - 1)(b_2 - 1)$$
$$A_1B_2 = (a_2 - 1)(b_2 - 2b + 1)$$
$$A_2B_1 = (a_2 - 2a + 1)(b_2 - 1)$$
$$A_2B_2 = (a_2 - 2a + 1)(b_2 - 2b + 1)$$

These eight expressions, when expanded, will form a set of orthogonal contrasts, each with a single degree of freedom. Instead of lining the nine treatment combinations into a single file (see exercises), Table 22.4 gives these eight contrasts in a "folded" $3 \times 3$ arrangement. Note that the coefficients for the interactions are actually products of corresponding coefficients of the main effects involved.

**Table 22.4  The eight contrasts among $3 \times 3$ treatment combinations**

|       | $b_2$ | $b_1$ | $b_0$ |   | $B_1$ |   |   |   | $B_2$ |   |   |
|-------|-------|-------|-------|---|-------|---|---|---|-------|---|---|
| $a_2$ | · | · | · |   | +1 | 0 | −1 |   | +1 | −2 | +1 |
| $a_1$ | · | · | · |   | +1 | 0 | −1 |   | +1 | −2 | +1 |
| $a_0$ | · | · | · |   | +1 | 0 | −1 |   | +1 | −2 | +1 |

|  | $A_1$ |  |   | | $A_1B_1$ | |   | | $A_1B_2$ | |
|----|----|----|---|----|----|----|---|----|----|----|
| +1 | +1 | +1 |   | +1 | 0 | −1 |   | +1 | −2 | +1 |
| 0 | 0 | 0 |   | 0 | 0 | 0 |   | 0 | 0 | 0 |
| −1 | −1 | −1 |   | −1 | 0 | +1 |   | −1 | +2 | −1 |

|  | $A_2$ |  |   | | $A_2B_1$ | |   | | $A_2B_2$ | |
|----|----|----|---|----|----|----|---|----|----|----|
| +1 | +1 | +1 |   | +1 | 0 | −1 |   | +1 | −2 | +1 |
| −2 | −2 | −2 |   | −2 | 0 | +2 |   | −2 | +4 | −2 |
| +1 | +1 | +1 |   | +1 | 0 | −1 |   | +1 | −2 | +1 |

## 4. Numerical calculation

The numerical value of the eight contrasts are obtained from the $3 \times 3$ table of treatment totals according to the coefficients of the contrast under consideration. It is convenient to enter the treatment totals at the upper left corner of Table 22.4. The main effects $A_1$ and $A_2$ are based on row totals, and the main effects $B_1$ and $B_2$ are based on column totals. The four interactions are based on the body of the table. To avoid arithmetic errors, it is also convenient to collect and add all the positive terms first and then all the negative terms. We shall calculate the value of $A_2 B_2$ for illustration. The central and the four corner terms are positive, while the four middle side terms are negative. Thus,

$$4(84) + 100 + 88 + 80 + 56 = \quad 660$$
$$2(112 + 84 + 60 + 56) = \quad \underline{624}$$
$$z = \quad 36$$
$$z^2 = 1{,}296$$

$D =$ sum of squares of coefficients of the contrast
$\quad = 1^2 + (-2)^2 + 1^2 + (-2)^2 + 4^2 + (-2)^2 + 1^2 + (-2)^2 + 1^2 = 36$
$r =$ number of replications of treatment combinations $\qquad = 4$

and

$$ssq = \frac{z^2}{Dr} = \frac{1296}{144} = 9 \qquad \text{with 1 } df$$

This and all the others have been listed in Table 22.5. Each component may then be tested against $s^2 = 28.33$ for significance. $F$ has 1 and 24 $df$.

*Table 22.5* **The value and sum of squares of the eight contrasts (Table 22.4)**

| Contrast | Value | $z$ | Divisor $D \times r$ | $ssq = z^2/Dr$ |
|----------|-------|-----|----------------------|-----------------|
| $A_1$ | $300 - 192 =$ | $108$ | $6 \times 4$ | $486$ |
| $A_2$ | $492 - 456 =$ | $36$ | $18 \times 4$ | $18$ |
| $B_1$ | $264 - 204 =$ | $60$ | $6 \times 4$ | $150$ |
| $B_2$ | $468 - 504 =$ | $-36$ | $18 \times 4$ | $18$ |
| $A_1 B_1$ | $156 - 168 =$ | $-12$ | $4 \times 4$ | $9$ |
| $A_1 B_2$ | $300 - 360 =$ | $-60$ | $12 \times 4$ | $75$ |
| $A_2 B_1$ | $300 - 312 =$ | $-12$ | $12 \times 4$ | $3$ |
| $A_2 B_2$ | $660 - 624 =$ | $36$ | $36 \times 4$ | $9$ |
| | | | Total treatment $ssq$ | $768$ |

*Table 22.6* Calculation of contrast values $z$ for $3 \times 3$ factorial combinations from treatment totals

| | Value of | | | Value of | |
|---|---|---|---|---|---|
| | $(b_2 - b_0)$ | $(b_2 - 2b_1 + b_0)$ | | $(a_2 - a_0)$ | $(a_2 - 2a_1 + a_0)$ |
| Given $a_2$ | 12 | −36 | Given $b_2$ | 20 | 12 |
| Given $a_1$ | 24 | +24 | Given $b_1$ | 56 | 0 |
| Given $a_0$ | 24 | 24 | Given $b_0$ | 32 | 24 |
| Sum | $B_1$ | $B_2$ | Sum | $A_1$ | $A_2$ |
| $(a_2 + a_1 + a_0)$ | 60 | −36 | $(b_2 + b_1 + b_0)$ | 108 | 36 |
| Difference | $A_1B_1$ | $A_1B_2$ | Difference | $A_1B_1$ | $A_2B_1$ |
| $(a_2 - a_0)$ | −12 | −60 | $(b_2 - b_0)$ | −12 | −12 |
| Difference | $A_2B_1$ | $A_2B_2$ | Difference | $A_1B_2$ | $A_2B_2$ |
| $(a_2 - 2a_1 + a_0)$ | −12 | +36 | $(b_2 - 2b_1 + b_0)$ | −60 | +36 |

Another method of numerical calculation is shown in Table 22.6. From the $3 \times 3$ table of treatment totals we calculate the values of $b_2 - b_0$ and $b_2 - 2b_1 + b_0$ for each level of factor $a$ separately. Likewise, the values of $a_2 - a_0$ and $a_2 - 2a_1 + a_0$ are obtained for each level of factor $b$. These are the 12 basic numbers from which the main effects and interactions are to be calculated. The sums of these numbers over all the levels of the other factor give the four main effects. The differences give the interactions. Note that the interactions (bottom of Table 22.6) have been calculated twice, and these provide an arithmetic check. It also shows that the interactions are symmetrical, that is, $A_1B_2 = B_2A_1$, etc. After these contrast values have been obtained, the calculation of the sum of squares of each component is the same as in Table 22.5.

## 5. Routine tabular procedure

Analogously to Table 21.3 and 21.5 for factors at two levels, a table of treatment totals for factors at three levels may also be set up systematically in such a way that the various effects may be obtained by repeated operations of sum and difference. The order of arranging the nine treatment combinations and subsequent calculations are shown in Table 22.7. The rules of "operation" are, briefly:

First three numbers are simple sums like: $\qquad a_2 + a_1 + a_0$
Second three numbers are differences of type: $\qquad a_2 - a_0$
Third three numbers are quantities of type: $\qquad a_2 - 2a + a_0$

*Table* 22.7   Calculation of eight contrast values (z) for 3 × 3 factorial combinations from treatment totals

| Levels of $a$ | $b$ | Short notation | Treatment totals | First operation | Second operation | Contrast $z$ |
|:---:|:---:|:---:|:---:|:---:|:---:|:---:|
| 0 | 0 | (1) | 56 | 204 | 720 | $Y$ |
| 1 | 0 | $a$ | 60 | 252 | 108 | $A_1$ |
| 2 | 0 | $a_2$ | 88 | 264 | 36 | $A_2$ |
| | | | | | | |
| 0 | 1 | $b$ | 56 | 32 | 60 | $B_1$ |
| 1 | 1 | $ab$ | 84 | 56 | $-12$ | $A_1B_1$ |
| 2 | 1 | $a_2b$ | 112 | 20 | $-12$ | $A_2B_1$ |
| | | | | | | |
| 0 | 2 | $b_2$ | 80 | 24 | $-36$ | $B_2$ |
| 1 | 2 | $ab_2$ | 84 | 0 | $-60$ | $A_1B_2$ |
| 2 | 2 | $a_2b_2$ | 100 | 12 | 36 | $A_2B_2$ |

of the numbers in the preceding column.   Thus, the three segments of the "first operation" column are obtained as follows:

$$56 + 60 + 88 = 204 \qquad 88 - 56 = 32 \qquad 88 - 2(60) + 56 = 24$$
$$56 + 84 + 112 = 252 \qquad 112 - 56 = 56 \qquad 112 - 2(84) + 56 = 0$$
$$80 + 84 + 100 = 264 \qquad 100 - 80 = 20 \qquad 100 - 2(84) + 80 = 12$$

This is one of the many instances in mathematics and statistics when it takes longer to explain than do.   The numbers in the "second operation" column, Table 22.7, are obtained exactly the same way from the preceding column.   The very top number, 720, is the grand total of all observations in the experiment.   The remaining eight numbers are the contrast values, so that $A_1$ appears in the same row with treatment $a$, and so on.   The reader should check that these $z$ values are identical with those obtained in Table 22.5, although in a slightly different order. The rest of the calculation is the same as before.

For 3 × 3 factorials this mechanical method has little advantage over the direct method using the contrast expressions.   When there are more than two factors, however, the mechanical method will be simpler if all of the single components are desired.   It is also suitable for programming.

## 6. Large interactions

It has been said in the preceding chapter that one of the advantages in factorial experiments is to be able to study interactions between factors. In the example considered above, the main effects are large and the interactions are small.   When interactions are large, however, the interpretation of factor effects needs careful examination of the various treatment totals (or means) and a mere presentation of an analysis-of-variance table may be wholly inadequate or even misleading. As an illustration, let us suppose that the treatment totals for four replications are as given here. These are the same nine numbers as before, but their treatment

|  | $b_2$ | $b_1$ | $b_0$ |  |
|---|---|---|---|---|
| $a_2$ | 56 | 84 | 100 | 240 |
| $a_1$ | 80 | 88 | 84 | 252 |
| $a_0$ | 112 | 60 | 56 | 228 |
|  | 248 | 232 | 240 | 720 |

labels have been changed. The total *ssq* for the nine treatment combinations remains the same: 768. A preliminary subdivision of the treatment effects is shown in the accompanying table. From this analy-

| Effects | df | ssq | msy | F |
|---|---|---|---|---|
| Main effects, $A$ | 2 | 24.00 | 12.00 | |
| Main effects, $B$ | 2 | 10.67 | 5.33 | |
| Interactions, $AB$ | 4 | 733.33 | 183.33 | 6.47 |
| Error | 24 | 680.00 | 28.83 | |

sis as well as from the marginal totals of treatments, one may conclude that factors $a$ and $b$ have no effects. On the other hand, an examination of the $3 \times 3$ treatment totals will reveal that both factors have strong effects, each depending upon the level of the other factor. For instance, at the given $a_0$ level, the effect of factor $b$ may be seen from the totals 112, 60, 56. At the given $a_2$ level, the effect of factor $b$ is reflected by 56, 84, 100, the changing trend having been reversed. And this is what we mean by interaction. The term *main* effect would be misleading if we forgot its definition. The main effect of one factor is defined as the *average* effect of that factor over all the levels of the other factor(s). Now, because of the opposite changing trends, the average effect becomes negligible, but this does not mean that the factors have no effects. Hence, when interactions are large, we must examine the effects at different levels separately.

Let us first subdivide the total interaction *ssq* into four single components according to the contrasts shown in Table 22.4 or by any of the other methods described in preceding sections. The results are listed in the accompanying table. The only significant component of the

| Component | $z$ | $z^2$ | $D \times r$ | $z^2/Dr$ |
|---|---|---|---|---|
| $A_1B_1$ | $-100$ | 10,000 | $4 \times 4$ | 625.00 |
| $A_1B_2$ | $-60$ | 3,600 | $12 \times 4$ | 75.00 |
| $A_2B_1$ | 20 | 400 | $12 \times 4$ | 8.33 |
| $A_2B_2$ | 60 | 3,600 | $36 \times 4$ | 25.00 |
| | | Total interaction *ssq* | | 733.33 |

interactions is

$$A_1B_1 = a_2b_2 - a_2b_0 - a_0b_2 + a_0b_0$$
$$= a_2b_2 - a_2 - b_2 + (1)$$

This component is equivalent to the interaction for factors at two levels, omitting the middle level. It reflects the opposing increasing and decreasing trend of the factor effects.

Now, in view of the nature of the interactions, perhaps a more descriptive analysis is to find the effects of one factor at each level of the other factor. For illustration, let us find the effects of factor $b$ at fixed levels of factor $a$. For example, at $a_2$, the *ssq* due to factor $b$ is

$$\frac{56^2 + 84^2 + 100^2}{4} - \frac{240^2}{12} = 248$$

The *ssq*'s due to $b$ at given $a_1$ and $a_0$ levels are calculated the same way. The results are summarized in Table 22.8. Obviously, factor $b$ has no effect at the $a_1$ level, but its effects at $a_0$ and $a_2$ levels are significant. Each of the components shown in Table 22.8 has 2 *df*. This can be further subdivided into single components by the contrasts $(1, 0, -1)$ and $(1, -2, 1)$ if needed, but it is evident that most of the effects are linear. Description of this nature is much more accurate than the mere

*Table* 22.8 Analysis of a 3 × 3 factorial experiment; subdivision of treatment effects at separate levels

| Source | *df* | *ssq* | Subdivision | *df* | *ssq* | *msq* | *F* |
|---|---|---|---|---|---|---|---|
| Blocks | 3 | 180 | | | | | |
| Treatments | 8 | 768 | Between $a$ rows $(A)$ | 2 | 24 | 12 | |
| | | | At $a_2$, between $b$'s | 2 | 248 | 124 | 4.38 |
| | | | At $a_1$, between $b$'s | 2 | 8 | 4 | |
| | | | At $a_0$, between $b$'s | 2 | 488 | 244 | 8.61 |
| Error | 24 | 680 | | | | 28.33 | |
| Total | 35 | 1,628 | | | | | |

statement that there are no main effects. A similar analysis may also be made at fixed levels of factor $b$. The conclusion is more or less the same.

## 7. Three factors, each at three levels

The principles involved in the analysis of a $3 \times 3 \times 3$ factorial experiment are the same as those for a $3 \times 3$ factorial, although the arithmetic labor will in general be much longer. For the sake of simplicity let us assume that there is only one replication ($r = 1$) of the 27 treatment combinations (Table 22.9 top). The problem of single-replication experiments will be discussed later. For the time being we are merely con-

*Table 22.9* Treatment combination totals of a $3 \times 3 \times 3$ factorial and preliminary subdivisions of treatment sum of squares

| | $b_2$ | | | $b_1$ | | | $b_0$ | | |
|---|---|---|---|---|---|---|---|---|---|
| | $a_2$ | $a_1$ | $a_0$ | $a_2$ | $a_1$ | $a_0$ | $a_2$ | $a_1$ | $a_0$ |
| $c_2$ | 26 | 33 | 28 | 14 | 18 | 21 | 28 | 10 | 11 |
| $c_1$ | 20 | 18 | 26 | 24 | 31 | 28 | 30 | 16 | 14 |
| $c_0$ | 8 | 24 | 24 | 13 | 23 | 20 | 8 | 13 | 11 |

$$ssq_{a \times b \times c} = 26^2 + 33^2 + \cdots - 540^2/27 = 1{,}456$$

| | $a_2$ | $a_1$ | $a_0$ |
|---|---|---|---|
| $b_2$ | 54 | 75 | 78 |
| $b_1$ | 51 | 72 | 69 |
| $b_0$ | 66 | 39 | 36 |
| | 171 | 186 | 183 |

$$ssq_{a \times b} = 648.0$$
$$ssq_a = 14.0$$

| | $b_2$ | $b_1$ | $b_0$ |
|---|---|---|---|
| $c_2$ | 87 | 53 | 49 |
| $c_1$ | 75 | 83 | 60 |
| $c_0$ | 56 | 56 | 32 |
| | 207 | 192 | 141 |

$$ssq_{b \times c} = 753.3$$
$$ssq_b = 266.0$$

| | $c_2$ | $c_1$ | $c_0$ |
|---|---|---|---|
| $a_2$ | 68 | 74 | 29 |
| $a_1$ | 61 | 65 | 60 |
| $a_0$ | 60 | 68 | 55 |
| | 189 | 207 | 144 | 540 |

$$ssq_{c \times a} = 445.3$$
$$ssq_c = 234.0$$

| Effects | $df$ | $ssq$ | $msq$ |
|---|---|---|---|
| $A$ | 2 | 14.00 | 7.00 |
| $B$ | 2 | 266.00 | 133.00 |
| $AB$ | 4 | 368.00 | 92.00 |
| $C$ | 2 | 234.00 | 117.00 |
| $AC$ | 4 | 197.33 | 49.33 |
| $BC$ | 4 | 253.33 | 63.33 |
| $ABC$ | 8 | 123.34 | 15.42 |
| Total | 26 | 1,456.00 | |

cerned with the procedure of subdividing the total treatment *ssq* into various components. This procedure is independent of the number of replications because we are using treatment totals in computations.

The degree of subdivision, as in the case of $3 \times 3$, is flexible. The preliminary and easiest subdivision is shown in Table 22.9, the top portion of which is really a three-dimensional classification equivalent to that for $2 \times 2 \times 2$ (Table 21.6). The sum of squares for these 27 numbers is calculated and denoted by $ssq_{a \times b \times c}$, where the subscript $a \times b \times c$ indicates the factors included in this sum of squares. The next step is to construct three two-way tables, each for a pair of factors, as shown in the middle portion of the table. For the $a \times c$ table, the factor $b$ has been telescoped. Thus, the entry for $a_2c_2$ is

$$26 + 14 + 28 = 68$$

etc. That is the sum of the three $b$ layers. For each of these two-way tables, the sum of squares is calculated. For the $a \times c$ table,

$$ssq_{c \times a} = \frac{1}{3}(68^2 + 74^2 + \cdots) - \frac{540^2}{27} = 445.33$$

Finally, from the marginal totals, the sum of squares for single factors is calculated. For example,

$$ssq_c = \frac{1}{9}(189^2 + \cdots) - \frac{540^2}{27} = 234.0$$

The interaction *ssq* is obtained by subtraction. For the interaction $AC$, its *ssq* is

$$ssq_{c \times a} - ssq_c - ssq_a = 445.3 - 234.0 - 14.0 = 197.3$$

The *ssq* for three-factor interaction $(ABC)$ is obtained by subtraction of all components from the total 1,450.0. These components have been summarized at the bottom of Table 22.9, which completes the preliminary subdivision.

In large experiments involving many factors and thus many treatment combinations, experience shows that interaction of high order is often small and is of the order of random variation. If so, we may assume that the variation for $ABC$ is due to random error rather than any meaningful and systematic effects. Under this assumption, $123.34/8 = 15.42$ may be taken as an estimate of $s^2$, against which the significance of other effects may be tested.

### 8. Single components of $3 \times 3 \times 3$

Each of the main effects $(A, B, C)$ may be further subdivided into single components by the method described previously. All the two-factor interactions may be subdivided into single components by applying the method for $3 \times 3$. The eight single components for three-factor interactions may be obtained by an extension of the same principle. It may seem very complicated to enumerate the 26 orthogonal contrasts for the 27 treatment combinations. This is where the succinct symbolic expressions are at their very best; without them the task of enumerating 26 orthogonal expressions would indeed be prohibitive. The formal expressions for the single components are given in Table 22.10, from which any single component may be calculated separately. The shorthand notation system adopted in that table should be made clear by considering a few examples.

$$C_1 = (c_2 - c_0)(a_2 + a_1 + a_0)(b_2 + b_1 + b_0)$$
$$B_2 = (b_2 - 2b_1 + b_0)(a_2 + a_1 + a_0)(c_2 + c_1 + c_0)$$
$$B_2C_1 = (b_2 - 2b_1 + b_0)(c_2 - c_0)(a_2 + a_1 + a_0)$$
$$A_2B_1C_2 = (a_2 - 2a_1 + a_0)(b_2 - b_0)(c_2 - 2c_1 + c_0)$$

When these expressions are formally expanded, we obtain the appropriate contrasts in terms of the treatment combination totals. The corresponding values of $D$, the sum of squares of coefficients of the treatment totals, are indicated in the last column of Table 22.10. These $D$ values may, of course, be actually calculated from the coefficients after the expansion of the symbolic product. There is, however, a shortcut. The marginal sum expression $(1, 1, 1)$ contributes a factor of 3 to $D$. The linear contrast expression $(1, 0, -1)$ contributes a factor 2, and the quadratic

**Table 22.10** Symbolic expressions for single components of a $3 \times 3 \times 3$ factorial

| | | | | |
|---|---|---|---|---|
| $(1,0,-1) \equiv (a_2 - a_0)$ | or | $(b_2 - b_0)$ | or | $(c_2 - c_0)$ |
| $(1,-2,1) \equiv (a_2 - 2a_1 + a_0)$ | or | $(b_2 - 2b_1 + b_0)$ | or | $(c_2 - 2c_1 + c_0)$ |
| $(1,1,1) \equiv (a_2 + a_1 + a_0)$ | or | $(b_2 + b_1 + b_0)$ | or | $(c_2 + c_1 + c_0)$ |

| Effects | | | Number | Type of expression | | | $D$ |
|---|---|---|---|---|---|---|---|
| $A_1$ | $B_1$ | $C_1$ | 3 | $(1, 0, -1)$ | $(1, 1, 1)$ | $(1, 1, 1)$ | 18 |
| $A_2$ | $B_2$ | $C_2$ | 3 | $(1, -2, 1)$ | $(1, 1, 1)$ | $(1, 1, 1)$ | 54 |
| $A_1B_1$ | $A_1C_1$ | $B_1C_1$ | 3 | $(1, 0, -1)$ | $(1, 0, -1)$ | $(1, 1, 1)$ | 12 |
| $A_1B_2$ | $A_1C_2$ | $B_1C_2$ | 6 | $(1, 0, -1)$ | $(1, -2, 1)$ | $(1, 1, 1)$ | 36 |
| $A_2B_1$ | $A_2C_1$ | $B_2C_1$ | | | | | |
| $A_2B_2$ | $A_2C_2$ | $B_2C_2$ | 3 | $(1, -2, 1)$ | $(1, -2, 1)$ | $(1, 1, 1)$ | 108 |
| | $A_1B_1C_1$ | | 1 | $(1, 0, -1)$ | $(1, 0, -1)$ | $(1, 0, -1)$ | 8 |
| $A_1B_1C_2$ | $A_1B_2C_1$ | $A_2B_1C_1$ | 3 | $(1, 0, -1)$ | $(1, 0, -1)$ | $(1, -2, 1)$ | 24 |
| $A_1B_2C_2$ | $A_2B_1C_2$ | $A_2B_2C_1$ | 3 | $(1, 0, -1)$ | $(1, -2, 1)$ | $(1, -2, 1)$ | 72 |
| | $A_2B_2C_2$ | | 1 | $(1, -2, 1)$ | $(1, -2, 1)$ | $(1, -2, 1)$ | 216 |
| | Total | | 26 | | | | |

*Table* 22.11 Routine tabular method of calculating single effects and sum of squares

| Levels of $a$ $b$ $c$ | Treatment total | Operations 1st | 2d | 3d | Effect (contrast) | Divisor $D$ | $ssq = z^2/D$ |
|---|---|---|---|---|---|---|---|
| 0 0 0 | 11 | 32 | 144 | 540 | $Y$ | | |
| 1 0 0 | 13 | 56 | 207 | $-12$ | $A_1$ | 18 | 8.00 |
| 2 0 0 | 8 | 56 | 189 | $-18$ | $A_2$ | 54 | 6.00 |
| 0 1 0 | 20 | 60 | $-26$ | 66 | $B_1$ | 18 | 242.00 |
| 1 1 0 | 23 | 83 | 6 | $-54$ | $A_1B_1$ | 12 | 243.00 |
| 2 1 0 | 13 | 64 | 8 | $-42$ | $A_2B_1$ | 36 | 49.00 |
| 0 2 0 | 24 | 49 | $-36$ | $-36$ | $B_2$ | 54 | 24.00 |
| 1 2 0 | 24 | 53 | 12 | 42 | $A_1B_2$ | 36 | 49.00 |
| 2 2 0 | 8 | 87 | 6 | 54 | $A_2B_2$ | 108 | 27.00 |
| 0 0 1 | 14 | $-3$ | 24 | 45 | $C_1$ | 18 | 112.50 |
| 1 0 1 | 16 | $-7$ | 4 | 34 | $A_1C_1$ | 12 | 96.33 |
| 2 0 1 | 30 | $-16$ | 38 | 42 | $A_2C_1$ | 36 | 49.00 |
| 0 1 1 | 28 | 16 | $-13$ | 14 | $B_1C_1$ | 12 | 16.33 |
| 1 1 1 | 31 | $-4$ | $-22$ | $-6$ | $A_1B_1C_1$ | 8 | 4.50 |
| 2 1 1 | 24 | $-6$ | $-19$ | $-22$ | $A_2B_1C_1$ | 24 | 20.17 |
| 0 2 1 | 26 | 17 | $-9$ | 54 | $B_2C_1$ | 36 | 81.00 |
| 1 2 1 | 18 | $-7$ | $-2$ | 34 | $A_1B_2C_1$ | 24 | 48.17 |
| 2 2 1 | 20 | $-2$ | $-31$ | 6 | $A_2B_2C_1$ | 72 | 0.50 |
| 0 0 2 | 11 | $-7$ | $-24$ | $-81$ | $C_2$ | 54 | 121.50 |
| 1 0 2 | 10 | $-13$ | $-42$ | $-30$ | $A_1C_2$ | 36 | 25.00 |
| 2 0 2 | 28 | $-16$ | 30 | $-54$ | $A_2C_2$ | 108 | 27.00 |
| 0 1 2 | 21 | 12 | $-5$ | 54 | $B_1C_2$ | 36 | 81.00 |
| 1 1 2 | 18 | $-10$ | 18 | 12 | $A_1B_1C_2$ | 24 | 6.00 |
| 2 1 2 | 14 | 10 | 29 | $-36$ | $A_2B_1C_2$ | 72 | 18.00 |
| 0 2 2 | 28 | 19 | 3 | 90 | $B_2C_2$ | 108 | 75.00 |
| 1 2 2 | 33 | $-1$ | 42 | $-12$ | $A_1B_2C_2$ | 72 | 2.00 |
| 2 2 2 | 26 | $-12$ | 9 | $-72$ | $A_2B_2C_2$ | 216 | 24.00 |
| Totals | 540 | | | | | | 1,456.00 |

expression $(1, -2, 1)$ contributes a factor 6 to $D$. Thus, for contrast $A_2$, the divisor is $D = 6 \times 3 \times 3 = 54$; for the contrast $A_2B_2$, the divisor is $D = 6 \times 6 \times 3 = 108$; and for the contrast $A_2B_2C_1$, the divisor is $D = 6 \times 6 \times 2 = 72$, etc. By means of Table 22.10 any and all of the components may be calculated. If the treatment totals are from $r$ replications, the final divisor is $D \times r$.

Table 22.11 gives the routine tabular method of calculating each single component and its sum of squares. Being an extension of Table 22.7, it requires very little additional explanation. The top nine numbers of "first operation" column are the sums of triplets of the preceding column (treatment total). The middle nine numbers are differences of the type $(1, 0, -1)$. Thus, $8 - 11 = -3$, $13 - 20 = -7$, etc. The bottom nine numbers are deviations of the type $(1, -2, 1)$. For instance, $8 - 2(13) + 11 = -7$, etc. This operation is repeated three times, yielding the value of contrasts, $z$. The *ssq* is then calculated the usual way. Note that not only do the 26 components add up to the total treatment *ssq* 1,456.0 but the components of each type of contrasts add up to those obtained in the preliminary subdivision shown in Table 22.9. For example, the components $A_1$ and $A_2$ yield $8.00 + 6.00 = 14.00$, and the eight components of $ABC$ give

$$4.50 + 20.17 + 48.17 + 0.50 + 6.00 + 18.00 + 2.00 + 24.00 = 123.34$$

correctly. In practice, it is not always necessary to calculate all the single components. However, if a certain type of interaction is large, careful examination of the three-way and two-way tables of treatment totals is necessary for meaningful interpretation.

## EXERCISES

**22.1** The following are results of a $3 \times 3$ factorial experiment in four randomized blocks. Give a complete analysis of the eight components of the treatment effects and test their significance. In order to check your answers, use different methods of subdivision.

|  | \multicolumn{3}{c}{$a_2$} | \multicolumn{3}{c}{$a_1$} | \multicolumn{3}{c}{$a_0$} | Block total |
|---|---|---|---|---|---|---|---|---|---|---|
|  | $b_2$ | $b_1$ | $b_0$ | $b_2$ | $b_1$ | $b_0$ | $b_2$ | $b_1$ | $b_0$ |  |
| I | 52 | 28 | 56 | 66 | 36 | 20 | 56 | 42 | 22 | 378 |
| II | 48 | 44 | 48 | 54 | 24 | 38 | 34 | 16 | 36 | 342 |
| III | 60 | 48 | 40 | 56 | 62 | 32 | 52 | 36 | 28 | 414 |
| IV | 40 | 48 | 16 | 48 | 46 | 22 | 34 | 26 | 26 | 306 |
| Treatment total | 200 | 168 | 160 | 224 | 168 | 112 | 176 | 120 | 112 | 1,440 |

*Ans.:* Consult Table 22.2. Factors $a$ and $b$ are interchanged and each observed value has been multiplied by 2. Check your answers with those of the text after appropriate modifications have been made.

**22.2** The eight contrasts shown in Table 22.4, when written out in single file, will look like the accompanying table. We take the trouble to write all these out just to convince the reader that the subdivision of *ssq* is, as always, based on orthogonal contrasts of the type with which we have been familiar since Chap. 12. Table 22.4 is merely a more succinct way of writing such contrasts.

| Contrast | | $a_2$ | | | $a_1$ | | | $a_0$ | |
|---|---|:---:|:---:|:---:|:---:|:---:|:---:|:---:|:---:|:---:|
| | | $b_2$ | $b_1$ | $b_0$ | $b_2$ | $b_1$ | $b_0$ | $b_2$ | $b_1$ | $b_0$ |
| $A$ | $A_1$ | +1 | +1 | +1 | 0 | 0 | 0 | −1 | −1 | −1 |
| | $A_2$ | +1 | +1 | +1 | −2 | −2 | −2 | +1 | +1 | +1 |
| $B$ | $B_1$ | +1 | 0 | −1 | +1 | 0 | −1 | +1 | 0 | −1 |
| | $B_2$ | +1 | −2 | +1 | +1 | −2 | +1 | +1 | −2 | +1 |
| $AB$ | $A_1B_1$ | +1 | 0 | −1 | 0 | 0 | 0 | −1 | 0 | +1 |
| | $A_1B_2$ | +1 | −2 | +1 | 0 | 0 | 0 | −1 | +2 | −1 |
| | $A_2B_1$ | +1 | 0 | −1 | −2 | 0 | +2 | +1 | 0 | −1 |
| | $A_2B_2$ | +1 | −2 | +1 | −2 | +4 | −2 | +1 | −2 | +1 |

**22.3** In Table 22.4 the main effects $A_1$, $A_2$ and $B_1$, $B_2$ are placed in the margins and the four interactions are placed inside the table body. A kind of reversed arrangement is possible, so that the main effects are in the table body and the interactions are the marginal products. For brevity let us simply write $A_1 = (a_2 - 1)$, etc., and omit the factor $(b_2 + b + 1)$, etc. The arrangement is as shown in the accompanying table. Write out the corresponding coefficients of treatment totals for each contrast.

| $A_1$ <br> $(a_2 - 1)$ | $B_1$ <br> $(b_2 - 1)$ | $A_1B_1$ <br> $(a_2 - 1)(b_2 - 1)$ |
|---|---|---|
| $B_2$ <br> $(b_2 - 2b + 1)$ | $A_2$ <br> $(a_2 - 2a + 1)$ | $A_2B_2$ <br> $(a_2 - 2a + 1)(b_2 - 2b + 1)$ |
| $A_1B_2$ <br> $(a_2 - 1)(b_2 - 2b + 1)$ | $A_2B_1$ <br> $(a_2 - 2a + 1)(b_2 - 1)$ | Interactions |

**22.4** For a $3 \times 3$ factorial, the four contrasts for interactions are not uniquely determined. Some other contrasts may be used, and which is selected depends on the nature and interpretation of the factor effects. Another set of orthogonal contrasts for interactions is given in the accompanying tables. All of these are orthogonal to main effects because the coefficients add to zero for each row as well as for each column. Interchange of rows and/or of columns or turning these tables 90° will generate other sets of contrasts.

|       | $b_2$ | $b_1$ | $b_0$ |
|-------|-------|-------|-------|
| $a_2$ | 0     | 0     | 0     |
| $a_1$ | +1    | −1    | 0     |
| $a_0$ | −1    | +1    | 0     |

|       | $b_2$ | $b_1$ | $b_0$ |
|-------|-------|-------|-------|
| $a_2$ | 0     | 0     | 0     |
| $a_1$ | +1    | +1    | −2    |
| $a_0$ | −1    | −1    | 2     |

|       | $b_2$ | $b_1$ | $b_0$ |
|-------|-------|-------|-------|
| $a_2$ | −2    | +2    | 0     |
| $a_1$ | +1    | −1    | 0     |
| $a_0$ | +1    | −1    | 0     |

|       | $b_2$ | $b_1$ | $b_0$ |
|-------|-------|-------|-------|
| $a_2$ | −2    | −2    | +4    |
| $a_1$ | +1    | +1    | −2    |
| $a_0$ | +1    | +1    | −2    |

**22.5**   To get familiar with the routine tabular method of calculating the contrast values from treatment totals, we arrange the data in a different order.   See if you get results identical with Table 22.7.

| Levels of $a$ $b$ | Short notation | Treatment totals | First operation | Second operation | Contrast $z$ |
|---|---|---|---|---|---|
| 0   0 | (1) | 56 | | | $Y$ |
| 0   1 | $b$ | 56 | | | $B_1$ |
| 0   2 | $b_2$ | 80 | | | $B_2$ |
| 1   0 | $a$ | 60 | | | $A_1$ |
| 1   1 | $ab$ | 84 | | | $A_1B_1$ |
| 1   2 | $ab_2$ | 84 | | | $A_1B_2$ |
| 2   0 | $a_2$ | 88 | | | $A_2$ |
| 2   1 | $a_2b$ | 112 | | | $A_2B_1$ |
| 2   2 | $a_2b_2$ | 100 | | | $A_2B_2$ |

**22.6**   Consider the treatment totals of the $3 \times 3$ factorial and the preliminary subdivision of treatment effects once more.   The $AB$ $ssq$ may be subdivided in various ways.   The method given in Table 22.4 gives concrete physical meaning and therefore is usually useful for interpretation of data.   However, we shall here give another

|       | $b_2$ | $b_1$ | $b_0$ |     |
|-------|-------|-------|-------|-----|
| $a_2$ | 100   | 112   | 88    | 300 |
| $a_1$ | 84    | 84    | 60    | 228 |
| $a_0$ | 80    | 56    | 56    | 192 |
|       | 264   | 252   | 204   | 720 |

| Effects | $df$ | $ssq$ |
|---------|------|-------|
| $A$     | 2    | 504   |
| $B$     | 2    | 168   |
| $AB$    | 4    | 96    |
| Treatments | 8 | 768   |

method of subdivision, not because it is useful in isolating certain treatment effects, but because it is useful in achieving some other designs. In Chap. 17 we have mentioned Graeco-Latin square arrangement, and there is only one $3 \times 3$ Graeco-Latin square. When we take the sums of Greek letters, we discover that they are the diago-

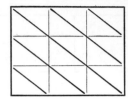 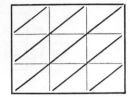

| $A\alpha$ | $B\beta$ | $C\gamma$ |
| $B\gamma$ | $C\alpha$ | $A\beta$ |
| $C\beta$ | $A\gamma$ | $B\alpha$ |

nal sums running from upper left to lower right. Similarly, when we take the sums of the Latin letters, we discover that they are the diagonal sums running from upper right to lower left. These two sets are mutually orthogonal, and both are orthogonal to rows and to columns. Hence the sum of squares for such sets of totals must be due to interaction, whatever their physical meaning. These sums are listed below ($A$, $B$, $C$ are not to be confused with main effects).

$$
\begin{array}{ll}
\alpha \quad 240 & A \quad 216 \\
\beta \quad 252 & B \quad 252 \\
\gamma \quad 228 & C \quad 252
\end{array}
$$

The sum of squares for the three Greek letter totals is found to be

$$\frac{240^2 + 252^2 + 228^2}{12} - \frac{720^2}{36} = 24 \text{ with } 2 \text{ } df$$

and the sum of squares for the three Latin letter totals is found to be 72 with 2 $df$. We note that $72 + 24 = 96 = ssq$ for $AB$ with 4 $df$.

It is important to note that this subdivision of interactions into two components, each with 2 $df$, is *not* equivalent to a certain pooling of the four single components (Table 22.5):

$$
A_1B_1 \quad 9 \qquad A_1B_2 \quad 75 \qquad A_2B_1 \quad 3 \qquad A_2B_2 \quad 9
$$

The subdivision into 72 and 24 is an entirely different subdivision of the total interaction sum of squares.

# 23

# Other types of factorials

In the preceding two chapters we have dealt with the most basic type of analysis for factors which are actually administered at levels (amounts) with equal intervals, such as 5, 10, and 15 grams. The method may be extended to factors with more than three levels. There are, however, occasions on which one or more of the factors involved are not quantitative, but, rather, qualitative in nature. There are also occasions on which certain treatment combinations are indistinguishable in practice. The analysis procedure of such cases requires slight modification. The present chapter extends the method to factors with unequal numbers of levels and also gives some examples of qualitative factors to show the flexibility of the analysis. In dealing with modified cases we shall by way of emphasis occasionally modify our notations as well.

## 1. The 2 $\times$ 3 factorial

When one factor (say, $a$) is administered at two levels ($a_1$, $a_2$) and another factor (say, $b$) at three levels with equal intervals ($b_0$, $b_1$, $b_2$), we say it

*Table* 23.1 Breakdown of treatment effects of a 2 × 3 factorial (based on data of Table 17.1)

| | $a_2$ | | | $a_1$ | | | Effect $z$ | Divisor $D \times r$ | ssq $z^2/Dr$ |
|---|---|---|---|---|---|---|---|---|---|
| | $b_2$ 144 | $b_1$ 138 | $b_0$ 108 | $b_2$ 162 | $b_1$ 90 | $b_0$ 78 | | | |
| $A$ | +1 | +1 | +1 | −1 | −1 | −1 | 60 | 6 × 6 | 100 |
| $B_1$ | +1 | 0 | −1 | +1 | 0 | −1 | 120 | 4 × 6 | 600 |
| $B_2$ | +1 | −2 | +1 | +1 | −2 | +1 | 36 | 12 × 6 | 18 |
| $AB_1$ | +1 | 0 | −1 | −1 | 0 | +1 | −48 | 4 × 6 | 96 |
| $AB_2$ | +1 | −2 | +1 | −1 | +2 | −1 | −84 | 12 × 6 | 98 |
| | | | | | Table 17.2 | | | Total treatment ssq | 912 |

is a 2 × 3 factorial experiment. It is the simplest example of factors with mixed levels. The analysis hardly needs explanation, because we merely combine the techniques of the preceding two chapters. The five contrasts among the six treatment combinations are, symbolically,

$$A = (a_2 - a_1)(b_2 + b_1 + b_0)$$
$$B_1 = (a_2 + a_1)(b_2 - b_0)$$
$$B_2 = (a_2 + a_1)(b_2 - 2b_1 + b_0)$$
$$AB_1 = (a_2 - a_1)(b_2 - b_0)$$
$$AB_2 = (a_2 - a_1)(b_2 - 2b_1 + b_0)$$

Table 23.1 gives the coefficients of the treatment combinations after these symbolic expressions have been expanded; this is a numerical workout from a 6 × 6 Latin-square data (Table 17.1). It is seen that the sum of the five-component ssq is 912, as was found in Table 17.2. Each component may then be tested against $s^2 = 14.2$ with 20 *df*.

## 2. Two kinds, each at three levels

When the six treatment combinations are six different mixtures of two ingredients of various amounts, the interpretation of the contrasts shown in Table 23.1 is no different from what we have said in the preceding two chapters. However, as repeatedly emphasized, the interpretation depends on the nature of the factors under consideration. Now, suppose that $a_1$ and $a_2$ are two drugs rather than two different amounts of the same material. It may be imagined that $a_1$ is the "old" or "standard" drug and $a_2$ is the new drug, each being administered at three levels $(b_0, b_1, b_2)$. For the old $a_1$ drug, the slope or linear effect is $a_1(b_2 - b_0)$; for the new $a_2$ drug, the response slope is $a_2(b_2 - b_0)$. If these two slopes are equal, that is, the two response lines are parallel, we should have

$$a_2(b_2 - b_0) - a_1(b_2 - b_0) = 0$$

But this is the contrast $AB_1$ in Table 23.1. Hence, the meaning of this contrast is to compare the two slopes of the response line; in other words, it tests the parallelism of the two lines. The treatment totals of Table 23.1 are plotted in Fig. 23.1, from which we see that the response curves to the two drugs ($a_1$ and $a_2$) are entirely different. $F = 96/14.2 = 6.76$, significant at the 5 per cent level.

Figure 23.1 also shows that the curvature of the two response lines is likewise very different, one being concave and the other convex. For the $a_1$ drug, the curvature is measured by

$$a_1(b_2 - 2b_1 + b_0) = 162 - 2(90) + 78 = +60$$

while for the $a_2$ drug it is $a_2(b_2 - 2b_1 + b_0) = 144 - 2(138) + 108 = -24$. The opposite signs ($+60$ and $-24$) indicate the opposite nature of the curvature for the two drugs. The contrast corresponding to this difference is $AB_2 = -84$, which is also significant. It is due to this opposite curvature of the two lines that the overall $B_2$ effect is very small,

$$F = \frac{18}{14.2} = 1.27$$

To summarize, the interpretation of the contrasts shown in Table 23.1 should be made in the light of Fig. 23.1 when $a_1$ and $a_2$ represent two test materials.

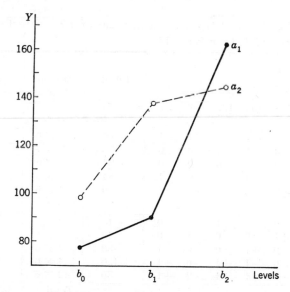

**Fig. 23.1** Responses to two drugs ($a_1$ and $a_2$) at three levels.

## 3. Other types of comparison

It should be reiterated that the set of five contrasts shown in Table 23.1 is not the only valid set that may be employed. For instance, the physician may want to compare the two drugs at each level separately. Or, he may want to study the linear and quadratic effects of each drug separately. Finally, the experiment may involve three factors (drugs $b_0$, $b_1$, $b_2$), each administered at two levels ($a_1$ and $a_2$). This is, of course, also a $2 \times 3$ factorial design.

Two additional simple sets of contrasts are shown in Table 23.2. The meaning of these comparisons is at once clear in terms of Fig. 23.1. For instance, the second contrast in the lower set of the table,

$$a_2 b_1 - a_1 b_1 = 138 - 90 = 48$$

gives the difference between the two drugs at the $b_1$ level. The component $ssq$'s of each set add up to the total treatment $ssq = 912$ with 5 $df$. Each component may be tested against $s^2 = 14.2$ with 20 $df$. Note that some components are the same as those in Table 23.1, while others are not.

Many other sets of comparisons may be readily constructed. When there is a reason to do so, we may first perform a $2 \times 2$ analysis for two of the three levels ($b_0$, $b_1$ or $b_1$, $b_2$), which takes up three degrees of freedom, and the remaining two degrees of freedom may be used for comparisons involving the third level. Something like this becomes advisable in the case to be described in the next section.

*Table* 23.2  Possible types of comparisons of a $2 \times 3$ factorial (treatment totals from Table 17.1)

| $a_2$ | | | $a_1$ | | | Effect | Divisor | $ssq$ |
|---|---|---|---|---|---|---|---|---|
| $b_2$ | $b_1$ | $b_0$ | $b_2$ | $b_1$ | $b_0$ | $z$ | $D \times r$ | $z^2/Dr$ |
| 144 | 138 | 108 | 162 | 90 | 78 | | | |
| +1 | +1 | +1 | −1 | −1 | −1 | 60 | 6 × 6 | 100 |
| +1 | 0 | −1 | 0 | 0 | 0 | 36 | 2 × 6 | 108 |
| +1 | −2 | +1 | 0 | 0 | 0 | −24 | 6 × 6 | 16 |
| 0 | 0 | 0 | +1 | 0 | −1 | 84 | 2 × 6 | 588 |
| 0 | 0 | 0 | +1 | −2 | +1 | 60 | 6 × 6 | 100 |
| | | | | | | | Total | 912 |
| +1 | 0 | 0 | −1 | 0 | 0 | −18 | 2 × 6 | 27 |
| 0 | +1 | 0 | 0 | −1 | 0 | 48 | 2 × 6 | 192 |
| 0 | 0 | +1 | 0 | 0 | −1 | 30 | 2 × 6 | 75 |
| +1 | 0 | −1 | +1 | 0 | −1 | 120 | 4 × 6 | 600 |
| +1 | −2 | +1 | +1 | −2 | +1 | 36 | 12 × 6 | 18 |
| | | | | | | | Total | 912 |

## 4. Incomplete treatment combinations

When factor $a$ has $l_a$ levels and factor $b$ has $l_b$ levels, the complete factorial design requires all $l_a \times l_b$ treatment combinations. However, this design is not always feasible either because of the large number of combinations or because of prior knowledge that certain combinations are of no interest. Hence the investigator may choose to test only those combinations which he thinks are useful for his purpose. The general subject of incomplete combinations is a difficult one because of the nonorthogonality of the factor effects, and usually it requires long and complicated calculations (necessary for nonorthogonal data). In this section we shall consider an extremely simple case which could be handled with the elementary tools in our possession.

Now, let us suppose that the factor $a$ has two levels or two kinds of material and that factor $b$ has four levels. The six (not eight) combinations tested in the experiment and the treatment totals ($r = 6$) are as given in the accompanying table. In such a case we may still obtain

|       | $b_3$ | $b_2$ | $b_1$ | $b_0$ |
|-------|-------|-------|-------|-------|
| $a_1$ | ...   | 162   | 90    | 78    |
| $a_2$ | 144   | 138   | 108   | ...   |

the effects of factor $b$ separately for each given kind or level of $a$. This takes up 2 $df$ for the $a_1$ material and 2 $df$ for the $a_2$ material. The remaining 1 $df$ is for between $a_1$ and $a_2$, which, however, does not represent the pure effect of factor $a$ but is also influenced by factor $b$. This method is, in fact, the same as the first set of contrasts in Table 23.2.

Alternatively, it may be noted that the four middle treatment combinations ($a_2b_2$, $a_2b_1$, $a_1b_2$, $a_1b_1$) constitute a complete $2 \times 2$ factorial by themselves and may be analyzed as such independently. This is done in Table 23.3, the first three comparisons of which are the same as those in Table 21.1. The fourth contrast is between the two extreme treatments ($a_2b_3 - a_1b_0$). The last contrast gives an overall test of parallelism for the two response lines, because its value is zero when the treatment totals (or means) have equal increments over levels of $b$ for both $a_1$ and $a_2$ materials, as may be seen in the following scheme, where $g$, $g'$, and $h$ are

|       | $b_3$     | $b_2$     | $b_1$   | $b_0$ |
|-------|-----------|-----------|---------|-------|
| $a_1$ | ...       | $g + 2h$  | $g + h$ | $g$   |
| $a_2$ | $g' + 2h$ | $g' + h$  | $g'$    | ...   |

*Table* 23.3 A possible subdivision of treatment $ssq$ for an incomplete $2 \times 4$ factorial experiment

| $a_2$ | | | $a_1$ | | | Effect | Divisor | $ssq$ |
|---|---|---|---|---|---|---|---|---|
| $b_3$ | $b_2$ | $b_1$ | $b_2$ | $b_1$ | $b_0$ | $z$ | $D \times r$ | $z^2/Dr$ |
| 144 | 138 | 108 | 162 | 90 | 78 | | | |
| 0 | +1 | +1 | −1 | −1 | 0 | − 6 | 4 × 6 | 1.50 |
| 0 | +1 | −1 | +1 | −1 | 0 | 102 | 4 × 6 | 433.50 |
| 0 | +1 | −1 | −1 | +1 | 0 | −42 | 4 × 6 | 73.50 |
| +1 | 0 | 0 | 0 | 0 | −1 | 66 | 2 × 6 | 363.00 |
| −2 | +1 | +1 | +1 | +1 | −2 | 54 | 12 × 6 | 40.50 |
| | | | | | | | Total | 912.00 |

certain constants. When this is true, twice the sum of the two extreme combinations is equal to the sum of the middle four. These five contrasts are orthogonal to each other, and the five individual $ssq$'s add up to the total treatment $ssq = 912$. Then each component may be tested against $s^2 = 14.2$ with 20 $df$. A similar analysis may be made when $a_1$ and $a_2$ are interchanged.

### 5. Zero level and dummy treatments

When the lowest level of application of the factors is actually zero (that is, no application at all as an absolute control) and the factors involved are different kinds of material such as different derivatives of a basic chemical compound, different preparations of a vaccine or antiserum, different varieties or strains of organisms, different forms of an active ingredient, etc., then the three treatment "combinations" at the zero level are all identical, receiving no active ingredients at all. For convenience, let us change the notations slightly and denote the three different drugs by $a$, $b$, $c$ (aspirin, Bufferin, Coricidin; or Ajax, Babo, Comet, if you like) each administered at three levels—0 (none given), 1 (10 grains per day), 2 (20 grains per day). The nine combinations are then as tabulated here. But the three combinations at the zero level ($a_0$, $b_0$, $c_0$)

|  |  | Material | | |
|---|---|---|---|---|
| | | $a$ | $b$ | $c$ |
| | 0 | $a_0$ | $b_0$ | $c_0$ |
| Quantity | 1 | $a_1$ | $b_1$ | $c_1$ |
| | 2 | $a_2$ | $b_2$ | $c_2$ |

really represent the same condition (placebo): the patient receives no drug of any kind. There are actually only seven (not nine) distinct treatments. The factorial may also be regarded as a 2 × 3 plus an extra control, instead of the superficial 3 × 3.

The investigator may, of course, conduct an experiment with just seven treatments, each replicated a certain number of times. In such a case, the number of observations at the zero level is only one-third of that at the other two levels and the quantitative effect of the drugs will not be as accurately determined as when there is an equal number of replications at all levels. In many instances, it is desirable to preserve the superficial 3 × 3 structure with equal number of placebos and treatments at 1 and 2 levels. Then the treatments $a_0$, $b_0$, $c_0$ are called dummy treatments. They are assigned at random to patients as if they were different. The analysis of the treatment effects, however, requires a slight modification.

As a numerical example, let us use the same treatment totals of the 3 × 3 experiment with $r = 4$ replications considered in the preceding chapter (Table 22.2) assuming now that $b_2$, $b_1$, $b_0$ represent three different materials $a$, $b$, $c$ and that $a_0$, $a_1$, $a_2$ represent the three quantities. The treatment totals may be rewritten as in the accompanying table. The

|  |  | Material | | | |
|---|---|---|---|---|---|
|  |  | $a$ | $b$ | $c$ | |
| Quantity | 0 | (pooled) | | | 192 |
|  | 1 | 84 | 84 | 60 | 228 |
|  | 2 | 100 | 112 | 88 | 300 |
|  |  |  |  |  | 720 |
| Levels (2 + 1) | | 184 | 196 | 148 | 528 |
| Levels (2 − 1) | | 16 | 28 | 28 | 72 |

chief modification is pooling the three dummy treatment totals together to yield a single total for the placebos. The total treatment *ssq* is now

$$\frac{192^2}{12} + \frac{84^2 + \cdots + 88^2}{4} - \frac{720^2}{36} = 672 \qquad \text{with 6 } df$$

The *ssq* due to quantity (0, 1, 2) is based on the respective marginal totals, and it is calculated the usual way:

$$\frac{192^2 + 228^2 + 300^2}{12} - \frac{720^2}{36} = 504 \qquad \text{with 2 } df$$

The differences between the materials are based on the 2 × 3 table, because the placebos ($a_0$, $b_0$, $c_0$) give no information on the quality of the drugs. The *ssq* due to the different forms of material is

$$\frac{184^2 + 196^2 + 148^2}{8} - \frac{528^2}{24} = 156 \qquad \text{with 2 } df$$

The interaction is based on the differences between the two levels of application for the various materials:

$$\frac{16^2 + 28^2 + 28^2}{8} - \frac{72^2}{24} = 12 \qquad \text{with 2 } df$$

The *ssq* for the latter three numbers may be further subdivided into single components by the usual method. Note that

$$504 + 156 + 12 = 672$$

The error *ssq* is obtained by subtraction.

*Table* 23.4   Analysis of a 3 × 3 factorial with dummy treatments (data from Table 22.2)

| Source | Table 22.3 *df* | Table 22.3 *ssq* | With dummies *df* | With dummies *ssq* | Subdivision | *df* | *ssq* | *msq* |
|---|---|---|---|---|---|---|---|---|
| Blocks | 3 | 180 | 3 | 180 | | | | |
| Treatments | 8 | 768 | 6 | 672 | Quantity (0, 1, 2) | 2 | 504 | 252 |
| | | | | | Material (*a*, *b*, *c*) | 2 | 156 | 78 |
| | | | | | Interaction | 2 | 12 | 6 |
| Error | 24 | 680 | 26 | 776 | | | | 29.85 |
| Total | 35 | 1,628 | 35 | 1,628 | | | | |

The analysis of variance is given in Table 23.4, in which the original results for nine treatment combinations are also included for comparison. It is clear that the treatment *ssq* loses 2 *df* because of the dummies and the error *ssq* gains 2 *df* because the variations among the three dummies are considered to be random error. That this is actually the case may be seen from the *ssq* for the dummy totals:

$$\frac{80^2 + 56^2 + 56^2}{4} - \frac{192^2}{12} = 96 \qquad \text{with 2 } df$$

and

$$\text{Treatment } ssq = 768 - 96 = 672 \qquad \text{with } 8 - 2 = \ 6 \ df$$
$$\text{Error } ssq = 680 + 96 = 776 \qquad \text{with } 24 + 2 = 26 \ df$$

The estimate of the error variance is now $s^2 = {}^{776}\!/_{26} = 29.85$, against which the significance of the various effects may be tested.

## 6. Time of application

Combinations involving time of application with some factors at zero level also result in dummy treatments. Suppose that a drug $b$ is either given or not given and, when given, that it may be given early ($t_1$) or late ($t_2$). The "four" combinations are:

| Drug | Time |
|---|---|
| Without $b$: | Early, late |
| With $b$: | Early, late |

It is seen that the first two "combinations" are identical, since no drug has been given, and the early-late classification is irrelevant. In such a case, we may either preserve the superficial $2 \times 2$ structure and have equal numbers of observations for patients receiving and not receiving drug $b$, or simply conduct the experiment with three treatments: 0 (no drug), $bt_1$ (drug early), and $bt_2$ (drug late), with equal numbers of replications. The choice between the two schemes depends upon the main purpose of the investigation. If the effect of the presence or absence of the drug is the chief problem, then the $2 \times 2$ structure with dummies will be preferable. On the other hand, if time of application is equally important, then the three-treatment scheme will be preferable, because two-thirds (instead of one-half) of the data are useful in evaluating the effect of time.

We shall conclude this section by considering a numerical example of a $2 \times 2 \times 2$ experiment with dummy treatments. The treatment totals shown in Table 23.5 are the same as those in Table 21.6, but now we change the factor $a$ into $t$, time of application with respect to drug $b$. It is seen that $t_1$ and $t_2$ without $b$ are identical, and the treatment totals are pooled to yield a single total. Factor $c$ is applied at two levels or represents two different ways of handling the patients. There are in reality six distinct treatments, and their total $ssq$ is

$$\frac{87^2 + 114^2}{6} + \frac{72^2 + 69^2 + 60^2 + 78^2}{3} - \frac{480^2}{24} = 370.50 \qquad \text{with 5 } df$$

The five single components are easily calculated. From the $b \times c$ two-way table, we can calculate the $ssq$ for $B$, $C$, and $BC$ by the usual method (Chap. 21). To obtain the $ssq$ for time, and for its interaction with factor $c$, we can only use half of the data (right half, with $b$). Thus,

$$ssq \text{ (time)} = \frac{(147 - 132)^2}{12} = 18.75$$

$$ssq(TC) = \frac{(72 + 78 - 69 - 60)^2}{12} = 36.75$$

*Table 23.5* Data and analysis of a $2 \times 2 \times 2$ factorial with dummy treatments (reference Table 21.6; number of replications $r = 3$)

|  | Without $b$ | | With $b$ | |
|---|---|---|---|---|
|  | $t_1$ | $t_2$ | $t_1$ | $t_2$ |
| $c_1$ | $(36 + 51) = 87$ | | 72 | 69 |
| $c_2$ | $(51 + 63) = 114$ | | 60 | 78 |
|  |  | 201 | 132 | 147 |

|  | Two-way table | | |
|---|---|---|---|
|  | without $b$ | with $b$ | |
| $c_1$ | 87 | 141 | 228 |
| $c_2$ | 114 | 138 | 252 |
|  | 201 | 279 | 480 |

The analysis of variance

| Source | Table 21.5 | | With dummies | | Subdivision | $df$ | $ssq$ |
|---|---|---|---|---|---|---|---|
|  | $df$ | $ssq$ | $df$ | $ssq$ |  |  |  |
| Blocks | 2 | 16 | 2 | 16.00 |  |  |  |
| Treatments | 7 | 432 | 5 | 370.50 | $B$ | 1 | 253.50 |
|  |  |  |  |  | $C$ | 1 | 24.00 |
|  |  |  |  |  | $BC$ | 1 | 37.50 |
|  |  |  |  |  | $T$ (time) | 1 | 18.75 |
|  |  |  |  |  | $TC$ | 2 | 36.75 |
| Error | 14 | 276 | 16 | 337.50 |  | $s^2 = 21.09$ | |
| Total | 23 | 724 | 23 | 724.00 |  |  |  |

These five components are orthogonal and add up to 370.50. The error $ssq$ is obtained by subtraction. The analysis of variance is shown in Table 23.5. The $ssq = 432.00 - 370.50 = 61.50$ with $2\ df$ lost from treatments and gained by error is the $ssq$ for the dummy totals 36, 51, and that for 51, 63, each with a single $df$.

## 7. Factor at four levels

When factor $a$ is applied at four levels with equal intervals, there are $3\ df$ for the main effects which may be further analyzed into single components (Chap. 12, Example 4):

| | |
|---|---|
| Linear effect: | $A_1 = 3a_3 + a_2 - a_1 - 3a_0$ |
| Quadratic effect: | $A_2 = a_3 - a_2 - a_1 + a_0$ |
| Cubic effect: | $A_3 = a_3 - 3a_2 + 3a_1 - a_0$ |

Usually the linear effect is more important (larger) than the other two. Note that $A_1$ is, except for a constant multiplier, the regression coefficient of response on the level of the factor, as it is for factors at two or three levels. Also note the similarity of the expression $a_3 - a_2 - a_1 + a_0$ here with $a_2 - 2a_1 + a_0$ for three levels. In both cases, it is the difference between the ends and the middle, and hence it measures the curvature of the response line.

If a second factor $b$ is applied at three levels, there will be $4 \times 3 = 12$ treatment combinations. The preliminary subdivision of the treatment $ssq$ is obtained the usual way from the $4 \times 3$ two-way table for treatment totals.

|  | df |
|---|---|
| $A$ | 3 |
| $B$ | 2 |
| $AB$ | 6 |
| All treatments | 11 |

Any single component of the interactions, if desired, may be obtained by writing out the corresponding symbolic expression. For example,

$$A_3B_2 = (a_3 - 3a_2 + 3a_1 - a_0)(b_2 - 2b_1 + b_0)$$

The most important component of $AB$ is probably the "linear $\times$ linear" interaction,

$$A_1B_1 = (3a_3 + a_2 - a_1 - 3a_0)(b_2 - b_0)$$

and it is the one worthwhile isolating in practice. It is unnecessary to break down all effects into single components. We leave the numerical calculations as an exercise.

To summarize, whatever the number of factors and whatever the number of levels, as long as all possible treatment combinations occur in complete blocks, the various components can always be calculated by the use of orthogonal contrasts.

## 8. Analysis of $3 \times 2 \times 2$

We close this chapter with an analysis of a three-factor experiment, the first (factor $a$) having three levels and the other two (factors $b$ and $c$) having two levels each. The data and the preliminary analysis are shown in Table 23.6. Although no new concept is involved, the reader should follow every phase of the calculation both as a review of the last few chapters and as a basis for comparison with the analysis of the same set of data, assuming a different design in the next chapter (Exercises 24.3 to 24.5).

The calculations presented in Table 23.6, being those for an ordinary randomized block, need no comment. We then proceed to break down the treatment $ssq = 945.33$ with 11 $df$ into more specific components by constructing three two-way tables of the treatment totals as shown in Table 23.7. The $ssq$ due to main effects and interactions of the factors

*Table* 23.6  Data of a $3 \times 2 \times 2$ factorial experiment and preliminary analysis of variance

| | $a_0$ | | | | $a_1$ | | | | $a_2$ | | | | Replication total |
|---|---|---|---|---|---|---|---|---|---|---|---|---|---|
| | (1) | $b$ | $c$ | $bc$ | (1) | $b$ | $c$ | $bc$ | (1) | $b$ | $c$ | $bc$ | |
| I | 12 | 8 | 11 | 14 | 13 | 28 | 13 | 24 | 26 | 21 | 33 | 31 | 234 |
| II | 8 | 26 | 18 | 16 | 11 | 20 | 22 | 30 | 28 | 27 | 28 | 24 | 258 |
| III | 19 | 24 | 10 | 20 | 17 | 18 | 14 | 18 | 24 | 24 | 23 | 17 | 228 |
| Treatment total | 39 | 58 | 39 | 50 | 41 | 66 | 49 | 72 | 78 | 72 | 84 | 72 | 720 |

The four basic quantities

(area) $A = 12^2 + 8^2 + \cdots + 17^2 = 16{,}028$   $R = \frac{1}{12}(234^2 + 258^2 + 228^2) = 14{,}442$

(area) $C = \frac{1}{36}(720^2) = 14{,}400$   $T = \frac{1}{3}(39^2 + 58^2 + \cdots + 72^2) = 15{,}345.33$

Analysis of variance

| Source | df | ssq | msq | F |
|---|---|---|---|---|
| Replications | 2 | 42.00 | | |
| Treatments | 11 | 945.33 | 85.94 | 2.95 |
| Error | 22 | 640.67 | $s^2 = 29.12$ | |
| Total | 35 | 1,628.00 | | $F_{.05} = 2.26$ |

may then be calculated from these tables in the usual way.   For example, from the $a \times b$ table, we first calculate the total sum of squares

$$ssq_{a \times b} = \frac{78^2 + \cdots + 144^2}{6} - \frac{720^2}{36} = 15{,}312 - 14{,}400 = 912$$

Similar quantities are calculated for the $a \times c$ and $b \times c$ tables.   The *ssq* due to main effects are calculated from marginal totals.   Thus,

$$ssq_a = \frac{186^2 + 228^2 + 306^2}{12} - \frac{720^2}{36} = 15{,}018 - 14{,}400 = 618$$

$$ssq_b = \frac{330^2 + 390^2}{18} - \frac{720^2}{36} = 14{,}500 - 14{,}400 = 100$$

Then the sum of squares due to interaction $AB$ is

$$ssq_{a \times b} - ssq_a - ssq_b = 912 - 618 - 100 = 194$$

All the other quantities are calculated the same way.   The $ssq = 1.55$ due to $ABC$ is obtained by subtracting from 945.33 all the other sums of squares.   The results are listed at the bottom of Table 23.7.   This is as far as we shall go in this example.   If components with a single degree of freedom (such as $A_1B$ and $A_2B$) are desired, the reader can always write out the symbolic expressions and the corresponding sums of squares by the method already described.

*Table* 23.7  Subdivision of treatment $ssq$ (data from Table 23.6)

|  | $a_0$ | $a_1$ | $a_2$ |  |
|---|---|---|---|---|
| $b_0$ | 78 | 90 | 162 | 330 |
| $b_1$ | 108 | 138 | 144 | 390 |
|  | 186 | 228 | 306 | 720 |

|  | $a_0$ | $a_1$ | $a_2$ |  |
|---|---|---|---|---|
| $c_0$ | 97 | 107 | 150 | 354 |
| $c_1$ | 89 | 121 | 156 | 366 |
|  | 186 | 228 | 306 | 720 |

|  | $c_0$ | $c_1$ |  |
|---|---|---|---|
| $b_0$ | 158 | 172 | 330 |
| $b_1$ | 196 | 194 | 390 |
|  | 354 | 366 | 720 |

$\frac{1}{6}(78^2 + \cdots + 144^2) = 15,312$    $\frac{1}{6}(97^2 + \cdots + 156^2) = 15,042.67$    $\frac{1}{6}(158^2 + \cdots + 194^2) = 14,511.11$

$\frac{1}{12}(186^2 + 228^2 + 306^2) = 15,018$    $\frac{1}{18}(330^2 + 390^2) = 14,500$    $\frac{1}{18}(354^2 + 366^2) = 14,404$

$ssq_{a \times b} = 912$      $ssq_{a \times c} = 642.67$      $ssq_{b \times c} = 111.11$

| Effect | $df$ | $ssq$ | $msq$ | $F$ |
|---|---|---|---|---|
| $A$ | 2 | 618.00 | 309.00 | 10.6 |
| $B$ | 1 | 100.00 | 100.00 | 3.4 |
| $AB$ | 2 | 194.00 | 97.00 | 3.3 |
| $C$ | 1 | 4.00 | 4.00 | |
| $AC$ | 2 | 20.67 | 10.33 | |
| $BC$ | 1 | 7.11 | 7.11 | |
| $ABC$ | 2 | 1.55 | 0.78 | |
| Total | 11 | 945.33 | $s^2 = 29.12$ | |

*Table* 23.8  An alternative method of obtaining main effects and interactions of a $3 \times 2 \times 2$ factorial experiment (data from Table 23.6)

|  | Treatment totals | | | | Pair sum and difference | | | | | | | | |
|---|---|---|---|---|---|---|---|---|---|---|---|---|---|
|  | $a_0$ | $a_1$ | $a_2$ | Total | $a_0$ | $a_1$ | $a_2$ | Total | $a_0$ | $a_1$ | $a_2$ | Total | |
| (1) | 39 | 41 | 78 | 158 | 97 | 107 | 150 | 354 | 186 | 228 | 306 | 720 | $Y$ |
| $b$ | 58 | 66 | 72 | 196 | 89 | 121 | 156 | 366 | 30 | 48 | $-18$ | 60 | $B$ |
| $c$ | 39 | 49 | 84 | 172 | 19 | 25 | $-6$ | 38 | $-8$ | 14 | 6 | 12 | $C$ |
| $bc$ | 50 | 72 | 72 | 194 | 11 | 23 | $-12$ | 22 | $-8$ | $-2$ | $-6$ | $-16$ | $BC$ |

|  | $ssq$, each with 1 $df$ |  | $ssq$, each with 2 $df$ |
|---|---|---|---|
| Corr. | $\dfrac{(720)^2}{36} = 14,400$ | $A$ | $\dfrac{186^2 + 228^2 + 306^2}{12} - \dfrac{720^2}{36} = 618$ |
| $B$ | $\dfrac{(60)^2}{36} = 100$ | $AB$ | $\dfrac{30^2 + 48^2 + 18^2}{12} - \dfrac{60^2}{36} = 194$ |
| $C$ | $\dfrac{(12)^2}{36} = 4$ | $AC$ | $\dfrac{8^2 + 14^2 + 6^2}{12} - \dfrac{12^2}{36} = 20.67$ |
| $BC$ | $\dfrac{(-16)^2}{36} = 7.11$ | $ABC$ | $\dfrac{8^2 + 2^2 + 6^2}{12} - \dfrac{16^2}{36} = 1.55$ |

We would like, however, to take this opportunity to give an alternative numerical procedure for analyzing the $3 \times 2 \times 2$ components. The new procedure outlined in Table 23.8 is to take the place of Table 23.7. The four combinations of factors $b$ and $c$ are denoted by the simplified notations (1), $b$, $c$, $bc$. Each of these is combined with $a_0$, $a_1$, $a_2$. The 12 treatment totals are arranged as shown at the left side of Table 23.8, with an extra "total" column added. Then the method of pair sum and difference (Chap. 21) is applied twice to each of the four columns. The "total" column provides a check at each stage of the procedure. The $ssq$'s for $B$, $C$, $BC$, each with 1 $df$, are then calculated the usual way as for factors with two levels each. The $ssq$ for the main effect of factor $a$ is calculated from the three numbers 186, 228, 306. That for interaction $AB$ is calculated from the next three numbers, 30, 48, $-18$, and so on. The full result is shown in the lower portion of Table 23.8. Note that the $ssq$'s thus obtained are identical with those in Table 23.7, and they thereby provide a complete check of the accuracy of the arithmetic. We shall refer to this table once more in the exercises for the next chapter.

**EXERCISES**

**23.1** The 36 numbers in Table 16.1 (also in Tables 17.1, 18.4, and 22.2) may be used as data for various factorial designs, as shown here. When the chief aim is to establish a response curve, the number of levels of a factor may be as few as 5 or 6 or as many as 9 or 10 or more. The 36 numbers may also be used as $9 \times 4$, $12 \times 3$, and $18 \times 2$ designs with a single replication. It is suggested that the student try some of the analyses by arranging the data appropriately. The fact that the single-component $ssq$'s add up to the total-treatment $ssq$ should provide an automatic check of the correctness of the arithmetic. The calculations should be easy, because the total $ssq$ for the entire experiment is known. The same 36 numbers will also be used in later chapters in connection with other subjects.

| Treatment combinations | No. of replications | Treatment combinations | No. of replications |
|:---:|:---:|:---:|:---:|
| $2 \times 2$ | 9 | $6 \times 6$ | 1 |
| $3 \times 2$ | 6 | $9 \times 2$ | 2 |
| $3 \times 3$ | 4 | $3 \times 2 \times 2$ | 3 |
| $4 \times 3$ | 3 | $3 \times 3 \times 2$ | 2 |
| $6 \times 2$ | 3 | $4 \times 3 \times 3$ | 1 |
| $6 \times 3$ | 2 | $3 \times 3 \times 2 \times 2$ | 1 |

**23.2** Since the reader is familiar with factors at two and three levels, it is advisable to work out an example involving four levels. Suppose that the data are as shown in the accompanying table. Verify the preliminary analysis. It is seen that most of the variation is due to blocks and that the treatments are nonsignificant. However, as an exercise, we shall proceed to find the three single components of $A$ and the six components of $AB$.

| | Block I | | | Block II | | | Block III | | | Treatment totals | | | |
|---|---|---|---|---|---|---|---|---|---|---|---|---|---|---|
| | $b_2$ | $b_1$ | $b_0$ | $b_2$ | $b_1$ | $b_0$ | $b_2$ | $b_1$ | $b_0$ | | $b_2$ | $b_1$ | $b_0$ | |
| $a_0$ | 20 | 24 | 17 | 24 | 23 | 13 | 8 | 11 | 13 | | 52 | 58 | 43 | 153 |
| $a_1$ | 24 | 27 | 17 | 22 | 12 | 8 | 24 | 19 | 18 | | 70 | 58 | 43 | 171 |
| $a_2$ | 26 | 33 | 28 | 14 | 18 | 21 | 28 | 10 | 11 | | 68 | 61 | 60 | 189 |
| $a_3$ | 30 | 28 | 26 | 24 | 31 | 18 | 20 | 16 | 14 | | 74 | 75 | 58 | 207 |
| Total | | 300 | | | 228 | | | 192 | | | 264 | 252 | 204 | 720 |

| Source | $df$ | $ssq$ | Effects | $df$ | $ssq$ | $msq$ |
|---|---|---|---|---|---|---|
| Blocks | 2 | 504.00 | | | | |
| Treatments | 11 | 413.33 | $A$ | 3 | 180.00 | 60.00 |
| | | | $B$ | 2 | 168.00 | 84.00 |
| | | | $AB$ | 6 | 65.33 | 10.89 |
| Error | 22 | 710.67 | | | | 32.30 |
| Total | 35 | 1,628.00 | | | | |

It will be found that $A_2$ and $A_3$ are zero and the whole $ssq = 180$ is due to linear effect ($A_1$) with 1 $df$. Plot the four totals 153, 171, 189, 207 against the levels of factor $a$. What do you notice?

The six components of $AB$ interactions are calculated from the contrasts supplied in the accompanying table. Fill in the coefficients.

| | | $b_2 - b_0$ | | | | $b_2 - b_0$ | | $b_2 - b_0$ | |
|---|---|---|---|---|---|---|---|---|---|
| $a_0$ | $-3$ | $-3$ | $0$ | $+3$ | $+1$ | $0$ | $-1$ | $0$ | |
| $a_1$ | $-1$ | $-1$ | $0$ | $+1$ | $-1$ | $0$ | $+3$ | $0$ | |
| $a_2$ | $+1$ | $+1$ | $0$ | $1$ | $-1$ | $0$ | $-3$ | $0$ | |
| $a_3$ | $+3$ | $+3$ | $0$ | $-3$ | $+1$ | $0$ | $+1$ | $0$ | |

| | | $b_2 - 2b_1 + b_0$ | | | $b_2 - 2b_1 + b_0$ | | $b_2 - 2b_1 + b_0$ | | |
|---|---|---|---|---|---|---|---|---|---|
| $a_0$ | $-3$ | $\mid 6$ | $+1$ | | $-1$ | $-1$ | $+2$ | $-1$ | |
| $a_1$ | $-1$ | $+2$ | $-1$ | | $+3$ | $+3$ | $-6$ | $+3$ | |
| $a_2$ | $+1$ | $-2$ | $-1$ | | $-3$ | $-3$ | $+6$ | $-3$ | |
| $a_3$ | $+3$ | $-6$ | $+1$ | | $+1$ | $+1$ | $-2$ | $+1$ | |

| Contrasts | $z$ | $z^2$ | $D \times r$ | $z^2/Dr$ |
|---|---|---|---|---|
| $A_1B_1$ | 2 | 4 | $40 \times 3$ | 0.03 |
| $A_1B_2$ | 18 | 324 | $120 \times 3$ | 0.90 |
| $A_2B_1$ | $-10$ | 100 | $8 \times 3$ | 4.17 |
| $A_2B_2$ | $-42$ | 1,764 | $24 \times 3$ | 24.50 |
| $A_3B_1$ | 64 | 4,096 | $40 \times 3$ | 34.13 |
| $A_3B_2$ | $-24$ | 576 | $120 \times 3$ | 1.60 |
| | | Total interaction $ssq$ | | 65.33 |

**23.3** Consider the treatment totals of 3 materials $\times$ 3 quantities once more.

Material

|  | | $a$ | $b$ | $c$ | |
|---|---|---|---|---|---|
| | $q_0$ | (pooled) | | | 192 |
| Quantity | $q_1$ | 84 | 84 | 60 | 228 |
| | $q_2$ | 100 | 112 | 88 | 300 |
| $2q_2 + q_1$ | | 284 | 308 | 236 | 828 |
| $2q_1 - q_2$ | | 68 | 56 | 32 | 156 |

The sums of squares for the different materials and their interactions with quantities 1 and 2 may be calculated in many different ways according to the nature of the factors and the hypothesis in the investigator's mind. For instance, if we think that the differences between the materials at the $q_2$ level should be twice as large as those at the $q_1$ level, then instead of taking the simple sum $q_2 + q_1$, we may take the value $2q_2 + q_1$ to calculate the *ssq* for the three materials. For interaction, we take $2q_1 - q_2$ which is orthogonal to the sum. The denominator is $D \times r$, where $D = 1^2 + 2^2 = 5$ Thus,

$$\left.\begin{array}{l} \dfrac{284^2 + 308^2 + 236^2}{5 \times 4} - \dfrac{828^2}{60} = 134.40 \\[2mm] \dfrac{68^2 + 56^2 + 32^2}{5 \times 4} - \dfrac{156^2}{60} = 33.60 \end{array}\right\} 168.00$$

The total is equal to $156 + 12 = 168$, as before (Table 23.4).

**23.4** There is another type of study which may sound like factorial at first but actually is not; it is the so-called "concentration time study." For instance, four concentrations of certain material may be combined with four periods of time in the following manner:

| Concentration (for example, ppm): | 1 | 2 | 5 | 10 |
|---|---|---|---|---|
| Time (for example, minutes): | 10 | 5 | 2 | 1 |
| Dosage (concentration $\times$ time): | 10 | 10 | 10 | 10 |

In studies of this type, the total dosage remains the same but the rate of its delivery varies. This is *not* a study of dosage; nor is it a study of concentration or time per se. It is a study of the difference between "slow delivery" and "quick delivery" of a fixed amount of material. The scheme above has four "treatments" (that is, four ways of giving the dosage).

The study of the effect of low-level radiation over a long period of time vs. that of acute radiation over a short period is of this type.

Of course, the concentration $\times$ time study may be combined with time or dosage studies by increasing the number of treatments. For instance, to the four treatments above we may add the following four:

| Concentration: | 1 | 2 | 5 | 10 |
|---|---|---|---|---|
| Time: | 20 | 10 | 4 | 2 |

so that the dosage is 20 in each combination.

# 24

# Simple
# confounding systems

To most students the use of the principle of confounding in experimental designs is novel, and yet it is one of the most important achievements in the art of designing experiments and statistical analysis. We shall first explain what confounding means by considering some very simple examples, and we shall postpone general discussion until later in the chapter. The principle may best be explained by the following special case.

## 1. Confounding in 2 × 2 × 2

If three factors ($a$, $b$, $c$) are involved and each is administered at two levels, there are eight different treatment combinations. When all experimental units are homogeneous, we simply assign the treatments at random to the units and there is no further problem. If blocking of the units is needed, however, a block must contain eight homogeneous units, to which the eight treatments are to be assigned at random; and this seemingly simple requirement is not always easy to fulfill in practice. (It is certainly a great deal more difficult to find 16 homogeneous units for a $2 \times 2 \times 2 \times 2$ factorial experiment, but we shall use 8 treatments as an example.)   A smaller block, that is, a smaller number of homogeneous units, may be readily available.   Suppose that blocks of "size" four (that is, blocks of four units each) are obtainable.   The problem now is how to conduct a valid experiment for eight treatments in blocks of only four units each.   Which four of the eight treatments should be included in the same block?

We have learned (Chap. 21) that there are seven orthogonal contrasts among the eight treatments, each having certain physical meaning.   Of these, the three-factor interaction $ABC$ is probably the least important one, and experience shows that this effect is usually smaller than that of the other six comparisons.   We also recall that for each comparison the expression consists of four positive terms and four negative terms (Table 21.4).   In particular,

$$ABC = (abc + a + b + c) - (ab + ac + bc + 1)$$

Now, if we put the four treatments $abc$, $a$, $b$, $c$ in one block and the other four treatments $(ab, ac, bc, 1)$ in another block, then the difference between the two blocks has dual meanings: it represents the difference between the units of these two blocks as well as the contrast $ABC$. These two kinds of differences cannot be separated by any statistical manipulation; they are inherently mixed up.   In such a situation, we say that the block difference and interaction $ABC$ are *confounded;* or, more briefly, $ABC$ is confounded with blocks.

When each block contains all treatments as it does in preceding chapters, we use the terms *replications* and *blocks* synonymously; but with confounding designs, these two terms must be kept explicitly distinct. In the example under consideration, one block contains four treatments and two complementary blocks form a replication of all eight treatments. If in an experiment with $r$ replications (that is, $2r$ blocks) half of the blocks are of the type $(abc, a, b, c)$ and half are $(ab, ac, bc, 1)$, we say that $ABC$ has been *completely* confounded in the experiment, because the same contrast $(ABC)$ is confounded with blocks in every replication.

In analyzing the experimental results, the sum of squares due to blocks is calculated as in the case of complete blocks.   By so calculating it we do not necessarily assume that $ABC = 0$, but, whatever it is, we incorporate it into the block differences.   The information on $ABC$ as a separate effect has been sacrificed in the interest of smaller blocks.   The smaller block with more homogeneous units will make the experimental error smaller than it otherwise would be. In brief, at the cost of $ABC$, we gain a smaller experimental error so that the other comparisons may be made more accurately.

At this stage the student invariably wonders how this would affect the other six comparisons. With blocks of four units each, how are we to evaluate and compare the other effects? The answer may be as surprising as it is simple: there is no disturbance whatsoever of the other six comparisons. The various effects and their sums of squares are to be calculated the usual way as if there were no confounding of $ABC$ with the blocks. Why? Because $ABC$ is orthogonal to every one of the other six comparisons. This gives the student an additional insight into the meaning and application of orthogonality. Take, for instance, the main effect of factor $a$:

$$ A = (abc + a - b - c) - (-ab - ac + bc + 1) $$

the terms being written in the same order as in $ABC$ to preserve the block structure. We note that in the first block there are two positive and two negative terms, so that their difference is independent of the block effect. The same thing is true for the four treatments in the second block. Taking the expression $A$ as a whole, we see that it is orthogonal to the blocks (that is, unaffected by the block differences) and it is calculated in the same way as in complete blocks. The reader may write out all the other expressions in the same manner and satisfy himself that for each comparison there are two positive and two negative terms in each block. This is merely another way of exhibiting the fact that $ABC$ is orthogonal to all the other comparisons.

## 2. Analysis of 2 × 2 × 2 with complete confounding

The difference and similarity in analysis between an ordinary complete randomized blocks experiment and one with $ABC$ completely confounded with blocks may be exhibited by analyzing the same set of data in two different ways, so that we can see the change in degrees of freedom and in sums of squares. For this purpose the data of Exercise 16.6, having been used in Table 21.5 for an ordinary 2 × 2 × 2, have been reproduced here in Table 24.1 with new blocks of four units each. The four treatments in each block have been assigned to the units at random. The *ssq* for the six blocks is calculated the usual way. The analyses of variance for three complete replications and for six incomplete blocks (confounding $ABC$) are given side by side in Table 24.1 for comparison. Some brief comments on each *ssq* (blocks, treatments, error) are in order.

For convenience the blocks containing ($abc$, $a$, $b$, $c$) are tabulated

*Table* 24.1 Data and analysis of a 2 × 2 × 2 factorial in blocks of four units with interaction $ABC$ completely confounded with blocks

| | *abc* | *a* | *b* | *c* | Block total | | *ab* | *ac* | *bc* | 1 | Block total |
|---|---|---|---|---|---|---|---|---|---|---|---|
| I | 30 | 16 | 24 | 19 | 89 | | 28 | 16 | 27 | 8 | 79 |
| II | 23 | 16 | 28 | 16 | 83 | | 18 | 25 | 16 | 10 | 69 |
| III | 25 | 19 | 20 | 16 | 80 | | 23 | 22 | 17 | 18 | 80 |
| Treatment total | 78 | 51 | 72 | 51 | 252 | | 69 | 63 | 60 | 36 | 228 |

Analysis of variance

| Variation | "Old" replications (Table 21.5) | | "New" blocks | | Components | | Source identification |
|---|---|---|---|---|---|---|---|
| | *df* | *ssq* | *df* | *ssq* | *df* | *ssq* | |
| Replications or blocks | 2 | 16 | 5 | 53 | 2 | 16 | Replications |
| | | | | | 1 | 24 | Interaction $ABC$ |
| | | | | | 2 | 13 | "Old" error |
| Treatments | 7 | 432 | 6 | 408† | 3 | 198 | *abc, a, b, c* totals |
| | | | | | 3 | 210 | *ab, ac, bc,* (1) totals |
| Error | 14 | 276 | 12 | 263 | 6 | 59.5 | (*abc, a, b, c*) expt. |
| | | | | | 6 | 203.5 | (*ab, ac, bc,* 1) expt. |
| Total | 23 | 724 | 23 | 724 | 23 | 724 | |

† The six single components (*A, B, AB, C, AC, BC*) remain the same as before (Table 21.5).

together, and those containing (*ab, ac, bc,* 1) are likewise grouped together. If the six block totals are arranged in a two-way table, we see that

| | | |
|---|---|---|
| 89 | 79 | 168 |
| 83 | 69 | 152 |
| 80 | 80 | 160 |
| 252 | 228 | 480 |

$ssq = 53$ consists of three components: (1) The $ssq$ for the marginal totals 168, 152, 160 is the $ssq = 16$ for replications with 2 $df$. (2) The $ssq$ for the totals 252 and 228 is the $ssq = 24$ for interaction $ABC$ with 1 $df$ if there were no confounding. (3) The remaining portion, $53 - 16 - 24 = 13$ with 2 $df$, would belong to the error term if we used complete replications without confounding. The composition of the treatment *ssq* = 408 with 6 *df* is also easily seen. It is the old treatment *ssq* 432 based on all eight treatment totals minus the *ssq* for *ABC*. Thus, $432 - 24 = 408$. We may, however, see it another way:

$$\text{Old treatment } ssq = \frac{78^2 + \cdots + 36^2}{3} - \frac{480^2}{24} = 432$$

$$\text{Old } ABC \; ssq = \frac{252^2 + 228^2}{12} - \frac{480^2}{24} = 24$$

Subtracting the latter from the former, we obtain the

$$\text{New treatment } ssq = \frac{78^2 + 51^2 + 72^2 + 51^2}{3} - \frac{252^2}{12}$$
$$+ \frac{69^2 + 63^2 + 60^2 + 36^2}{3} - \frac{228^2}{12}$$

$$= 198 + 210 = 408$$

It is clear that the new treatment *ssq* = 408 with 6 *df* consists of two major components, one from the four treatment totals of the (*abc, a, b, c*) blocks with 3 *df* and one from the four treatment totals of the (*ab, ac, bc,* 1) blocks with 3 *df*. In other words, each half of the experiment is regarded as a whole experiment for treatment *ssq*.

Finally, we come to the error *ssq*. In practice it is always obtained by subtraction, of course. However, it is instructive to see how it may be obtained some other way. Exactly analogously to the treatment *ssq*, the error *ssq* may be calculated from the separate halves of the experiment. Taking the three (*abc, a, b, c*) blocks by themselves, we may calculate an error *ssq* by the usual method. The reader will find that it is equal to 59.50 with 6 *df*. Similarly, taking the three (*ab, ac, bc,* 1) blocks by themselves, we find that the error *ssq* is 203.50 with 6 *df*. The sum of these two is $59.50 + 203.50 = 263.00$ with 12 *df*. Alternatively, the old error *ssq* minus that which goes to block differences gives $276 - 13 = 263$.

Since we shall not be able to go into details of the many different types of complicated confounding systems in this introductory text, we explain this simple case in detail to show some of the general features of confounding designs. The block differences always contain certain treatment effects, although in complicated cases we cannot identify them as explicitly as we did here with *ABC*. Some of the error *ssq*'s also go into the block differences. With complete confounding, the treatments lose degrees of freedom as well as sums of squares.

## 3. Missing value

From the analysis above it is clear that in a completely confounded experiment the two halves are regarded pretty much like independent whole experiments. Each half contributes independently to the error *ssq*. Now, if one observation is missing, it affects only half of the experiment. For example, if the observation on treatment *abc* in the first replication (see Table 24.1) is missing, it affects the half with (*abc*, *a*, *b*, *c*) blocks only. The incomplete data would be as in the accompanying table, where *y* is the unknown and the dots are known numbers.

|  | abc | a | b | c |  |
|---|---|---|---|---|---|
| I | y | · | · | · | 59 |
| II | · | · | · | · | · |
| III | · | · | · | · | · |
|  | 48 | · | · | · | 222 |

The formula for missing value (Chap. 19) for ordinary randomized blocks is then applicable to this half of the experiment. Thus

$$y = \frac{4(48) + 3(59) - 222}{(4 - 1)(3 - 1)} = \frac{147}{6} = 24.50$$

If another observation in the other half of the experiment (say, on treatment *ab*) is missing, it may be estimated again as a single missing value by the same procedure. Only when two observations are missing in the same half of the experiment do we use the formula for two missing values.

## 4. Partial confounding

In replicated experiments it is neither necessary nor advisable to confound the same interaction with blocks in all replications. Indeed, if some information in all interactions is desired, different interactions may be confounded with blocks in different replications. To be concrete, let us consider the $2 \times 2 \times 2$ factorial once more. In the preceding section the interaction $ABC$ has been confounded with blocks of four units each. In other replications, the interactions $AB$, $AC$, $BC$ may be similarly confounded (with blocks). Table 24.2 (based on Table 21.4)

*Table* 24.2    Confounding of interaction with blocks for a
2 × 2 × 2 factorial experiment

| Replication | Two complementary blocks | | Interaction confounded |
|---|---|---|---|

| I | $abc$  $a$  $b$  $c$ | $ab$  $ac$  $bc$  $1$ | $ABC$ |
| II | $abc$  $ab$  $c$  $1$ | $ac$  $bc$  $a$  $b$ | $AB$ |
| III | $abc$  $ac$  $b$  $1$ | $ab$  $bc$  $a$  $c$ | $AC$ |
| IV | $abc$  $bc$  $a$  $1$ | $ac$  $ab$  $b$  $c$ | $BC$ |

gives a plan for four replications, each interaction being confounded once in one replication. A confounding design of this nature is called partial confounding.

In practice it is unnecessary to have exactly four replications, nor is it necessary to confound all interactions the same number of times. Knowledge of the factors and their reactions will determine which interaction is the least important one and hence to be confounded with the blocks. Even when there are four replications, we do not have to adopt the plan of Table 24.2 but may decide to confound $ABC$ in two replications and confound $BC$, say, in the other two. There is considerable freedom of choice in confounding designs.

The analysis of a partial confounding experiment is usually longer than that with complete confounding. To illustrate, let us suppose that the experiment has four replications ($r = 4$) according to the plan shown in Table 24.2. The calculation of the main effects ($A$, $B$, $C$) and their sums of squares is the same as before, since they are orthogonal with the blocks and unaffected by the confounding design. To calculate interaction $ABC$ and its *ssq*, however, we omit replication I (in which it has been confounded) and the calculation proceeds the same way as before (Chap. 21) with the data of replications II to IV. In other words, as far as interaction $ABC$ is concerned, the experiment consists of only three replications ($r = 3$), instead of four. Similarly, the calculation of $AB$ and its *ssq* is based on replications I, III, and IV, also with $r = 3$, replication II being omitted. Partial confounding leads to partial information. We say that $ABC$, etc., have three-fourths information relative to that of main effects, because only three-fourths of the experimental observations have been used to evaluate the interactions. The analysis of the experiment is as follows:

|  | $df$ |
|---|---|
| Blocks | 7 |
| $A$ | 1 |
| $B$ | 1 |
| $C$ | 1 |
| $AB$ | $1'$ |
| $AC$ | $1'$ |
| $BC$ | $1'$ |
| $ABC$ | $1'$ |
| Error | 17 |
| Total | 31 |

The prime serves as an indicator to remind us of the fact that the information is partial because of the partial confounding. The error *ssq* is obtained by subtraction.

The formula for a missing value in a partially confounded experiment is necessarily very long and will not be given here. Even the formulas found in more comprehensive texts are for special cases. This and other considerations lead to the conclusion that if a certain high-order interaction is known to be unimportant, it is simplest to use complete confounding (same confounding in all replications).

## 5. Further confounding

We shall not develop this subject in any detail but just mention enough about it to show the reader the many possibilities of confounding. To conduct a four-factor factorial $2 \times 2 \times 2 \times 2 = 2^4$ experiment, we need blocks of "size" 16 for complete replications. Exactly analogously to the cases of $2^3$ discussed above, the interaction (Exercise 21.7)

$$ABCD = (abcd + ab + ac + bc + ad + bd + cd + 1)$$
$$- (abc + abd + acd + bcd + a + b + c + d)$$

may be confounded with blocks of 8 units each. In such a design all the other contrasts are still orthogonal to the (incomplete) blocks and their *ssq*'s are calculated the usual way. Again, different interactions (for example, $BCD$) may be confounded with blocks in different replications, resulting in partial confounding. So far, the situation here is the same as that for $2 \times 2 \times 2$.

A new problem arises when blocks of 8 units each are still too large to achieve homogeneity, and we prefer to conduct the experiment in smaller blocks, say, of 4 units each. Then the 16 treatment combinations are to be allocated to 4 (incomplete) blocks, each containing only 4 treatments:

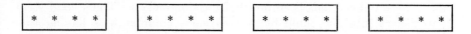

The problem is how this is to be done. Which four should go to the same block? The comparison among the four blocks has 3 *df*. This implies that a certain 3 contrasts among the treatments must have been confounded with the blocks. The allocation of the 16 treatments to the four blocks depends entirely on the particular three contrasts to be confounded with the blocks. Hence, the problem reduces to that of how to choose the three contrasts whose information is to be sacrificed in the interest of small (homogeneous) blocks.

The design for simultaneous confounding of three contrasts is not as straightforward as the simple confounding of one contrast at a time. For factors at two levels each, however, there is a very simple rule by which the confounding arrangement may be achieved. Of the three contrasts to be confounded, two may be written down at will (that is, arbitrarily) and the third one is automatically determined. For example, if we wish to confound $ABCD$ and $BCD$, then the third confounded contrast is, symbolically, the product $(ABCD)(BCD) = AB^2C^2D^2 = A$, where the "square" of any letter is taken as unity (or simply drops out). This so-called rule arises from the combinatorial fact of the contrasts concerned. The method of writing down the four treatments of each block will be illustrated later. For the time being, all we know is that the three contrasts confounded with blocks are $ABCD$, $BCD$, $A$.

The scheme exemplified above is hardly practical, because it involves the sacrifice of information about the main effect of a factor. A better confounding design is of the type $ABCD$, $AB$, $CD$; although this scheme does not involve any main effect, it does sacrifice two primary interactions ($AB$ and $CD$) which are often important and necessary for interpretation. Probably the best compromise and most practical scheme is of the type $AB$, $ACD$, $BCD$, shown in Table 24.3. Note that this confounding scheme does not involve $ABCD$ but confounds only one of the six primary interactions.

The method of allocating treatments to blocks is also shown in Table 24.3. First, write out the three contrasts $AB$, $ACD$, $BCD$ among the 16 treatment combinations. We need work out only two contrasts, since the third is simply the "product" of the other two. Thus, $(AB)(BCD) = B^2ACD = ACD$; $(AB)(ACD) = A^2BCD = BCD$; or

$$(ACD)(BCD) = ABC^2D^2 = AB$$

(The "product" property is in the coefficients of the contrasts:

$$(+1)(-1) = -1$$

etc. The number 1 has been omitted for simplicity.) The second step is to collect into the same block treatment combinations having the same sign sequence patterns. Thus, the four treatments (1), *abc, abd, cd* belong to the same block because they all have the same sign sequence $+, -, -$. The resulting four blocks are shown at the bottom of Table 24.3. The numbering of the blocks (I, II, III, IV) is, of course, arbitrary. That the comparisons among the four blocks are actually the contrasts $AB$, $ACD$, $BCD$ may be confirmed by observing the following:

$$AB = IV - III - II + I$$
$$ACD = IV - III + II - I$$
$$BCD = IV + III - II - I$$

This is the confounding scheme for any one replication. If the same scheme is adopted in all replications, the result is complete confounding of these contrasts. Alternatively, different confounding schemes may be used in different replications, resulting in partial confounding. There are five other confounding schemes similar to the one given above, and they may be obtained by simple permutation of letters. Thus, corre-

**Table 24.3** Confounding for a $2^4$ factorial in blocks of four units each

| Inter-action | Treatment combinations | | | | | | | | | | | | | | | |
|---|---|---|---|---|---|---|---|---|---|---|---|---|---|---|---|---|
| | (1) | *a* | *b* | *ab* | *c* | *ac* | *bc* | *abc* | *d* | *ad* | *bd* | *abd* | *cd* | *acd* | *bcd* | *abcd* |
| $AB$ | + | − | − | + | + | − | − | + | + | − | − | + | + | − | − | + |
| $ACD$ | − | + | − | + | + | − | + | − | + | − | + | − | − | + | − | + |
| $BCD$ | − | − | + | + | + | + | − | − | + | + | − | − | − | − | + | + |
| Block | I | II | III | IV | IV | III | II | I | IV | III | II | I | I | II | III | IV |

| Block I | (1) | *abc* | *abd* | *cd* |
|---|---|---|---|---|

| Block II | *a* | *bc* | *bd* | *acd* |
|---|---|---|---|---|

| Block III | *b* | *ac* | *ad* | *bcd* |
|---|---|---|---|---|

| Block IV | *ab* | *c* | *d* | *abcd* |
|---|---|---|---|---|

sponding to $AB$, we may have $AC$, $AD$, $BC$, $BD$, $CD$. If there are six replications in the experiment, we may adopt one system for each replication, achieving balanced partial confounding for all interactions concerned.

The principle of confounding outlined above may be extended to the general case $2^n$ (that is, $n$ factors, each at two levels). Extensive tables of the confounding schemes are available. Apparently, the construction of confounding designs has become a small branch of mathematics by itself. In this section I think we have gained enough basic understanding of the subject.

## 6. Confounding for factors at three levels

Although the basic reason for and result of confounding for factors at three levels are the same as those for factors at two levels, the technique involved is much more sophisticated. Consider the $3 \times 3 \times 3$ factorial with 27 treatment combinations. Obviously it is impossible to confound only one contrast (as we did for $2 \times 2 \times 2$), using blocks of size $13\frac{1}{2}$. It is, however, possible to confound certain *two* contrasts with three blocks, each of size 9:

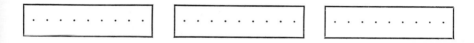

These three complementary blocks make a complete replication of the 27 treatment combinations. The problem now is one of allocation: which 9 of the 27 should go to the same block?

The allocation of treatments to blocks depends on what contrasts are to be sacrificed. From the experimental viewpoint we naturally wish to conserve all the main effects ($A$, $B$, $C$). This implies that the nine treatment combinations in any block should contain levels 0, 1, 2, each exactly three times, for each of the three factors ($a$, $b$, $c$). That is, if we concentrate on any single factor, say, $c$, the nine treatments in each of the three blocks are $c_0$, $c_0$, $c_0$, $c_1$, $c_1$, $c_1$, $c_2$, $c_2$, $c_2$, so that the main effects of the factor are orthogonal to the blocks.

Next, we would also wish to conserve all the primary interactions ($AB$, $AC$, $BC$) within the blocks. If so, the nine treatments in each block should contain the combinations of levels (0, 1, 2) of one factor with those (0, 1, 2) of another. In other words, when we concentrate on any pair of factors ($a,b$; $a,c$; $b,c$), each block contains the nine combinations 00, 01, 02, 10, 11, 12, 20, 21, 22. Note that this condition covers the previous one; when two-factor interactions are conserved, the main effects are automatically conserved. When these conditions are satisfied, then we know that the two contrasts confounded with the blocks are necessarily certain components of the three-factor interaction $ABC$ (which has 8 $df$, Chap. 22). This is the most desirable design for practical purposes.

Having been told of the many restrictions in allocation, the reader may wonder if it is possible to find such a design at all. The fact is that it is not only possible but surprisingly simple, thanks to an ingenious application of $3 \times 3$ Latin squares by Fisher (1935 and 1960). The reader should check carefully that the following nine treatment combinations in a block do satisfy all the conditions required for conserving two-factor interactions. The three rows of the square are the three levels

|        | $b_0$ | $b_1$ | $b_2$ |
|--------|-------|-------|-------|
| $a_0$  | $c_0$ | $c_1$ | $c_2$ |
| $a_1$  | $c_1$ | $c_2$ | $c_0$ |
| $a_2$  | $c_2$ | $c_0$ | $c_1$ |

or

Levels of

| $a$ | $b$ | $c$ |
|-----|-----|-----|
| 0 | 0 | 0 |
| 0 | 1 | 1 |
| 0 | 2 | 2 |
| 1 | 0 | 1 |
| 1 | 1 | 2 |
| 1 | 2 | 0 |
| 2 | 0 | 2 |
| 2 | 1 | 0 |
| 2 | 2 | 1 |

of factor $a$, and the three columns are the three levels of factor $b$. Then the three levels of factor $c$ are arranged into a $3 \times 3$ Latin square, so that for any pair of factors, the nine combinations $(0, 1, 2)(0, 1, 2)$ are all present, and these nine should go to the same block. The side table of levels of $a$, $b$, $c$ is just another way of writing the Latin square.

The treatment combinations shown above represent only 9 of the total 27. The second group of 9 combinations may be obtained by "cyclic substitution," replacing level 0 by 1, 1 by 2, and 2 by 0 for any one of the factors. This has been done in Table 24.4 for the levels of factor $c$. The third group of 9 combinations is obtained from the second by another round of cyclic substitution. The three groups of 9 thus obtained (top row of Table 24.4) are precisely the 27 treatment combinations divided into 3 complementary blocks conserving primary interactions. The three Latin squares of system I (top row of Table 24.4) constitute a "cyclic group."

The first square of system I is the familiar "standard" square. Interchange of levels 1 and 2 gives the first square of system II, from which another cyclic group may be generated. Interchange of the second and third rows of the standard square gives system III, and a similar inter-change of columns gives system IV. Each system is a suitable confound-ing design for the 27 treatment combinations. Thus, we see there are actually four ways of constructing the required confounding blocks.

Finally, it should be pointed out that the four systems of confounding are orthogonal to each other. To help see this, the treatment combina-tions are given shorthand symbols for ease of tracing. The nine combi-nations in the first square (block) of system I are denoted by $x_1, \ldots, x_9$,

*Table 24.4* **Confounding for a $3 \times 3 \times 3$ factorial in blocks of nine units each. Rows are $a_0, a_1, a_2$; columns are $b_0, b_1, b_2$. Body of Latin square gives the levels of factor $c$; arrow indicates cyclic substitution**

System             Three complementary blocks

I

| 0 | 1 | 2 |
|---|---|---|
| 1 | 2 | 0 |
| 2 | 0 | 1 |

$\rightarrow$

| 1 | 2 | 0 |
|---|---|---|
| 2 | 0 | 1 |
| 0 | 1 | 2 |

$\rightarrow$

| 2 | 0 | 1 |
|---|---|---|
| 0 | 1 | 2 |
| 1 | 2 | 0 |

II

| 0 | 2 | 1 |
|---|---|---|
| 2 | 1 | 0 |
| 1 | 0 | 2 |

$\rightarrow$

| 1 | 0 | 2 |
|---|---|---|
| 0 | 2 | 1 |
| 2 | 1 | 0 |

$\rightarrow$

| 2 | 1 | 0 |
|---|---|---|
| 1 | 0 | 2 |
| 0 | 2 | 1 |

III

| 0 | 1 | 2 |
|---|---|---|
| 2 | 0 | 1 |
| 1 | 2 | 0 |

$\rightarrow$

| 1 | 2 | 0 |
|---|---|---|
| 0 | 1 | 2 |
| 2 | 0 | 1 |

$\rightarrow$

| 2 | 0 | 1 |
|---|---|---|
| 1 | 2 | 0 |
| 0 | 1 | 2 |

IV

| 0 | 2 | 1 |
|---|---|---|
| 1 | 0 | 2 |
| 2 | 1 | 0 |

$\rightarrow$

| 1 | 0 | 2 |
|---|---|---|
| 2 | 1 | 0 |
| 0 | 2 | 1 |

$\rightarrow$

| 2 | 1 | 0 |
|---|---|---|
| 0 | 2 | 1 |
| 1 | 0 | 2 |

Orthogonality of the systems

I

| $x_1$ | $x_2$ | $x_3$ |
|---|---|---|
| $x_4$ | $x_5$ | $x_6$ |
| $x_7$ | $x_8$ | $x_9$ |

| $y_1$ | $y_2$ | $y_3$ |
|---|---|---|
| $y_4$ | $y_5$ | $y_6$ |
| $y_7$ | $y_8$ | $y_9$ |

| $z_1$ | $z_2$ | $z_3$ |
|---|---|---|
| $z_4$ | $z_5$ | $z_6$ |
| $z_7$ | $z_8$ | $z_9$ |

II

| $x_1$ | $y_2$ | $z_3$ |
|---|---|---|
| $y_4$ | $z_5$ | $x_6$ |
| $z_7$ | $x_8$ | $y_9$ |

| $y_1$ | $z_2$ | $x_3$ |
|---|---|---|
| $z_4$ | $x_5$ | $y_6$ |
| $x_7$ | $y_8$ | $z_9$ |

| $z_1$ | $x_2$ | $y_3$ |
|---|---|---|
| $x_4$ | $y_5$ | $z_6$ |
| $y_7$ | $z_8$ | $x_9$ |

those of the second square by $y_1, \ldots, y_9$, and of the third, $z_1, \ldots, z_9$ (see lower portion of Table 24.4). The treatment combinations of system II are then written in terms of these shorthand symbols at the bottom of Table 24.4. We note immediately that, of the nine treatments in each block of system II, three are from the first block, three from the second block, and three from the third block of system I. Hence systems I and II are orthogonal to each other. The reader may write out the blocks of systems III and IV in terms of the same symbols and satisfy himself that any two of the four systems are orthogonal to each other. In summary, these four systems form an orthogonal set of subdividing the 8 *df* of interaction $ABC$ into four pairs of degrees of freedom.

## 7. Example of confounding in 3 × 3 × 3

To reduce the arithmetic labor to a minimum, let us suppose that there is only a single replication of the 27 treatments (Table 22.9), tested in three blocks according to confounding system I (Table 24.5). This is a particular case of complete confounding. If there are *r* replications using the same confounding system, the procedure described below applies to treatment totals of the *r* replications. In fact, we shall use the term *treatment totals* although there is only one replication in this particular example.

The sums of squares due to main effects ($A$, $B$, $C$) and primary interactions ($AB$, $AC$, $BC$) remain the same as those obtained previously (Table 22.9), because these comparisons are unaffected by the confounding blocks. Hence we shall be concerned here with the interaction $ABC$ only.

The top row of Table 24.5 gives the treatment combinations of the three blocks, the corresponding measurements, and the block totals. The *ssq* due to blocks is calculated the usual way:

$$\text{Blocks } ssq = \frac{186^2 + 168^2 + 186^2}{9} - \frac{540^2}{27} = 24.00 \qquad \text{with 2 } df$$

These 2 *df*, originally part of $ABC$, are now confounded with the blocks. The next step is to calculate the *ssq* due to the unconfounded components of $ABC$ which still have 6 *df*. If the experiment is conducted in whole replications without confounding, the total $ABC$ *ssq* is found to be 123.34 with 8 *df* as shown in Table 22.9. Thus the remaining unconfounded component of $ABC$ should have the *ssq*.

$$123.34 - 24.00 = 99.34 \qquad \text{with 6 } df$$

*Table 24.5* Confounding of a $3 \times 3 \times 3$ factorial in blocks of nine units each (data from Table 22.9, reproduced in Table 24.6). Rows are $a_0$, $a_1$, $a_2$; columns are $b_0$, $b_1$, $b_2$. Levels of factor $c$ are indicated by (0), (1), (2) in the body of the tables

Treatments in three complementary blocks

I

| (0) 11 | (1) 28 | (2) 28 |
| (1) 16 | (2) 18 | (0) 24 |
| (2) 28 | (0) 13 | (1) 20 |

186

| (1) 14 | (2) 21 | (0) 24 |
| (2) 10 | (0) 23 | (1) 18 |
| (0)  8 | (1) 24 | (2) 26 |

168

| (2) 11 | (0) 20 | (1) 26 |
| (0) 13 | (1) 31 | (2) 33 |
| (1) 30 | (2) 14 | (0)  8 |

186

Calculation of $ABC$ components

II

| (0) 11 | (2) 21 | (1) 26 |
| (2) 10 | (1) 31 | (0) 24 |
| (1) 30 | (0) 13 | (2) 26 |

192

| (1) 14 | (0) 20 | (2) 28 |
| (0) 13 | (2) 18 | (1) 18 |
| (2) 28 | (1) 24 | (0)  8 |

171

| (2) 11 | (1) 28 | (0) 24 |
| (1) 16 | (0) 23 | (2) 33 |
| (0)  8 | (2) 14 | (1) 20 |

177

III

| (0) 11 | (1) 28 | (2) 28 |
| (2) 10 | (0) 23 | (1) 18 |
| (1) 30 | (2) 14 | (0)  8 |

170

| (1) 14 | (2) 21 | (0) 24 |
| (0) 13 | (1) 31 | (2) 33 |
| (2) 28 | (0) 13 | (1) 20 |

197

| (2) 11 | (0) 20 | (1) 26 |
| (1) 16 | (2) 18 | (0) 24 |
| (0)  8 | (1) 24 | (2) 26 |

173

IV

| (0) 11 | (2) 21 | (1) 26 |
| (1) 16 | (0) 23 | (2) 33 |
| (2) 28 | (1) 24 | (0)  8 |

190

| (1) 14 | (0) 20 | (2) 28 |
| (2) 10 | (1) 31 | (0) 24 |
| (0)  8 | (2) 14 | (1) 20 |

169

| (2) 11 | (1) 28 | (0) 24 |
| (0) 13 | (2) 18 | (1) 18 |
| (1) 30 | (0) 13 | (2) 26 |

181

In the case of complete confounding, this is the simplest computation procedure, and it is entirely analogous to the case of complete confounding in $2 \times 2 \times 2$.

To lay a broader foundation for partial confounding experiments, let us calculate the conserved component of $ABC$ independently (instead of by subtraction). The experiment was carried out according to confounding system I of Table 24.4. Now, the treatment totals may be collected into three groups according to system II, then recollected according to system III, and finally recollected according to system IV. These groupings, as noted before, are orthogonal to the blocks as well as among themselves. The results are shown in Table 24.5. Each system of grouping yields three group totals, from which a sum of squares may be calculated. These $ssq$'s are conserved components of $ABC$.

$$ABC\text{(II)}: \qquad \frac{192^2 + 171^2 + 177^2}{9} - \frac{540^2}{27} = 26.00 \qquad \text{with } 2 \; df$$

$$ABC\text{(III)}: \qquad \frac{170^2 + 197^2 + 173^2}{9} - \frac{540^2}{27} = 48.67 \qquad \text{with } 2 \; df$$

$$ABC\text{(IV)}: \qquad \frac{190^2 + 169^2 + 181^2}{9} - \frac{540^2}{27} = 24.67 \qquad \text{with } 2 \; df$$

$$\text{Total conserved } ABC = 99.34 \qquad \text{with } 6 \; df$$

The eight single components of $ABC$ (that is, $A_1B_1C_1$, $A_2B_1C_1$, etc.) have been given in Table 22.11, assuming no confounding. It is important to note that the confounded pair of contrasts and the three conserved pairs of $ABC$ shown above are *not* equivalent to any combination of interaction terms in Table 22.11. These are two entirely different subdivisions of the total $ABC$ *ssq*'s. The single components (Table 22.11) have definite physical meaning. The four grouping systems (Tables 24.4 and 24.5) are difficult to interpret in terms of particular effects, although we know that they conserve all primary interactions and confound certain components of $ABC$.

The reason for calculating the conserved $ABC$ *ssq*'s is twofold: one is to see if they are extraordinarily large and another is to enable us to obtain the error *ssq* by subtraction in the general case. Our example has a single replication. If the experiment had $r$ complete replications, using the same confounding system in each replication, the analysis would take the form given in the accompanying table.

|        | df         | When $r = 4$ | When $r = 1$ |
|--------|------------|--------------|--------------|
| Blocks | $3r - 1$   | 11           | 2            |
| $A$    | 2          | 2            | 2            |
| $B$    | 2          | 2            | 2            |
| $AB$   | 4          | 4            | 4            |
| $C$    | 2          | 2            | 2            |
| $AC$   | 4          | 4            | 4            |
| $BC$   | 4          | 4            | 4            |
| $ABC$  | 6          | 6            | (Error) 6    |
| Error  | $24(r-1)$  | 72           | 0            |
| Total  | $27r - 1$  | 107          | 26           |

When the confounding system is the same in all replications (complete confounding), the error *ssq* may be calculated independently, as in the case of $2 \times 2 \times 2$. The $r$ blocks with the same nine treatments may be considered as a randomized block experiment by themselves, giving an error *ssq* with $(9 - 1)(r - 1) = 8(r - 1) \; df$. The error *ssq* for the entire experiment is the sum of the three separate error *ssq*'s from the three smaller experiments, with $3 \times 8(r - 1) = 24(r - 1)$ *df*.

It should also be mentioned that the 27 treatment means (or totals) cannot be presented as unbiased estimates of the respective treatments because of the confounding design. Only 9 of the 27 treatments appear in a presumably homogeneous block, and there is heterogeneity among the blocks. Hence, the individual treatment means should be corrected according to blocks. We shall not give the details here, since in practice it is hardly necessary to do so unless the high-order interaction $ABC$ is very large (in comparison with $s^2$). For most practical purposes it is sufficient to present the three two-way tables ($a \times b$, $a \times c$, $b \times c$) as we did in Table 22.9. The means (or totals) for main effects and primary interactions need no correction by blocks, because they are unaffected by the confounding arrangement. In later chapters when we come to general incomplete block designs, however, the method of correcting treatment means by blocks will be given, because it is necessary for those designs.

## 8. Routine method of diagonal sums

To calculate the conserved components of $ABC$, we have in Table 24.5 adopted the long but easy-to-understand method of collecting and recollecting the treatment totals according to the orthogonal schemes of Table 24.4. The reader will notice (as shown in Table 24.4) that the nine treatments in any one block of a system are merely a certain three diagonal combinations, one from each block of another system. This combinatorial fact led Yates (1937) to develop a simple routine method of obtaining the final group totals necessary for calculating the $ABC$ *ssq*. The procedure is outlined in Table 24.6, beginning with a systematic tabulation of the 27 treatment totals. For the nine treatments involving $b_2$, we first take the principal (upper left to lower right) diagonal sums, which are called the "$I$ totals." Thus, $26 + 18 + 24 = 68$, $20 + 24 + 28 = 72$, and $8 + 33 + 26 = 67$. Similar sums are obtained for the next nine and the last nine treatments. Then the diagonal sums in the other direction (upper right to lower left), which are called the "$J$ totals," are obtained. Thus, for the first nine treatments, $28 + 18 + 8 = 54$, $26 + 24 + 26 = 76$, and $24 + 33 + 20 = 77$. Similar $J$ totals are calculated for the other sets of treatments. The nine $I$ totals and the nine $J$ totals thus obtained are given in the middle of Table 24.6.

The final step is to calculate again the $I$ and $J$ totals for each set of the nine $I$ and $J$ totals obtained previously. For example,

$$68 + 68 + 32 = 168, \text{ etc.}$$

*Table 24.6*  Treatment totals in a
3 × 3 × 3 factorial experiment
(confounded with blocks of nine units
each)

|       | $b_2$ | | | $b_1$ | | | $b_0$ | | |
|-------|-------|-------|-------|-------|-------|-------|-------|-------|-------|
|       | $a_2$ | $a_1$ | $a_0$ | $a_2$ | $a_1$ | $a_0$ | $a_2$ | $a_1$ | $a_0$ |
| $c_2$ | 26 | 33 | 28 | 14 | 18 | 21 | 28 | 10 | 11 |
| $c_1$ | 20 | 18 | 26 | 24 | 31 | 28 | 30 | 16 | 14 |
| $c_0$ | 8 | 24 | 24 | 13 | 23 | 20 | 13 | 8 | 11 |

Calculation of diagonal sums

| $I$ totals ($\searrow$) | | | $J$ totals ($\nearrow$) | | |
|------|------|------|------|------|------|
| 68 | 65 | 55 | 54 | 65 | 35 |
| 72 | 68 | 54 | 76 | 65 | 55 |
| 67 | 59 | 32 | 77 | 62 | 51 |

| $I$ totals | $J$ totals | $I$ totals | $J$ totals |
|------|------|------|------|
| 168 | 190 | 170 | 177 |
| 186 | 181 | 173 | 171 |
| 186 | 169 | 197 | 192 |

The results are four sets of final totals. Note that the first set (168, 186, 186) is identical with block totals. The remaining three sets are identical with the group totals obtained in Table 24.5 by the long method of collection. The order of appearance of these totals depends on the way the 27 treatment totals are arranged. (See exercises.)

Similarly to the case of factors at two levels, we may have partial confounding experiments, that is, experiments using different confounding systems (Table 24.4) in different replications. In particular, if there are four (or eight) replications, each confounding system may be used once (or twice), thus achieving a balanced partial confounding design. The arithmetic procedure of analysis is necessarily long, but the principle remains the same as in the case of 2 × 2 × 2. Each of the four components of $ABC$ may be calculated from the three replications in which they are not confounded with three-fourths information relative to that for main effects and primary interactions. We shall not give massive numerical examples here; realistic data may be found in more comprehensive textbooks, which the reader must consult when he has an experiment of that particular type.

## 9. Confounding in mixed series

When some of the factors have two levels and some have three levels, it is also possible to have confounding designs, but, in general, the analysis is more complicated. The difficulties involved may be seen from the simple factorial of $3 \times 3 \times 2$ (in the order of $a$, $b$, $c$). First of all, we wish to conserve all main effects. Hence, the block size must be a multiple of 2, so that $c_0$ and $c_1$ may occur the same number of times. By the same token, the block size must be a multiple of 3, so that $a_0$, $a_1$, $a_2$ and $b_0$, $b_1$, $b_2$ may occur the same number of times. It is seen that the only feasible confounding design is one with block size 6, whereby 3 blocks make up a complete replication of the 18 treatments. With blocks of size 6, we strive to include the 6 combinations of factors $a$ and $c$, as well as the 6 combinations of $b$ and $c$, thus conserving the interactions $AC$ and $BC$. Such designs are shown in Table 24.7; they are derived from Table 24.4 by deleting the third column of each square. The two columns now correspond to the two levels of factor $c$. The three rows are $a_0$, $a_1$, $a_2$. The levels of factor $b$ are indicated in the body of the rectangles.

We note that with blocks of size 6 conserving $AC$ and $BC$, the interaction $AB$ has been confounded to a certain extent, because only 6 of the 9 possible combinations between factors $a$ and $b$ appear in any one block. The confounding of certain primary interactions is a usual feature for factors with different numbers of level, and it is this fact that limits the utility of confounding designs in a mixed series. The triple interaction $ABC$ is also partially confounded with the blocks, although not in a clearcut way as in the case of $3 \times 3 \times 3$. In analysis, the sums of squares of the following effects are calculated in the usual way. But

$$
\begin{array}{c}
& & df \\
\text{Main effects} & \left\{ \begin{array}{cc} A & 2 \\ B & 2 \\ C & 1 \end{array} \right. \\
\text{Interactions} & \left\{ \begin{array}{cc} AC & 2 \\ BC & 2 \end{array} \right.
\end{array}
$$

the $ssq$'s for interactions $AB$ and $ABC$, each with 4 $df$, require special calculation. The detailed procedure is too long to be given here. If the triple interaction $ABC$ is unimportant, its $ssq$ may be pooled with that for error. The $ssq$ for $AB$ should be calculated by a special procedure.

Pretty much the same situation prevails for the mixed series $3 \times 2 \times 2$. It is possible to use blocks of size 6, containing all combinations of $a \times b$ and all combinations of $a \times c$, thus conserving interactions $AB$ and $AC$.

*Table 24.7* Confounding of a 3 × 3 × 2 factorial in blocks of six units each. Rows are $a_0$, $a_1$, $a_2$; columns are $c_0$, $c_1$; levels of factor $b$ are in the body of the rectangles. The design is derived from Table 24.4 by deleting the third column of each square; arrow indicates cyclic substitution

| System | Three blocks | | | System | Three blocks | | |
|---|---|---|---|---|---|---|---|
| I | $\begin{matrix}0 & 1\\1 & 2\\2 & 0\end{matrix}$ → | $\begin{matrix}1 & 2\\2 & 0\\0 & 1\end{matrix}$ → | $\begin{matrix}2 & 0\\0 & 1\\1 & 2\end{matrix}$ | II | $\begin{matrix}0 & 2\\2 & 1\\1 & 0\end{matrix}$ → | $\begin{matrix}1 & 0\\0 & 2\\2 & 1\end{matrix}$ → | $\begin{matrix}2 & 1\\1 & 0\\0 & 2\end{matrix}$ |
| IV | $\begin{matrix}0 & 2\\1 & 0\\2 & 1\end{matrix}$ → | $\begin{matrix}1 & 0\\2 & 1\\0 & 2\end{matrix}$ → | $\begin{matrix}2 & 1\\0 & 2\\1 & 0\end{matrix}$ | III | $\begin{matrix}0 & 1\\2 & 0\\1 & 2\end{matrix}$ → | $\begin{matrix}1 & 2\\0 & 1\\2 & 0\end{matrix}$ → | $\begin{matrix}2 & 0\\1 & 2\\0 & 1\end{matrix}$ |

**Confounding of a 3 × 2 × 2 factorial in blocks of six units each**

Rows: $a_0$, $a_1$, $a_2$    Columns: $b_0$, $b_1$    Body: levels of factor $c$

| Replication I | | Replication II | | Replication III | |
|---|---|---|---|---|---|
| Two blocks | | Two blocks | | Two blocks | |
| $\begin{matrix}0 & 1\\1 & 0\\1 & 0\end{matrix}$ | $\begin{matrix}1 & 0\\0 & 1\\0 & 1\end{matrix}$ | $\begin{matrix}1 & 0\\0 & 1\\1 & 0\end{matrix}$ | $\begin{matrix}0 & 1\\1 & 0\\0 & 1\end{matrix}$ | $\begin{matrix}1 & 0\\1 & 0\\0 & 1\end{matrix}$ | $\begin{matrix}0 & 1\\0 & 1\\1 & 0\end{matrix}$ |

NOTE: The first column of each replication is a permutation of 0, 1, 1. The second column is then automatically determined. The second block is the "mirror image" of the first.

Two blocks make up a complete replication of the 12 treatment combinations. The design is simple, as shown in the lower portion of Table 24.7. The primary interaction $BC$, however, is partially confounded with blocks, since its conservation requires a multiple of 4 units in a block. The triple interaction $ABC$ is, of course, also partially confounded. A numerical example of 3 × 2 × 2 in blocks of 6 units each is given in the exercises.

Since the historical monograph of Yates (1937), many other types of confounding designs have been worked out by various workers. One of the earlier extensions may be found in Jerome Li [Design and Statistical Analysis of Some Confounded Factorial Experiments, *Iowa State Univ. Eng. Exp. Sta., Bull.* 333 (1944)].

This section does not give a detailed numerical procedure for calculating the partially confounded interactions. Instead, the purpose is to make the reader aware of the existence of a number of confounded designs.

## 10. General comments

Confounding in factorial experiments is almost a necessity in agricultural field trials, because the homogeneity of land that can be used as a realistic block is limited by Nature, and no large area of land can be uniform in fertility. No wonder that most of the highly sophisticated confounding designs (such as the quasi-Latin square with two-dimensional confounding) are primarily designed for and thus readily applicable to field experiments, but they are too restricted in setup for general usage. It is for this reason that we have limited ourselves to simple confounding systems in this chapter. In experiments with mice, say, the need for confounding (in a $3 \times 3 \times 2$ factorial, for example) is less than in experiments with large animals or field trials, because fairly large numbers of inbred (genetically homogeneous) mice are available.

In medical and human experiments, the situation is not very clear in view of the limited availability of experimental individuals and the large variation from individual to individual. Frequently, even an ordinary blocking process is highly artificial and the so-called "matching" of experimental units according to some superficial characteristics is deceptive, because the characteristics employed for matching and the individual's innate ability to respond to the particular treatments under testing may or may not be associated at all. (In agriculture the fact that fertile soil produces higher yield is known.) Hence, in experiments with man, the very first consideration is the nature of the individuals available for the treatments.

On many occasions, according to my personal experience, not enough attention has been paid to proper randomization in human experiments. Randomization has been accepted by physicians (not without reluctance) as a valid procedure in clinical trials only recently. The art and technique of experimentation with man are still very young. At the present stage of development, the medical scientist should use simple designs with proper randomization and a sufficient number of replications, rather than delve into complicated confounding arrangements which not only introduce difficulties in calculation (which can be taken care of by a little extra training) but also introduce uncertainties in interpretation and understanding. In fact, if I may say so, many of the human experiments are already confounded with—sometimes known and sometimes unknown—other factors which impair the validity of the conclusion. Certainly we do not want to complicate the matter by introducing confounding of experimental factors.

On the other hand, there are good arguments for confounding designs in human experiments, precisely because of the limited uniformity of experimental individuals. In such a case, some a priori knowledge (or judgment, a poor substitute for knowledge) of the nature of the factors is needed to decide just which interactions are comparatively unimportant and can be sacrificed in the interest of small blocks.  As to the detailed design and analysis, the medical scientist will do well to consult a statistician at the very beginning of the planning stage of the experiment.   It is at this stage the statistician can contribute most to the experimental investigation.   Sometimes, it is advisable (albeit time-consuming) that the experimenter and statistician should have a "dry run" together—analyzing a set (or sets) of hypothetical data with the particular design agreed upon—to see if the information and conclusion are the kind the investigator really wants.   If they are not, some other design should be adopted.

Since many of the confounding designs for large numbers of treatments are not easily applicable to human experiments, and yet small blocks are of advantage even for a moderate number of treatments, we later give (especially in Chaps. 27 and 28) some general designs involving incomplete blocks, which are readily applicable to any type of experiment.

### EXERCISES

#### 24.1   Combining two experiments

One of the subjects that has not been developed in this book is the combined analysis of a series of experiments.   In order to combine the experimental results, the experiments must be done under very similar conditions and the error variance should be of the same order of magnitude in all the experiments.   We shall consider a very simple case here, using the data of Table 24.1 to illustrate the principle of joint analysis and its relationship with confounded experiments.

Suppose that four treatments (1, 2, 3, 4) are tested in two hospitals (VA and VB), each replicated three times, and that the data are as given in the accompanying tables.

|  | At VA Hospital | | | | Block total |  | At VB Hospital | | | | Block total |
|---|---|---|---|---|---|---|---|---|---|---|---|
|  | (1) | (2) | (3) | (4) |  |  | (1) | (2) | (3) | (4) |  |
| I | 30 | 16 | 24 | 19 | 89 |  | 28 | 16 | 27 | 8 | 79 |
| II | 23 | 16 | 28 | 16 | 83 |  | 18 | 25 | 16 | 10 | 69 |
| III | 25 | 19 | 20 | 16 | 80 |  | 23 | 22 | 17 | 18 | 80 |
| Treatment total | 78 | 51 | 72 | 51 | 252 |  | 69 | 63 | 60 | 36 | 228 |

If we ignore the hospitals, the two experiments may be taken as one with four treatments and *six* blocks, and therefore it may be analyzed by the simple method of randomized blocks into the following components:

|            | df | ssq |
|------------|----|-----|
| Blocks     | 5  | 53  |
| Treatments | 3  | 333 |
| "Error"    | 15 | 338 |
| Total      | 23 | 724 |

Further analysis concerns the subdivision of the blocks *ssq* into two components—one between the two hospitals (or in general, localities) and one within the hospitals. This is done the usual way and needs no explanation. The reader, however, should check the numerical results listed in Table 24.1.

The comparatively new feature in the analysis is the subdivision of the "error" *ssq* with 15 *df*. This includes two types of interactions; one is that for "treatments × hospitals" and the other is that for "treatments × blocks" (within hospitals). The latter is what we call "error" in the usual analysis of variance. The treatment × hospital interaction may be obtained from the accompanying tabulation of totals.

|      | (1) | (2) | (3) | (4) |     |
|------|-----|-----|-----|-----|-----|
| VA   | 78  | 51  | 72  | 51  | 252 |
| VB   | 69  | 63  | 60  | 36  | 228 |
|      | 147 | 114 | 132 | 87  | 480 |

The *ssq*'s between the hospitals and between the treatments are calculated the same way. The total *ssq* for the body of the table (eight numbers) is 432. Then the interaction *ssq* is $432 - 24 - 333 = 75$ with 3 *df*. These results have been given in Table 24.8. Check every component and compare with Table 24.1 and see where the similarities and differences are. In particular, note that $333 + 75 = 198 + 210 = 408$. What does this mean?

**Table 24.8　Combined analysis of two experiments**

| df | Components | df | ssq | Subcomponents | df | ssq |
|----|------------|----|-----|---------------|----|-----|
| 5  | { Blocks   | 4  | 29  | { VA blocks    | 2  | 10.5 |
|    |            |    |     | { VB blocks    | 2  | 18.5 |
|    | { Hospitals | 1 | 24  | Hospitals      | 1  | 24.0 |
| 3  | Treatments | 3  | 333 | Treatments     | 3  | 333.0 |
| 15 | { Treatment × hosp. | 3 | 75 | Treatment × hosp. | 3 | 75.0 |
|    | { Error (treat. × blocks) | 12 | 263 | { VA error | 6 | 59.5 |
|    |            |    |     | { VB error     | 6  | 203.5 |
| 23 | Total      | 23 | 724 |               | 23 | 724.0 |

**24.2**　The treatment totals (Table 24.6) may be arranged in any systematic order, and the successive *I-J* diagonal-sum method will yield the same final totals, from which the *ssq* for *ABC* components are calculated. For example, the 27 treatment combinations may be arranged as in the table on top of page 336. Then the nine *I* and

| | $c_0$ | | | $c_1$ | | | $c_2$ | | |
|---|---|---|---|---|---|---|---|---|---|
| | $b_0$ | $b_1$ | $b_2$ | $b_0$ | $b_1$ | $b_2$ | $b_0$ | $b_1$ | $b_2$ |
| $a_0$ | 11 | 20 | 24 | 14 | 28 | 26 | 11 | 21 | 28 |
| $a_1$ | 13 | 23 | 24 | 16 | 31 | 18 | 10 | 18 | 33 |
| $a_2$ | 8 | 13 | 8 | 30 | 24 | 20 | 28 | 14 | 26 |

$J$ totals for the three groups of nine treatments are as tabulated. Note that these final totals are the same as those shown in Table 24.6, but the order of appearance has changed.

| $I$ totals | | |
|---|---|---|
| 42 | 65 | 55 |
| 50 | 66 | 52 |
| 52 | 76 | 82 |

| $J$ totals | | |
|---|---|---|
| 55 | 87 | 74 |
| 48 | 56 | 58 |
| 41 | 64 | 57 |

| $I$ totals | $J$ totals |
|---|---|
| 190 | 173 |
| 181 | 170 |
| 169 | 197 |

| $I$ totals | $J$ totals |
|---|---|
| 168 | 171 |
| 186 | 177 |
| 186 | 192 |

### 24.3  Confounding in 3 $\times$ 2 $\times$ 2 with blocks of six units

In each replication the twelve treatments are divided into two blocks of six each according to the setup shown in Table 24.7. The six treatments in each block are then randomized (that is, assigned to the units at random). Three complete replications make the design balanced and the analysis comparatively easy. The data are arranged in Table 24.9 by adopting the same observed values of treatments as in Table 23.6. Now, the sum of squares due to the six blocks is

$$\frac{121^2 + 140^2 + \cdots + 121^2}{6} - \frac{720^2}{36} = 14{,}504 - 14{,}440 = 104 \qquad \text{with 5 } df$$

The *ssq* due to the 12 treatments has 11 *df*. The adopted design conserves all the main effects ($A$, $B$, $C$) and interactions $AB$ and $AC$. Hence the *ssq*'s corresponding to these effects remain the same as those in orthogonal experiments (Table 23.7 or 23.8, lower portion). Only the interactions $BC$ and $ABC$ have been partially confounded with the blocks and need special adjustment and new calculation. It will be recalled that the *ssq*'s due to these two effects in an unconfounded experiment are calculated from the following values (Table 23.8, lower right):

| | $a_0$ | $a_1$ | $a_2$ | Total | |
|---|---|---|---|---|---|
| | . | . | . | . | $Y$ |
| | . | . | . | . | $B$ |
| | . | . | . | . | $C$ |
| | $-8$ | $-2$ | $-6$ | $-16$ | $BC$ |

*Table 24.9* Data of a 3 × 2 × 2 factorial in blocks of six units each, according to plan shown in Table 24.7, lower portion (data rearranged from Table 23.6)

|  | Rep. I | | | | Rep. II | | | | Rep. III | | | |
|---|---|---|---|---|---|---|---|---|---|---|---|---|
|  | $b_0$ | | $b_1$ | | $b_0$ | | $b_1$ | | $b_0$ | | $b_1$ | |
| $a_0$ | (0) | 12 | (1) | 14 | (1) | 18 | (0) | 26 | (1) | 10 | (0) | 24 |
| $a_1$ | (1) | 13 | (0) | 28 | (0) | 11 | (1) | 30 | (1) | 14 | (0) | 18 |
| $a_2$ | (1) | 33 | (0) | 21 | (1) | 28 | (0) | 27 | (0) | 24 | (1) | 17 |
|  | | 121 | | | | 140 | | | | 107 | | |
| $a_0$ | (1) | 11 | (0) | 8 | (0) | 8 | (1) | 16 | (0) | 19 | (1) | 20 |
| $a_1$ | (0) | 13 | (1) | 24 | (1) | 22 | (0) | 20 | (0) | 17 | (1) | 18 |
| $a_2$ | (0) | 26 | (1) | 31 | (0) | 28 | (1) | 24 | (1) | 23 | (0) | 24 |
|  | | 113 | | | | 118 | | | | 121 | | |

Analysis of variance

| Source | | df | ssq | msq |
|---|---|---|---|---|
| Blocks | | 5 | 104.00 | |
| Table 23.7 | $A$ | 2 | 618.00 | 309.00 |
| | $B$ | 1 | 100.00 | 100.00 |
| | $AB$ | 2 | 194.00 | 97.00 |
| | $C$ | 1 | 4.00 | 4.00 |
| | $AC$ | 2 | 20.67 | 10.33 |
| New | $BC$ | 1 | 3.56 | 3.56 |
| | $ABC$ | 2 | 34.44 | 17.22 |
| Error | | 19 | 549.33 | 28.91 |
| Total | | 35 | 1,628.00 | |

We need only adjust the value $-16$ for $BC$ and the values $-8$, $-2$, $-6$ for $ABC$. The method of adjustment is as follows: Let the block differences (upper minus lower) in each replication be

$$g_1 = 121 - 113 = 8 \qquad g_2 = 140 - 118 = 22 \qquad g_3 = 107 - 121 = -14$$

Then the adjusted value of $BC$ is

$$\overline{BC} = BC + \tfrac{1}{3}(g_1 + g_2 + g_3) = -16 + \frac{8 + 22 - 14}{3} = -10.7$$

or

$$3\overline{BC} = 3BC + g_1 + g_2 + g_3 = 3(-16) + 8 + 22 - 14 = -32$$

The corresponding sum of squares is

$$\frac{(\overline{BC})^2}{32} = \frac{(3\overline{BC})^2}{32 \times 9} = \frac{(-32)^2}{288} = 3.56$$

The adjustments for $ABC$ (originally based on $-8$, $-2$, $-6$) are

$$\left.\begin{array}{l}3\overline{a_0BC} = 3a_0BC - g_1 + g_2 + g_3 = 3(-8) - 8 + 22 - 14 = -24 \\ 3\overline{a_1BC} = 3a_1BC + g_1 - g_2 + g_3 = 3(-2) + 8 - 22 - 14 = -34 \\ 3\overline{a_2BC} = 3a_2BC + g_1 + g_2 - g_3 = 3(-6) + 8 + 22 + 14 = +26\end{array}\right\} -32$$

A check in arithmetic is provided by the fact that the sum of these three adjusted numbers is equal to $3\overline{BC} = -32$. The *ssq* due to the adjusted $ABC$ is one-sixtyth of the sum of squares of these three numbers; viz.,

$$\tfrac{1}{60}\left(24^2 + 34^2 + 26^2 - \frac{32^2}{3}\right) = \frac{2066.67}{60} = 34.44 \qquad \text{with 2 } df$$

Replacing the *ssq*'s for $BC$ and $ABC$ in Table 23.7 by these new *ssq*'s for adjusted $BC$ and $ABC$, we obtain the analysis of variance in the lower portion of Table 24.9.

### 24.4 Adjusted treatment totals (or means)

All the marginal totals for single factors and the two-way $a \times b$ and $a \times c$ tables in Table 23.7 are unaffected by the confounding design. When divided by the appropriate number of observations that goes into the total, we obtain the unbiased estimates of the mean values of the treatments. Since the interaction $BC$ has been partially confounded, the $b \times c$ table needs adjustment which is based on the adjusted $\overline{BC} = -10.7$. Adopting this value, we use the reverse computation (Exercise 21.3) to obtain treatment totals from main effects and interactions, as shown in the accompanying table. The treatment total is one-fourth of the preceding column. For

| Effects | | Pair sum and difference | | Treatment total | | |
|---|---|---|---|---|---|---|
| | | | | Adjusted | | (Table 23.8) |
| $\overline{BC}$ | $-10.7$ | 1.3 | 781.3 | 195.3 | $bc$ | 194 |
| $C$ | 12 | 780.0 | 682.7 | 170.7 | $c$ | 172 |
| $B$ | 60 | 22.7 | 778.7 | 194.7 | $b$ | 196 |
| $Y$ | 720 | 660.0 | 637.3 | 159.3 | (1) | 158 |

example, $\tfrac{1}{4}(781.3) = 195.3$, etc. It is seen that the adjusted treatment totals for $b \times c$ combinations are very close to the uncorrected ones. Note that the marginal totals for each single factor remain the same, corrected or uncorrected. The treatment totals for the three-way table $a \times b \times c$ also need correction, because the triple interaction $ABC$ has been partially confounded with the blocks. In practice, however, it is hardly necessary to correct them, because the correction offers little help to interpretation.

### 24.5 The loss of information due to confounding

Confounding implies the loss of certain information about the treatments. For factorials of the type $2^n$ with blocks of size $2^{n-1}$ or $2^{n-2}$ we allocate the treatment combinations into blocks in such a way that certain contrasts (but no others) are confounded with blocks.   In any one replication, the confounding is clear-cut.   The same is true in the confounding designs of 3 × 3 × 3 in blocks of 9.   Here, in any one replication, certain two degrees of freedom of $ABC$, but not others, are confounded with the blocks.   The situation is not so clear-cut for the mixed series, and this is easily seen from the example of 3 × 2 × 2.   When two blocks of six treatments each make up one complete replication (Table 24.7), it implies that one degree of freedom has been confounded with the blocks; but what contrast does this 1 $df$ represent?   It does not represent any one at all in the sense we usually use the term.   It turns out that partially $BC$ and partially $ABC$ have been confounded.   How much of each, then?   It may be shown (Yates, 1937) that in a balanced situation the loss of information on $BC$ (1 $df$) is $\frac{1}{9}$ and the loss on $ABC$ (2 $df$) is $\frac{4}{9}$.   It is important to note that

$$1 \times \tfrac{1}{9} + 2 \times \tfrac{4}{9} = 1$$

which is exactly the 1 $df$ that has been confounded with the blocks.   We usually think that a single degree of freedom is indivisible.   The confounding design for mixed series, however, leads to something close to the "fission" of a single degree of freedom!

# 25

# Split-plot design

Another type of design generally applicable to field trials as well as to laboratory experiments is called the split-plot design. It may be regarded as a confounding factorial design in one respect, and it resembles replicated hierarchial classifications in another.

## 1. The design

We shall first explain the design in terms of field trials, because that way it is easiest to comprehend. Suppose that three treatments ($A_1$, $A_2$, $A_3$, not necessarily three levels of an ingredient) are to be tested in randomized blocks as shown in Fig. 25.1, and each treatment is assigned to a plot of land (unit) at random within each block. For certain types of treatment (for example, method of plowing or irrigation), the plot of land cannot be too small. The comparatively large plot of land is the unit of experimentation with respect to the treatments $A$'s.

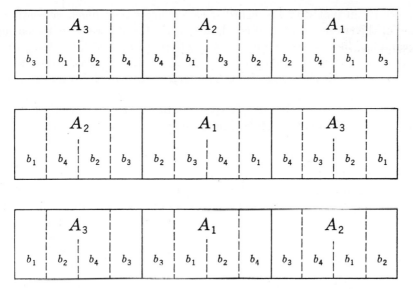

Fig. 25.1  A split-plot design with three treatments ($A_1$, $A_2$, $A_3$) for the whole units and four treatments ($b_1$, $b_2$, $b_3$, $b_4$) for the subunits within each whole unit.  Three replications are shown.

Suppose that at the same time there are four other treatments ($b_1$, $b_2$, $b_3$, $b_4$, not necessarily four levels of an ingredient) to be tested, each in combination with each of the $A$'s, so that there are $3 \times 4 = 12$ treatment combinations altogether.  In order to accommodate the $b$'s, each original plot (for the $A$'s) may be split into four subplots (subunits), and the four $b$'s are to be assigned to these subplots at random, as shown in Fig. 25.1. Hence the name "split-plot."

This type of design is to be clearly distinguished from the ordinary randomized blocks.  In the latter case, the 12 treatment combinations are assigned to the 12 units in a block at random.  In split-plot design, the $A$ treatments are assigned first, and then within each $A$, the four $b$'s are randomly assigned.  It is this crucial difference that leads to an entirely different method of analysis.

The plot (unit) and subplot (subunit) are merely historical terms to describe the procedure of allocating the treatment factors.  If we "promote" a subplot to the "rank" of a full plot, etc., we have the following terminology:

Subplot (subunit):     plot (unit)
Whole plot (unit):     block (incomplete)
Block:                 replication

Hence a full plot may be regarded as an incomplete block consisting of four of the twelve treatment combinations. The difference between such incomplete blocks also represents the difference between the $A$'s. From this point of view, the split-plot design may be described as a confounding system in which the main effects of the $A$'s have been confounded with the incomplete blocks.

In industrial and laboratory experiments, the treatments that cannot be administered in small scale should be the $A$'s for whole units, and the treatments that can be conveniently applied in small scale should be the $b$'s for the subunits. This aspect of the design is dictated by mechanical necessity rather than by choice. However, when there is a choice, the more important treatments should be the $b$'s for the subunits and the treatments of secondary importance should be the $A$'s for the whole units. The reason for this will be made clear by the method of analysis.

## 2. The analysis

In part 1 of Table 25.1 are listed the observations of a split-plot experiment. The analysis is best done in two stages. First of all, the whole unit totals are obtained, as shown in part 2. For example, $12 + 8 + 11 + 14 = 45$ for the first whole unit that receives treatment $A_1$. The two stages of

*Table* 25.1   Data from a split-plot design. $A_1$, $A_2$, $A_3$ are treatments for the whole units and $b_1$, $b_2$, $b_3$, $b_4$ are treatments for the subunits

Part 1   Observations of each subunit

| Whole unit: | $A_1$ | | | | $A_2$ | | | | $A_3$ | | | | Replication total |
|---|---|---|---|---|---|---|---|---|---|---|---|---|---|
| Subunit: | $b_1$ | $b_2$ | $b_3$ | $b_4$ | $b_1$ | $b_2$ | $b_3$ | $b_4$ | $b_1$ | $b_2$ | $b_3$ | $b_4$ | |
| I | 12 | 8 | 11 | 14 | 13 | 28 | 13 | 24 | 26 | 21 | 33 | 31 | 234 |
| II | 8 | 26 | 18 | 16 | 11 | 20 | 22 | 30 | 28 | 27 | 28 | 24 | 258 |
| III | 19 | 24 | 10 | 20 | 17 | 18 | 14 | 18 | 24 | 24 | 23 | 17 | 228 |
| Treatment total | 39 | 58 | 39 | 50 | 41 | 66 | 49 | 72 | 78 | 72 | 84 | 72 | 720 |

Part 2   Whole unit totals

| | $A_1$ | $A_2$ | $A_3$ | |
|---|---|---|---|---|
| I | 45 | 78 | 111 | 234 |
| II | 68 | 83 | 107 | 258 |
| III | 73 | 67 | 88 | 228 |
| $A$ totals | 186 | 228 | 306 | 720 |

Part 3   Treatment totals

| | $A_1$ | $A_2$ | $A_3$ | $b$ totals |
|---|---|---|---|---|
| $b_1$ | 39 | 41 | 78 | 158 |
| $b_2$ | 58 | 66 | 72 | 196 |
| $b_3$ | 39 | 49 | 84 | 172 |
| $b_4$ | 50 | 72 | 72 | 194 |
| $A$ totals | 186 | 228 | 306 | 720 |

analysis are based on a simple one-way classification: (1) *ssq*'s between whole units and (2) *ssq*'s within whole units. For this preliminary subdivision, we need the following three basic quantities:

$$A = 12^2 + 8^2 + \cdots + 23^2 + 17^2 = 16,028.00$$
$$B = \tfrac{1}{4}(45^2 + 78^2 + \cdots + 88^2) = 15,238.50$$
$$C = 720^2/36 = 14,400.00$$

The preliminary subdivisions are thus as tabulated here. The rest of the analysis consists of further subdivision of each of the two major components shown.

| Components | df | ssq |
|---|---|---|
| Between whole units | 8 | $B - C = $ 838.50 |
| Within whole units | 27 | $A - B = $ 789.50 |
| Total | 35 | $A - C = 1,628.00$ |

The whole unit totals, each being $b_1 + b_2 + b_3 + b_4$, contain no information on the $b$ treatments, but they show the replication differences as well as the main effects of the $A$'s. Thus, from part 2 of Table 25.1 we calculate the two additional quantities:

$$R = \tfrac{1}{12}(234^2 + 258^2 + 228^2) = 14,442.00$$
$$U = \tfrac{1}{12}(186^2 + 228^2 + 306^2) = 15,018.00$$

so that $R - C$ and $U - C$ are the sums of squares for replications and whole-unit treatments, respectively. The error *ssq* for between whole units is obtained by subtraction (or $B - R - U + C$). This completes the analysis for whole units as shown in the upper portion of Table 25.2.

*Table* 25.2  Analysis of a split-plot design experiment (data from Table 25.1)

| | Source | df | ssq | | msq |
|---|---|---|---|---|---|
| Between whole units | Replications | 2 | $R - C = $ | 42.00 | |
| | A treatments | 2 | $U - C = $ | 618.00 | 309.0 |
| | Error (whole) | 4 | (subt) | 178.50 | $44.6 = S^2$ |
| | Total | 8 | $B - C = $ | 838.50 | |
| Within whole units | b treatments | 3 | $V - C = $ | 111.11 | 37.0 |
| | Ab interactions | 6 | $T - U - V + C = $ | 216.22 | 36.0 |
| | Error (subunits) | 18 | (subt) | 462.17 | $25.7 = s^2$ |
| | Total | 27 | $A - B = $ | 789.50 | |
| | Grand total | 35 | $A - C = 1,628.00$ | | |

The information on the $b$ treatments and their interaction with the $A$'s can be sought only from a subdivision of the within-whole-unit sum of squares. The treatment totals (bottom line of part 1, Table 25.1) have been rearranged into the form of a two-way table in part 3, from which another two quantities may be calculated:

$$T = \tfrac{1}{3}(39^2 + 41^2 + \cdots + 72^2) = 15{,}345.33$$
$$V = \tfrac{1}{9}(158^2 + \cdots + 194^2) = 14{,}511.11$$

so that $T - C$ is the *ssq* for all treatments, $V - C$ is the *ssq* for the main effects of the $b$ treatments, and the remainder

$$(T - C) - (U - C) - (V - C) = T - U - V + C$$

is the *ssq* for interaction between the $A$'s and $b$'s. Again, the error *ssq* for the subunit treatments is obtained by subtraction, as shown in the lower portion of Table 25.2. The significance of the $A$'s is tested against $S^2 = 44.6$, and the significance of the $b$'s and interactions is tested against $s^2 = 25.7$. This completes the numerical procedure of the analysis. If desirable, the components for $A$ treatments, $b$ treatments, and their interactions may be further subdivided by the methods described in earlier chapters.

### 3. Comparison with randomized blocks

The most characteristic feature of the split-plot design is that there are two error variances: one for the whole-plot treatments and one for the subplot treatments and interactions. Ordinarily, the subunits within a whole unit are more homogeneous than the whole units themselves. Therefore, the error variance for whole units ($S^2 = 44.6$) is usually larger than the error variance for subunits ($s^2 = 25.7$), as our example shows. If the heterogeneity within and that between whole units are about the same, the two error variances should also be about the same.

If the twelve ($3 \times 4$) treatments are tested in ordinary randomized blocks, parts 1 and 3 of Table 25.1 will remain the same, and hence the *ssq* for replications and various components of the treatments will remain the same, but there will be no part 2 and hence no separate components for between and within whole units. It follows that the sum of the two error *ssq*'s in Table 25.2 must be the error *ssq* for an ordinary randomized

| Error | df | ssq | msq |
|---|---|---|---|
| Whole unit | 4 | 178.50 | 44.6 |
| Subunits | 18 | 462.17 | 25.7 |
| Randomized blocks | 22 | 640.67 | 29.1 |

blocks experiment. The error variance for the ordinary randomized blocks design is the average value of $S^2$ and $s^2$ for the split-plot design. Note that $29.1 = 640.67/22$ is exactly the error variance obtained in Table 23.6.

The fact that $S^2 > s^2$ for a split-plot design justifies our previous statement that, whenever possible, the more important treatments should be given to the subunits which have a smaller error variance and may be compared with greater accuracy.

When the heterogeneities within and between whole units are of about the same magnitude, the ordinary randomized blocks and split-plot designs will yield the same error variance and thus the same conclusions about treatment significance.

### 4. The general subdivision

Let $r$ be the number of replications, $a$ the number of treatments applied to the whole units, and $b$ the number of treatments applied to the subunits within each whole unit. The number of observations in each replication is $ab$, and there are $rab$ observations altogether. The total number of whole units is $ra$. Hence the analysis takes the general form of Table 25.3.

The split-plot technique may be applied to Latin squares also. The $a \times a$ Latin square is with respect to the whole-unit treatments. Each whole unit is then subdivided into $b$ subunits. The analysis is very similar to that for randomized blocks. The first stage is an analysis of the $a^2$ whole units, and the second stage is an analysis for within whole units. More details may be found in Exercise 25.4.

### 5. Variance of mean difference

The number of replications for each treatment is counted on the subunit basis. For an $A$ treatment, the error variance is $S^2$ and for a $b$ treatment

*Table 25.3*  General analysis of a split-plot design

| Major source | df | Components | df |
|---|---|---|---|
| Between whole units | $ra - 1$ | Replications<br>$A$ treatments<br>Error (whole) | $r - 1$<br>$a - 1$<br>$(r - 1)(a - 1)$ |
| Within whole units | $ra(b - 1)$ | $b$ treatments<br>Interactions<br>Error (subunits) | $b - 1$<br>$(a - 1)(b - 1)$<br>$(r - 1)a(b - 1)$ |
| Total | $rab - 1$ | Total | $rab - 1$ |

the error variance is $s^2$, as shown in Table 25.2. Hence the variance of the difference between two $A$ means and that between two $b$ means are respectively

$$V(A_1 - A_2) = \frac{2S^2}{rb}$$

$$V(b_1 - b_2) = \frac{2s^2}{ra}$$

If we consider two $b$ treatments in combination with a specified $A$ treatment, then there are only $r$ replications; hence,

$$V(A_1 b_1 - A_1 b_2) = \frac{2s^2}{r}$$

When two $A$ means in combination with a specified $b$ treatment are compared, the comparison involves both the main effect of $A$ and the interaction $Ab$. The error variance is a weighted mean of $S^2$ and $s^2$; thus,

$$V(A_1 b_1 - A_2 b_1) = \frac{2}{r}\left(\frac{S^2 + (b-1)s^2}{b}\right)$$

Finally, for the difference between two $A$ means in combination with different $b$ treatments, the variance formula above still applies, although the statistic $V(A_1 b_1 - A_2 b_3)$ does not exactly follow the $t$ distribution. For more details, see Cochran and Cox (1957).

### 6. Missing observation

Since the split-plot design is analogous to a complete confounding design in its structure, the method of estimating a missing observation is equally simple. Suppose that an observation $(y)$ is missing from replication II that is supposed to receive the $A_2 b_2$ treatment. The relevant data in Table 25.1 would be as shown in the accompanying table. The missing

|     | $A_2$ |   |   |   | |
|-----|-------|-------|-------|-------|-------|
|     | $b_1$ | $b_2$ | $b_3$ | $b_4$ | |
| I   | 13 | 28 | 13 | 24 | |
| II  | 11 | $y$ | 22 | 30 | $63 + y$ |
| III | 17 | 18 | 14 | 18 | |
|     | | $46 + y$ | | | $208 + y$ |

observation has no effect on treatments involving $A_1$ or $A_3$. As far as the data shown above are concerned, the design is the ordinary randomized blocks type with respect to the $b$'s (under a given $A$ treatment). Applying the estimation formula in Chap. 19, Sec. 1, we have

$$y = \frac{4(46) + 3(63) - 208}{(4 - 1)(3 - 1)} = \frac{165}{6} = 27.5$$

One degree of freedom should be subtracted from that for $s^2$, the error variance for the subunits.

If a second observation is missing and it belongs to another $A$ treatment, the same formula for a single missing value applies. Only when both missing values belong to the same $A$ treatment do we apply the formulas for two missing values or resort to iteration with one formula.

## 7. Sub-subunits

Analogously to the hierarchical classifications, each of the subunits may be further subdivided into a number of sub-subunits, to which another set of treatments ($c_1, c_2, \ldots$) may be assigned at random. The analysis follows the same general procedure. There will be three error variances: $S^2$ for the $A$ treatments of the whole units, $s^2$ for the $b$ treatments of the subunits, and $s_c^2$ for the $c$ treatments of the sub-subunits. Again, the three error $df$'s and $ssq$'s will add up to the $df$ and $ssq$ for the single error term if the experiment is conducted in an ordinary randomized blocks design of $abc$ units each.

### EXERCISES

**25.1** A split-plot experiment is done according to the plan shown in Fig. 25.2. $W, X, Y, Z$ are the treatments applied to the whole units, and $a, b, c$ are three treatments applied to the subunits. Study the general method of analysis outlined in Sec. 4 and compare the following subdivision with Table 25.2.

|  |  | $df$ |
|---|---|---|
| Between whole units | Replications | 2 |
|  | $W, X, Y, Z$ | 3 |
|  | Error (whole) | 6 |
|  | Total | 11 |
| Within whole units | $a, b, c$ | 2 |
|  | Interactions | 6 |
|  | Error (subunits) | 16 |
|  | Total | 24 |
|  | Grand total | 35 |

Note that the error $df$ $6 + 16 = 22 = 4 + 18$

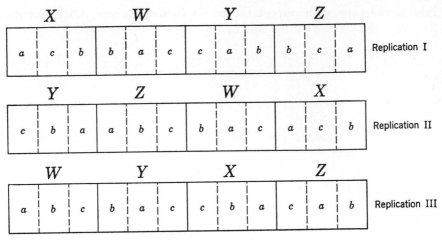

**Fig. 25.2** A split-plot design with four treatments for the whole units and three treatments for the subunits.

**25.2** Regarding the number of degrees of freedom for $S^2$ and $s^2$, show that

$$(r - 1)(a - 1) + (r - 1)a(b - 1) = (r - 1)(ab - 1)$$

If the $a \times b$ treatments were tested in ordinary randomized blocks, what would be the *df* for the error variance?

**25.3** The error sum of squares for whole units is

$$(B - C) - (R - C) - (U - C) = B - R - U + C$$

The error sum of squares for subunits is

$$(A - B) - (V - C) - (T - U - V + C) = A - B - T + U$$

Show that the sum of these two sums of squares is the error *ssq* for an ordinary randomized blocks experiment $(A - R - T + C)$.

**25.4** General analysis of an $a \times a$ Latin square, each whole unit consisting of $b$ subunits.

| Between whole units: $a^2 - 1$ | | |
|---|---|---|
| | Rows | $a - 1$ |
| | Columns | $a - 1$ |
| | $A$ treatments | $a - 1$ |
| | Error (whole) | $(a - 1)(a - 2)$ |
| Within whole units: $a^2(b - 1)$ | $b$ treatments | $b - 1$ |
| | Interactions | $(a - 1)(b - 1)$ |
| | Error (subunits) | $a(a - 1)(b - 1)$ |
| Total | $a^2b - 1$ | |
| | Total | $a^2b - 1$ |

**25.5** Suppose that six treatments ($A_1, \ldots, A_6$) are assigned to six units at random and then each unit is split into two halves for testing treatments $b_1$ and $b_2$.

There are three replications altogether and the data are as shown in the first of the

| | $A_1$ | | $A_2$ | | $A_3$ | | $A_4$ | | $A_5$ | | $A_6$ | |
|---|---|---|---|---|---|---|---|---|---|---|---|---|
| | $b_1$ | $b_2$ | $b_1$ | $b_2$ | $b_1$ | $b_2$ | $b_1$ | $b_2$ | $b_1$ | $b_2$ | $b_1$ | $b_2$ |
| I | 12 | 8 | 11 | 14 | 13 | 28 | 13 | 24 | 26 | 21 | 33 | 31 |
| II | 8 | 26 | 18 | 16 | 11 | 20 | 22 | 30 | 28 | 27 | 28 | 24 |
| III | 19 | 24 | 10 | 20 | 17 | 18 | 14 | 18 | 24 | 24 | 23 | 17 |
| | 39 | 58 | 39 | 50 | 41 | 66 | 49 | 72 | 78 | 72 | 84 | 72 |

Whole unit totals

| | $A_1$ | $A_2$ | $A_3$ | $A_4$ | $A_5$ | $A_6$ | |
|---|---|---|---|---|---|---|---|
| I | 20 | 25 | 41 | 37 | 47 | 64 | 234 |
| II | 34 | 34 | 31 | 52 | 55 | 52 | 258 |
| III | 43 | 30 | 35 | 32 | 48 | 40 | 228 |
| | 97 | 89 | 107 | 121 | 150 | 156 | 720 |

accompanying tables. The values of $A$, $C$, $R$, $T$ remain the same as in the example of Table 25.1. Why? Calculate the basic quantities $B$, $U$, $V$ and verify the results shown in the accompanying table of analysis of variance.

Analysis of Variance

| Source | df | ssq | Subdivision | df | ssq | msq |
|---|---|---|---|---|---|---|
| Between whole units | 17 | 1,094 | Replications | 2 | 42.00 | |
| | | | $A$ treatments | 5 | 642.67 | 128.5 |
| | | | Error (whole) | 10 | 409.33 | 40.9 |
| Within whole units | 18 | 534 | $b$ treatments | 1 | 100.00 | 100.0 |
| | | | Interactions | 5 | 202.67 | 40.5 |
| | | | Error (subunits) | 12 | 231.33 | 19.3 |
| Total | 35 | 1,628 | | 35 | 1,628.00 | |

# 26

# Incomplete Latin square

Up to now we have been dealing with orthogonal designs or designs with simple confounding, the analysis of which is quite straightforward, because the sum of squares due to various factors (or classifications) may be subdivided into clear-cut components and in a very easy manner. When the design is not orthogonal, the simple procedures of calculation we have learned breaks down, because the sum of squares can no longer be subdivided into clear-cut components. The reason for this will be made clear by studying a few examples. Whenever possible, especially in small experiments involving a few treatments, the experimenter is strongly advised to use simple orthogonal designs. However, there are circumstances that cannot be helped. Fortunately, not all unorthogonal designs are complicated. Since some of them are highly economical in experimental units and will prove very useful under a wide variety of conditions, we shall in this and the next two chapters present some of the most elementary types of unorthogonal designs which require only one or two extra steps of arithmetic in analysis. These designs are all originally due to F. Yates (1937). Since then, many other sophisticated designs have been developed by various authors, but we shall not deal with them in this volume.

## 1. Latin square with one row missing

Let us consider a $5 \times 5$ Latin-square experiment in which the five rows, say, represent the five replications and the columns represent groupings with respect to some other characteristics of the experimental units. Now suppose that one row (replication) is missing, either because of contamination or lack of response, shortage of experimental units or treatment agents, or shortage of time or of funds or because the replication has been done so poorly that the investigator wishes to exclude it in analysis. The latter point will be illustrated by an example in Exercise 26.2. The data in Table 26.1 represent a Latin square with one row missing. Whichever row is missing, there is no loss in generality in calling it the first row. The remaining data consist of five columns and four rows with five treatments. Our problem is how to analyze the data appropriately.

As a very crude approximation one may regard the incomplete square as a randomized block experiment, because each row still contains all five treatments. In doing so, however, the remaining restrictions imposed by

*Table* 26.1   A Latin square with one row missing

| (3) | (4) | (5) | (2) | (1) | Row total $R_i$ |
|-----|-----|-----|-----|-----|-----|
| | | | missing | | |
| (1) 9 | (5) 24 | (2) 11 | (4) 24 | (3) 17 | 85 |
| (2) 16 | (3) 23 | (1) 17 | (5) 19 | (4) 25 | 100 |
| (4) 22 | (2) 20 | (3) 20 | (1) 17 | (5) 26 | 105 |
| (5) 13 | (1) 9 | (4) 8 | (3) 20 | (2) 20 | 70 |
| $C_i$   60 | 76 | 56 | 80 | 88 | 360 |

| Treatments | (3) | (4) | (5) | (2) | (1) | Total |
|-----|-----|-----|-----|-----|-----|-----|
| $T_i$ | 80 | 79 | 82 | 67 | 52 | $360 = Y$ |
| $4T_i$ | 320 | 316 | 328 | 268 | 208 | $1{,}440 = 4Y$ |
| $4T_i + C_i$ | 380 | 392 | 384 | 348 | 296 | $1{,}800 = 5Y$ |

the column classification have been ignored and the resulting experimental error will thus be inflated. Now it is clear that an incomplete Latin square, though less accurate than a complete square, is yet something more than the randomized blocks. It may be regarded as an intermediate type of design between a true Latin square and an ordinary randomized block. Hence, an incomplete Latin square may be employed as a good experimental design in its own right. It does not necessarily arise from missing observations.

The reader may recall (Chap. 17) that a Latin square is an arrangement in which three classifications (rows, columns, treatments) are mutually orthogonal. When one row is missing, the remaining rows are still orthogonal to columns as well as to treatments, because each row still contains one unit from each column as well as one observation for each treatment. Hence the sum of squares due to rows is to be calculated in the usual manner. The orthogonality between columns and treatments has been destroyed by the missing row, however. In the $4 \times 5$ arrangement of Table 26.1, we see that each column no longer contains all treatments, nor does each treatment appear in all columns. It is due to this kind of entanglement that the *ssq* due to columns and treatments cannot be calculated in the ordinary way. In the following section we shall give only the arithmetic procedure of analysis with some justification based on intuition, and we shall leave the detailed explanations to later sections.

## 2. Procedure of analysis

It is best to do the easy and routine part of the calculation first. Let us first ignore the fact that columns are no longer orthogonal to treatments and calculate the following basic quantities in the usual way. Referring to Table 26.1 we have

$$A = 9^2 + 24^2 + \cdots = 7{,}066 \qquad R = \tfrac{1}{5}(85^2 + \cdots) = 6{,}630$$
$$C = \frac{360^2}{20} = 6{,}480 \qquad K = \tfrac{1}{4}(60^2 + \cdots) = 6{,}664$$

from which the *ssq*'s for total $(A - C)$, rows $(R - C)$, and columns $(K - C)$ may be obtained. The only new feature involved is in the calculation of the treatment *ssq*. Consider treatment (1), which appears in all columns except the last. The observed total for treatment (1) is

$$\begin{array}{ccccccccc} & Col.\ 1 & & Col.\ 2 & & Col.\ 3 & & Col.\ 4 \\ T_1 = & 9 & + & 9 & + & 17 & + & 17 & = 52 \end{array}$$

This is clearly a biased estimate of the true merit of treatment (1), because not all columns contributed to this total. If column 5 were a better column and if the treatment appeared in that column, the treatment total would be higher than that observed. In other words, a good column 5 would lead to an underestimate of treatment (1). Conversely, if column 5 were a poor column, the observed $T_1$ would be an overestimate of the merit of treatment (1). The average per unit value of column 5 is $\frac{1}{4}$ (column total) = $\frac{1}{4}(88)$ = 22. Hence, the "column adjusted" treatment (1) total is $52 + \frac{1}{4}(88) = 52 + 22 = 74$. For convenience and accuracy of arithmetic, we usually use totals instead of mean values obtained by division. Thus, multiplying the value above by 4, we obtain

$$4T_1 + C_5 = 4(52) + 88 = 208 + 88 = 296$$

This is the value shown at the very bottom line of Table 26.1, in which similarly adjusted values for all other treatments are also given. Note that for ease of calculation and to avoid possible confusion, the observed treatment totals have been arranged in the order (3), (4), (5), (2), (1), corresponding to the missing treatment in the columns. The symbol $T_j$ then simply means the total of the treatment that is missing in column $j$. When the treatment totals are arranged this way, the values of $4T_j + C_j$ may be obtained with a minimum risk of error. Thus, for the first column, we have $60 + 320 = 380$ for treatment (3), etc.

The final step is the calculation of the *adjusted* treatment sum of squares. Analogously to the usual procedure, we calculate an area $T^*$ and a correction term $C^*$, the adjusted *ssq* is then equal to $T^* - C^*$. In our example these values are

$$T^* = \frac{\Sigma(4T_j + C_j)^2}{5 \times 4 \times 3} = \frac{380^2 + \cdots + 296^2}{60} = 10{,}904$$

$$C^* = \frac{Y^2}{4 \times 3} = \frac{360^2}{12} = 10{,}800$$

so that the adjusted treatment *ssq* is $T^* - C^* = 104$. The analysis of variance is shown in Table 26.2. The error *ssq* is obtained by subtracting all the isolated components from the total. The significance of treatment

*Table* 26.2 Analysis of variance of an $n \times (n - 1)$ incomplete Latin square (data from Table 26.1, for which $n = 5$)

| Source | df | ssq | msq |
|---|---|---|---|
| Rows, orthogonal | $n - 2 = 3$ | $R - C = 150$ | |
| Columns, raw | $n - 1 = 4$ | $K - C = 184$ | |
| Treatments, adjusted | $n - 1 = 4$ | $T^* - C^* = 104$ | 26 |
| Error | $(n - 1)(n - 3) = 8$ | (subt) $= 148$ | 18.5 |
| Total | $n(n - 1) - 1 = 19$ | $A - C = 586$ | |

$\left.\begin{array}{c} 26 \\ 18.5 \end{array}\right\} F = 1.4$

effects should be tested against the adjusted treatment mean square. In this example, $F = 1.4$; the treatment effects are not significant at the 0.05 level.

### 3. Adjusted treatment means

In orthogonal experiments the treatment mean is simply the treatment total divided by the number of replications. In nonorthogonal experiments this is not the case. In our particular example the unbiased treatment mean must be calculated on the basis of the quantities $4T_j + C_j$ which are free from column effects. It may be shown that

$$\text{Adjusted treatment mean} = \frac{4T_j + C_j}{5 \times 3}$$

where the denominator $5 \times 3$ is $n(n - 2)$ as explained in a later section. Thus, for our example, the adjusted treatment means are as shown in Table 26.3. The variance between any two such adjusted treatment means is

$$\frac{2s^2}{n}\left(\frac{n-1}{n-2}\right) = \frac{2(18.5)}{5}\left(\frac{4}{3}\right) = 9.87$$

where $s^2 = 18.5$ is the error mean square given in Table 26.2. The standard error between two adjusted treatment means is $\sqrt{9.87} = 3.14$. We note that when $n$ is large, the variance given above approaches the usual expression $2s^2/n$ for orthogonal experiments. This completes the numerical analysis of the experiment.

### 4. General expressions

If $n$ is the dimension of the Latin square in which one row has been missing, the total number of observations is $N = n(n - 1)$. The ordinary correction term is $C = Y^2/n(n - 1)$, where $Y$ is the grand total of the $N$ observations. The adjusted treatment total is $(n - 1)T_j + C_j$, where $T_j$ is the observed total of the treatment that is missing in column $j$. Dividing such a quantity by $n(n - 2)$ gives the adjusted treatment mean.

*Table* 26.3   Adjusted treatment means for data in Table 26.1

|  | Treatments | | | | | Mean |
|---|---|---|---|---|---|---|
|  | (1) | (2) | (3) | (4) | (5) |  |
| $4T_j + C_j$ | 296 | 348 | 380 | 392 | 384 | $360 = Y$ |
| Adjusted mean | 19.73 | 23.20 | 25.33 | 26.13 | 25.60 | 24.00 |
| Effects, $t_j$ | $-4.27$ | $-0.80$ | $+1.33$ | $+2.13$ | $+1.60$ | 0 |

*Table 26.4* **Adjusted column totals and effects**

|  | | Treatments | | | | Total |
|---|---|---|---|---|---|---|
|  | (3) | (4) | (5) | (2) | (1) | |
| $C_j$ | 60 | 76 | 56 | 80 | 88 | $360 = Y$ |
| $4C_j + T_j$ | 320 | 383 | 306 | 387 | 404 | $1{,}800 = 5Y$ |
| Effects, $c_j$ | $-2.67$ | $+1.53$ | $-3.60$ | $+1.80$ | $+2.93$ | 0 |

The two new quantities required for calculating the adjusted treatment *ssq* are

$$T^* = \frac{\Sigma[(n-1)T_j + C_j]^2}{n(n-1)(n-2)} \qquad C^* = \frac{Y^2}{(n-1)(n-2)}$$

so that $T^* - C^*$ is the required treatment *ssq*. Test of significance should be made against the adjusted mean square for treatments. The general expressions for number of degrees of freedom are given in Table 26.2.

For a short introductory course or for those who do not care to go much into detail, the numerical procedure outlined above will probably be sufficient to enable handling such situations appropriately and the subsequent sections may be omitted. The following is intended to supply some additional explanations and point out some additional properties of the *ssq* due to columns and treatments.

## 5. Joint *ssq* due to treatments and columns

In the preceding analysis we have left observed column totals as they are and adjusted the treatment totals. Since the relationships between treatments and columns are mathematically symmetrical, we may also leave the observed treatment totals as they are and adjust the column totals by the same procedure through the same argument. Referring to the values of $C_j$ and $T_j$ given in the lower portion of Table 26.1, we see that the *raw* treatment *ssq* is

$$T - C = \frac{80^2 + \cdots + 52^2}{4} - \frac{360^2}{20} = 6{,}639.50 - 6{,}480 = 159.50$$

Now, instead of calculating $4T_j + C_j$, we calculate $4C_j + T_j$, the column totals adjusted for treatment effects. These values are given in Table 26.4.

Hence the adjusted column *ssq* is

$$K^* - C^* = \frac{\Sigma(4C_j + T_j)^2}{5 \times 4 \times 3} - \frac{Y^2}{4 \times 3} = \frac{655{,}710}{60} - \frac{129{,}600}{12} = 128.50$$

*Table 26.5* **Joint** *ssq* **due to treatments and columns**

| In this section | | In Table 26.2 | |
|---|---|---|---|
| Columns, adj. | $K^* - C^* = 128.50$ | Columns, raw | $K - C = 184$ |
| Treatments, raw | $T - C = 159.50$ | Treatments, adj. | $T^* - C^* = 104$ |
| | Joint $ssq = 288.00$ | | Joint $ssq = 288$ |

We summarize these and previous results in Table 26.5. We see that the joint $ssq$ due to columns and treatments remains the same whether we adjust columns and leave treatments as they are or adjust treatments and leave columns as they are. This is a general feature of nonorthogonal experiments in which the number of treatments equals that of blocks. For a test of significance of treatments, we do what we did in Table 26.2, that is, we adjust the treatments and leave the columns alone. The calculations made in this section, although not necessary for practical purposes, provide a thorough check on the accuracy of the arithmetic involved.

## 6. Mathematical notes on linear model

In presenting the arithmetical procedure in the preceding sections we have made a number of statements without rigorous justification. In this section we shall sketch the algebra involved as briefly as possible to see some of the underlying reasons. The starting point is, as usual, the linear model, which is the same as that for the ordinary Latin square (Chap. 17). That is, every observed value is the sum of a general constant $u$, a row effect $r_i$, a column effect $c_j$, a treatment effect $t_k$, and a random error $e$. For our present purpose, the rows being orthogonal to columns and to treatments, the row effects can be eliminated from further consideration, and so can the error terms. We shall concentrate on the treatments and columns which are now nonorthogonal to each other. As before, we still assume that

$$c_1 + \cdots + c_5 = 0 \qquad \text{so that} \qquad c_2 + c_3 + c_4 + c_5 = -c_1$$
$$t_1 + \cdots + t_5 = 0 \qquad \text{so that} \qquad t_1 + t_2 + t_4 + t_5 = -t_3$$

The relevant components of the observed values are listed in Table 26.6. We need only investigate column 1 and treatment (3), which is missing in that column, because other columns and their missing treatments have the same relationship. The totals of column 1 and treatment (3) are obtained by adding the appropriate single observations. Thus,

$$C_1 = 4u + 4c_1 - t_3 \qquad\qquad T_3 = 4u + 4t_3 - c_1$$

The observed column total contains a treatment effect, and the observed treatment total contains a column effect. There are five pairs of equations like this. Using the method of solving two equations with two unknowns, we see that the following expressions do not contain mixed effects:

$$4T_3 + C_1 = 20u + 15t_3 \qquad 4C_1 + T_3 = 20u + 15c_1$$

This is why we use the quantities $4T_j + C_j$ in numerical analysis to free the treatment totals of column effects. To obtain the general expression for an incomplete Latin square of dimension $n$, we need only replace 4 by $(n - 1)$ in the expressions sbove. Thus,

$$(n - 1)T_3 + C_1 = Y + n(n - 2)t_3$$
$$(n - 1)C_1 + T_3 = Y + n(n - 2)c_1$$

Dividing these expressions by $n(n - 2)$, we obtain the adjusted treatment and column means, respectively. The *effects* are estimated by

$$l_3 = \frac{(n - 1)T_3 + C_1 - Y}{n(n - 2)} \qquad c_1 = \frac{(n - 1)C_1 + T_3 - Y}{n(n - 2)}$$

The sum of squares for treatments and columns is a bit complicated in comparison with the usual orthogonal expressions, but it is still manageable. The joint sum of squares due to treatments and columns is

$$ssq(c,t) = c_1 C_1 + \cdots + c_5 C_5 + t_1 T_1 + \cdots + t_5 T_5$$

The column totals $C$ and treatment totals $T$ are those given in Table 26.1.

**Table 26.6** Linear components of each observation in an $n \times (n - 1)$ incomplete Latin square. (The arrangement is that of Table 26.1. Row effects, being orthogonal, are not shown here. Error is also omitted.)

| | | | | | |
|---|---|---|---|---|---|
| $u + c_1$ $+t_1$ | $u + c_2$ $+t_5$ | $u + c_3$ $+t_2$ | $u + c_4$ $+t_4$ | $u + c_5$ $+t_3$ | $5u$ |
| $u + c_1$ $+t_2$ | $u + c_2$ $+t_3$ | $u + c_3$ $+t_1$ | $u + c_4$ $+t_5$ | $u + c_5$ $+t_4$ | $5u$ |
| $u + c_1$ $+t_4$ | $u + c_2$ $+t_2$ | $u + c_3$ $+t_3$ | $u + c_4$ $+t_1$ | $u + c_5$ $+t_5$ | $5u$ |
| $u + c_1$ $+t_5$ | $u + c_2$ $+t_1$ | $u + c_3$ $+t_4$ | $u + c_4$ $+t_3$ | $u + c_5$ $+t_2$ | $5u$ |
| $C_1$ | $C_2$ | $C_3$ | $C_4$ | $C_5$ | $20u = Y$ |

The effects $c$ and $t$ have been calculated in Secs. 3 and 5. The reader should verify that $ssq(c,t) = \Sigma cC + \Sigma tT = 288$ as an exercise.

Finally, we come to the problem of splitting $ssq(c,t)$ into two components: one due to the $C$'s as they are and another due to $(n-1)T + C$. To see this, substitute the values of $c$ and $t$ in the expression above, remembering that $C_1$, $T_3$; $C_2$, $T_4$; etc., form pairs to be denoted by $C_j T_j$.

$$
ssq(c,t) = C_1 \left[ \frac{(n-1)C_1 + T_3 - Y}{n(n-2)} \right] + \cdots
$$

$$
+ T_3 \left[ \frac{(n-1)T_3 + C_1 - Y}{n(n-2)} \right] + \cdots
$$

$$
= \frac{1}{n(n-2)} \left[ (n-1)\Sigma C^2 + (n-1)\Sigma T^2 + 2\Sigma C_j T_j - 2Y^2 \right]
$$

If we subtract

$$
\text{Raw column } ssq = \frac{\Sigma C^2}{n-1} - \frac{Y^2}{n(n-1)}
$$

from the joint $ssq$ above, the remainder will be

$$
\text{Adjusted treatment } ssq = \frac{\Sigma[(n-1)T + C]^2}{n(n-1)(n-2)} - \frac{Y^2}{(n-1)(n-2)}
$$

This completes the proof that

$$
ssq(c,t) = \text{raw column } ssq + \text{adjusted treatment } ssq
$$

In practice, we calculate these two components as outlined in Sec. 2, because there is no shortcut to obtain the joint $ssq$. Since the expression for $ssq(c,t)$ is symmetrical with respect to $C$ and $T$, it follows that it may also be written as raw treatment $ssq$ + adjusted column $ssq$. The latter subdivision, however, is of no practical use except possibly for checking the arithmetic.

### 7. Some other cases

If one column, instead of one row, is missing from a Latin square, the analysis remains the same, substituting the row totals $R$ for the column totals $C$. Alternatively, the square may be turned 90° so that a missing column becomes a missing row.

If one of the treatments is missing, the remaining treatments are still orthogonal to columns as well as to rows, and hence the treatment means and $ssq$ are calculated the usual way and need no adjustment. But the columns and rows are no longer orthogonal to each other. In such a case, it is immaterial whether rows are adjusted and columns left alone or columns are adjusted and rows left alone. The purpose is simply to find the joint *ssq* due to columns and rows, so that the error *ssq* may be obtained by subtraction. One example of this kind is given as an exercise.

If one row and one column are missing, a more complicated analysis is available. When more than one row is missing, the analysis in general is very complicated and we shall not cover such cases here. However, certain particular cases known as Youden squares will be mentioned toward the end of the next chapter.

## EXERCISES

**26.1**  Verify: $ssq(c,t) = \Sigma cC + \Sigma tT = 288$ in our text example.

| $C$ | $c$ | $cC$ | $T$ | $t$ | $tT$ |
|-----|-----|------|-----|-----|------|
| 60 | $-2.67$ | $-160.00$ | 52 | $-4.27$ | $-221.87$ |
| 76 | $1.53$ | $116.53$ | 67 | $-0.80$ | $-53.60$ |
| 56 | $-3.60$ | $-201.60$ | 80 | $1.33$ | $106.67$ |
| 80 | $1.80$ | $144.00$ | 79 | $2.13$ | $168.53$ |
| 88 | $2.93$ | $258.13$ | 82 | $1.60$ | $131.20$ |

**26.2**  Consider once again the $6 \times 6$ Latin-square experiment in the exercises for Chap. 17. Suppose that, after the experiment was done, it was discovered that treatment (4) was applied with a wrong agent and the investigator wanted to exclude it from analysis. The data would be as shown here. Do an analysis of variance with the remaining data.

| | | | | | | $R$ | $5R$ | $C$ | $5R + C$ |
|---|---|---|---|---|---|---|---|---|---|
| (1) 11 | (6) 18 | (4) | (3) 19 | (2) 19 | (5) 3 | 70 | 350 | 70 | 420 |
| (5) 13 | (4) | (2) 23 | (1) 10 | (6) 28 | (3) 22 | 105 | 525 | 63 | 588 |
| (3) 19 | (2) 13 | (6) 21 | (5) 14 | (4) | (1) 21 | 88 | · | · | · |
| (6) 23 | (5) 5 | (3) 16 | (2) 13 | (1) 9 | (4) | 66 | · | · | · |
| (4) | (3) 12 | (1) 3 | (6) 19 | (5) 6 | (2) 15 | 55 | · | · | · |
| (2) 25 | (1) 15 | (5) 7 | (4) | (3) 26 | (6) 23 | 96 | · | · | · |
| $C$   91 | 63 | 70 | 84 | 88 | 84 | 480 | 2,400 | 480 | 2,880 |

| Treatment | (1) | (2) | (3) | (4) | (5) | (6) | Total |
|-----------|-----|-----|-----|-----|-----|-----|-------|
| Total | 78 | · | · | — | · | 132 | 480 |

The four ordinary basic quantities are

$$A = 11^2 + \cdots + 23^2 \qquad K = \tfrac{1}{5}(91^2 + \cdots + 84)^2$$
$$C = 480^2/30 \qquad T = \tfrac{1}{6}(78^2 + \cdots + 132^2)$$

The treatment is orthogonal to rows and to columns, and its *ssq* is calculated the usual way: $T - C$. The raw column *ssq* is $K - C$. To obtain the adjusted row *ssq*, we need the two adjusted quantities

$$R^* = \frac{\Sigma(5R + C)^2}{6 \times 5 \times 4} = \frac{420^2 + 588^2 + \cdots}{120} \qquad C^* = \frac{480^2}{5 \times 4}$$

To free the row totals from column effects, the $C$ value to be added to $5R$ is the total of the column missing in that row. For example, to the number $5R = 525$ for the second row in which the missing treatment (4) occurs in column 2 we add 63, the total of that column.

*Partial Ans.:* $ssq(c,t) = 471$. Complete the accompanying analysis-of-variance table.

| Source | df | ssq | msq |
|---|---|---|---|
| Rows, adjusted | . . . | 349.8 | |
| Columns, raw | . . . | . . . . . | |
| Treatment (orthogonal) | 4 | 732.0 | . . . . .⎫ |
| Error | 15 | 167.0 | 11.13⎭ $F = 16.4$ |
| Total | 29 | . . . . . | |

**26.3** For those who do not like heavy algebra and yet wish to follow the mathematical notes in Sec. 6, the best thing to do is work out the details of the smallest possible ($3 \times 3$) Latin square in which one row is missing. To further simplify the writing, we may let the treatments missing in columns 1, 2, 3 be treatments (1), (2), (3) without loss of generality. The setup is as follows, omitting the row effects.

| (1) | (2) | (3) |
|---|---|---|
| (2)<br>$u + c_1 + t_2$ | (3)<br>$u + c_2 + t_3$ | (1)<br>$u + c_3 + t_1$ |
| (3)<br>$u + c_1 + t_3$ | (1)<br>$u + c_2 + t_1$ | (2)<br>$u + c_3 + t_2$ |

$$C_1 = 2u + 2c_1 - t_1 \qquad C_2 = 2u + 2c_2 - t_2 \qquad C_3 = 2u + 2c_3 - t_3$$
$$T_1 = 2u + 2t_1 - c_1 \qquad T_2 = 2u + 2t_2 - c_2 \qquad T_3 = 2u + 2t_3 - c_3$$
$$2C_1 + T_1 = Y + 3c_1 \qquad 2C + T_2 = Y + 3c_2 \qquad 2C_3 + T_3 = Y + 3c_3$$
$$2T_1 + C_1 = Y + 3t_1 \qquad 2T + C_2 = Y + 3t_2 \qquad 2T_3 + C_3 = Y + 3t_3$$

where $Y = 6u = \Sigma C = \Sigma T$. The joint sum of squares due to columns and treatments is

$$ssq(c,t) = C_1 c_1 + C_2 c_2 + C_3 c_3 + T_1 t_1 + T_2 t_2 + T_3 t_3$$

Substituting the values of $c_j$ and $t_j$ given by the equations above,

$$ssq(c,t) = C_1 \left( \frac{2C_1 + T_1 - Y}{3} \right) + C_2 \left( \frac{2C_2 + T_2 - Y}{3} \right) + C_3 \left( \frac{2C_3 + T_3 - Y}{3} \right)$$
$$+ T_1 \left( \frac{2T_1 + C_1 - Y}{3} \right) + T_2 \left( \frac{2T_2 + C_2 - Y}{3} \right) + T_3 \left( \frac{2T_3 + C_3 - Y}{3} \right)$$
$$= \tfrac{1}{3}[2(C_1{}^2 + C_2{}^2 + C_3{}^2) + 2(T_1{}^2 + T_2{}^2 + T_3{}^2)$$
$$+ 2(C_1 T_1 + C_2 T_2 + C_3 T_3) - 2Y^2]$$

since $C_1 Y + C_2 Y + C_3 Y = Y^2$ and also $T_1 Y + T_2 Y + T_3 Y = Y^2$. Now, let us subtract the quantity (raw column $ssq$)

$$\tfrac{1}{2}(C_1{}^2 + C_2{}^2 + C_3{}^2) - \frac{Y^2}{6}$$

from the joint $ssq$ above. The remainder, known as the adjusted treatment $ssq$, is

$$\tfrac{1}{6}[(C_1{}^2 + C_2{}^2 + C_3{}^2) + 4(T_1{}^2 + T_2{}^2 + T_3{}^2) + 4(C_1 T_1 + C_2 T_2 + C_3 T_3)] - \frac{Y^2}{2}$$
$$= \tfrac{1}{6}[(2T_1 + C_1)^2 + (2T_2 + C_2)^2 + (2T_3 + C_3)^2] - \frac{Y^2}{2}$$

which is the desired form. Note that here $n = 3$, and hence $(n - 1) = 2$, $(n - 1)(n - 2) = 2$, and $n(n - 1)(n - 2) = 3 \times 2 \times 1 = 6$, in agreement with the general formula.

# 27

# Balanced
# incomplete blocks

The method of analyzing incomplete Latin squares opens the way to a much more general and flexible type of design originally called *incomplete randomized blocks* by Yates (1936) but later known as *balanced incomplete blocks*. Its potential usefulness in medical research, particularly when employing men as experimental units, has not been fully explored, although it has been in common use in agricultural and other experiments.

One of the first considerations and difficulties in testing several treatments is where and how to obtain homogeneous blocks. The larger the block, the more difficult they are to find. Suppose that seven treatments are to be compared. The ordinary randomized block design (Chap. 16) requires that each block contain seven more or less homogeneous experimental units. This requirement can frequently not be met with mouse litters, still less with human siblings or patients of the same characteristics. But it may be possible to find blocks of three homogeneous units each. Under certain circumstances the "size" of a block is limited by the capacity of laboratory equipments and instruments. The present chapter deals with the method of testing several treatments with small blocks. The extreme case—blocks of only two units—will be dealt with in the next chapter.

## 1. The balanced condition

Suppose that there are 7 treatments to be tested in blocks of 3 units each. One may take all possible 3's out of 7, and there are $\binom{7}{3} = 35$ such triplets of treatments, ranging from (1, 2, 3) to (5, 6, 7). In terms of experimentation, this means that 35 blocks of 3 units each are needed, and the total number of observations will be $35 \times 3 = 105$. Since there are only 7 treatments, this means that each treatment must be replicated $105/7 = 15$ times. Although the number of replications as well as the total size of the experiment in this particular case are still within practical range, the reader may readily see that the general solution of this kind is far from being practical because it involves too many replications. For instance, for 10 treatments in blocks of 4 units each, there will be $\binom{10}{4} = 210$ blocks with 840 observations, each treatment being replicated 84 times!

The so-called "balanced condition" we are seeking for experimental purposes is that every possible *pair* (not triplets, quadruplets, etc.) of treatments appears in the same block the same number of times. This will greatly reduce the number of blocks required. Let us return to the case of 7 treatments in blocks of 3 units each for illustration. It happens that only 7 blocks (instead of 35) are needed; Table 27.1 gives such a design. A detailed examination of the table will clarify the balanced nature. Concentrating on any one treatment, we find that each of the treatments appears three times in the entire experiment. Next, let us look at the pairs. There are $\binom{7}{2} = 21$ possible pairs of treatments, viz., (1, 2), (1,

*Table 27.1* Balanced incomplete blocks for $t = 7$ treatments tested in blocks of $k = 3$ units each, so that each pair of treatments occurs in the same block once ($\lambda = 1$). Upper number, in parentheses, indicates treatments; lower number is the corresponding measurement

| Block 1 | | | Block 2 | | | Block 3 | | | Block 4 | | |
|---|---|---|---|---|---|---|---|---|---|---|---|
| (2 | 1 | 4) | (5 | 3 | 2) | (3 | 6 | 4) | (5 | 4 | 7) |
| 15 | 10 | 11 | 15 | 12 | 4 | 5 | 10 | 14 | 19 | 14 | 19 |

| | Block 5 | | | Block 6 | | | Block 7 | | |
|---|---|---|---|---|---|---|---|---|---|
| (1 | 6 | 5) | (7 | 2 | 6) | (7 | 3 | 1) |
| 8 | 17 | 10 | 12 | 6 | 11 | 21 | 14 | 5 |

3), . . . , (6, 7).   The reader will find that each of these pairs appears in the same block once.   Each block provides three pairs.   In the first block, for instance, the three pairs are (1, 2), (1, 4), (2, 4).   The 7 blocks altogether provide precisely the 21 pairs, each occurring once in the same block and there being no duplication.   This is called a balanced incomplete blocks design, of which the example above is only a very special case.

When the number of experimental units per block is less than the number of treatments to be studied, the block is necessarily incomplete; hence the blocks and treatments are not orthogonal to each other.   The advantages of a balanced design are (1) the simplicity of the analysis in spite of the nonorthogonality, (2) equal precision for comparisons of all treatments, and (3) unbiased estimates of treatment effects readily available.

## 2. General relationship

Let

$t$ = number of treatments        $k$ = number of units per block
$r$ = number of replications       $b$ = number of blocks

It is understood that $k$ is smaller than $t$; for otherwise we would have an ordinary (complete) randomized block experiment and there would be no need for incomplete block designs.   By the very meaning of the symbols above, we see that the total number of units in the entire experiment is

$$N = tr = bk$$

For a balanced design it is required that each possible pair of treatments appear in the same block the same number of times.   Let $\lambda$ be the number of times each pair occurs in the same block.   Consider a fixed treatment, say, treatment (1).   There are $t - 1$ possible pairs involving (1).   If each pair occurs $\lambda$ times, there will be $\lambda(t - 1)$ pairs involving (1) in the entire experiment.   On the other hand, each block in which treatment (1) occurs provides $k - 1$ pairs involving (1), and (1) is replicated $r$ times in the experiment, that is, there are $r$ such blocks.   Hence

$$\lambda(t - 1) = r(k - 1) \qquad \lambda = \frac{r(k - 1)}{t - 1}$$

For a balanced design, the numbers $t$, $k$, $r$, $b$, and $\lambda$ must be integers and must satisfy the two conditions above.   For practical purposes of using it as experimental design, it should be further restricted that $r$ should not be too large.   In most practical designs, $r \leq 10$.

In terms of the general symbols, we see that the example in Table 27.1 is the case in which

$$t = b = 7 \qquad k = r = 3 \qquad \lambda = \frac{3(3 - 1)}{7 - 1} = 1$$

Each pair of treatments occurs in the same block once and once only.

## 3. Preliminary calculations

The analysis is very similar to that employed in the preceding chapter, except that there are no orthogonal rows (replications) here. The data in Table 27.1 are employed to illustrate the procedure. In practice the three treatments are allocated to the units of a block at random, just as in the case of ordinary randomized blocks, and this is why treatments shown in the blocks are not in any order. To facilitate the arithmetic, it is best to retabulate the data systematically as shown in the upper part of Table 27.2.

First, the block totals $B_i$, the treatment totals $T_j$, and the grand total $Y$ are obtained. Second, each treatment total is multiplied by $k = 3$. The third step is new and should be carefully taken. It is the calculation of the quantity denoted by $B_j^*$; it may best be remembered as the s.o.b. (*sum of blocks*) in which treatment $j$ appears. For example, treatment (1) appears in blocks 1, 5, 7. The sum of these three blocks is

$$B_1^* = B_1 + B_5 + B_7 = 36 + 35 + 40 = 111$$

Similarly, treatment (2) occurs in blocks 1, 2, 6; hence

$$B_2^* = B_1 + B_2 + B_6 = 36 + 31 + 29 = 96$$

and so on. When the data are arranged in the way shown in Table 27.2,

*Table 27.2* Tabulation of data and preliminary calculation for an experiment with seven treatments in balanced incomplete blocks of three units each ($k = 3$)

| Blocks | Treatments | | | | | | | Block total $B_i$ |
|---|---|---|---|---|---|---|---|---|
| | (1) | (2) | (3) | (4) | (5) | (6) | (7) | |
| 1 | 10 | 15 | | 11 | | | | 36 |
| 2 | | 4 | 12 | | 15 | | | 31 |
| 3 | | | 5 | 14 | | 10 | | 29 |
| 4 | | | | 14 | 19 | | 19 | 52 |
| 5 | 8 | | | | 10 | 17 | | 35 |
| 6 | | 6 | | | | 11 | 12 | 29 |
| 7 | 5 | | 14 | | | | 21 | 40 |
| $T_j$ | 23 | 25 | 31 | 39 | 44 | 38 | 52 | 252 = Y |
| $kT_j$ | 69 | 75 | 93 | 117 | 132 | 114 | 156 | 756 = kY |
| $B_j^*$ | 111 | 96 | 100 | 117 | 118 | 93 | 121 | 756 = kY |
| $Q_j$ | −42 | −21 | −7 | 0 | 14 | 21 | 35 | 0 |
| $Q_j^2$ | 1,764 | 441 | 49 | 0 | 196 | 441 | 1,225 | 4,116 = ΣQ² |

this sum may be obtained with little risk of confusion. For each treatment, it is simply the sum of the block totals corresponding to the single observations of that treatment. The final step shown in Table 27.2 is to calculate the quantity

$$Q_j = kT_j - B_j^*$$

for each treatment. Thus, for treatment (1), $Q_1 = 69 - 111 = -42$, etc. These $Q$ values reflect pure treatment effects and are free from block effects. An algebraic justification for this statement will be given in a later section. For the time being we may be satisfied with our intuitive reasoning. Consider treatment (1), which appears in blocks 1, 5, 7 only. If these are "good" blocks, the observed total of treatment (1), $T_1$, would be an overestimate of the merit of treatment (1). Conversely, if blocks 1, 5, 7 are "poor" blocks, treatment (1) is "unlucky" and $T_1$ would be an underestimate of the merit of treatment (1). Hence the observed $T_j$ should be adjusted according to the blocks in which treatment ($j$) occurs. Now $T_1$ is the sum of three observations, and so is each of the $B$'s. To put them on equal basis, we adjust $3T_1$ by the amount $B_1 + B_5 + B_7$. It should be noted that the $Q$ values add to zero.

## 4. Analysis of variance

In analyzing a small experiment with $t = b$, we ignore the treatment effects on blocks but adjust the treatments according to blocks. The sums of squares due to blocks are then calculated the usual way, in spite of the nonorthogonality. Thus, we need the basic quantities

(area) $A = \Sigma y^2 = 10^2 + 15^2 + \cdots = 3{,}486$

(area) $B = \frac{1}{3}\Sigma B_i^2 = \frac{1}{3}(36^2 + 31^2 + \cdots) = 3{,}156$

(area) $C = \dfrac{Y^2}{N} = \dfrac{252^2}{21} = 3{,}024$

so that $A - C =$ total *ssq* and $B - C =$ raw block *ssq*. The sum of squares due to treatments, *adjusted for blocks*, is

$$\frac{(t-1)\Sigma Q^2}{rt \cdot k(k-1)} = \frac{\Sigma Q^2}{kt\lambda} = \frac{4116}{21} = 196$$

The error *ssq* is obtained by subtraction. The analysis of variance is given in Table 27.3, in which $F = 1.95$, indicating nonsignificance of the treatment effects. The required $F$ value corresponding to 6 and 8 *df* is 3.58 at the 5 per cent significance level.

*Table 27.3* Analysis of variance of a balanced incomplete block experiment with $t = b = 7$ and $k = r = 3$ (data from Table 27.2)

| Source of variation | df | ssq | msq |
|---|---|---|---|
| Blocks, raw | $b - 1 = 6$ | $B - C = 132$ | |
| Treatments, adjusted | $t - 1 = 6$ | $\Sigma Q^2/kt\lambda = 196$ | $32.67\}$ $F = 1.95$ |
| Intrablock error | $rt - t - b + 1 = 8$ | (subt) 134 | $16.75\}$ |
| Total | $rt - 1 = 20$ | $A - C = 462$ | |

## 5. Adjusted treatment means; variance; efficiency

Since the $Q$ values are free from block influences, the estimates of treatment *effects* $t_j$ are based on them.

$$t_j = \frac{(t-1)Q_j}{tr(k-1)} = \frac{Q_j}{t\lambda} = \frac{Q_j}{7}$$

The sum of these $t$'s is zero. Let $\bar{y}$ be the general mean of the entire experiment. In our example, $\bar{y} = {}^{252}\!/_{21} = 12$. Then the

$$\text{Adjusted treatment mean} = \bar{y} + t_j$$

Referring to the $Q$ values in Table 27.2, we find the results listed in Table 27.4. The variance of the difference between two treatment means in an orthogonal experiment (wherein no adjustment is needed) is $2s^2/r$, where $s^2$ is the error variance and $r$ is the number of replications. For balanced incomplete block design, the variance of the difference between two adjusted treatment means is always larger than $2s^2/r$, being

$$\frac{2s^2}{r}\left[\frac{k(t-1)}{t(k-1)}\right] = \frac{2(16.75)}{3}\left[\frac{3 \times 6}{7 \times 2}\right] = 14.36$$

and the standard error is $\sqrt{14.36} = 3.79$. The ordinary $t$ test for the difference between two adjusted treatment means may be performed with this standard error. This amounts to saying that the error variance is

*Table 27.4* Treatment effects and adjusted means

| | Treatments | | | | | | | Mean |
|---|---|---|---|---|---|---|---|---|
| | (1) | (2) | (3) | (4) | (5) | (6) | (7) | |
| Effects, $t_j$ | $-6$ | $-3$ | $-1$ | $0$ | $2$ | $3$ | $5$ | $0$ |
| Adj. mean | $6$ | $9$ | $11$ | $12$ | $14$ | $15$ | $17$ | $12$ |

not simply $s^2 = 16.75$ for adjusted values but is what we would call the *effective* error variance:

$$s'^2 = s^2 \left[ \frac{k(t-1)}{t(k-1)} \right] = (16.75) \left[ \frac{3 \times 6}{7 \times 2} \right] = 21.54$$

The overall $F$ test for the seven treatments gives a nonsignificant result. If the seven treatments were seven levels of a factor, we see from the values of $t_j$ or adjusted means that the linear trend (with 1 $df$) would be highly significant. The linear trend should be tested against the effective error variance.

The value of $k$ being smaller than $t$, the adjusting factor for error variance given above is always greater than unity. In this example, the factor is equal to $(3 \times 6)/(7 \times 2) = {}^{18}\!/_{14} = 1.2857$. The reciprocal of this factor is called the *efficiency* of the incomplete block design. Thus

$$\text{Efficiency} = \frac{t(k-1)}{k(t-1)} = \frac{1 - 1/k}{1 - 1/t} = \frac{14}{18} = 0.78$$

The "efficiency" given above is based on the assumption that the error variance for complete blocks of size $t$ is of the same magnitude as that for incomplete blocks of size $k$. In practice, if the small incomplete blocks are more homogeneous than the large complete blocks, the efficiency is much higher than that shown above.

This completes the analysis of one type of balanced incomplete block experiments. We say one type because there are many other types available and the procedure of analysis requires modification when there are a large number of blocks. The following sections supply some explanatory notes to justify the arithmetic procedure adopted in the preceding sections and some notes on the construction of balanced incomplete blocks. These sections may be omitted by the uninterested reader, at least in his first reading.

## 6. The linear model and $Q$ value

The model for incomplete blocks is the same as that for complete blocks, that is, $y_{ij} = u + b_i + t_j + e_{ij}$, except that not all the $t$'s occur in the same block. We assume that $\Sigma t_j = 0$ over all treatments, $\Sigma b_i = 0$ over all blocks, and $\Sigma e_{ij} = 0$ for any one block as well as for any one treatment. Table 27.5 is a rewrite of Table 27.2 in algebraic form. For the sake of simplicity the general constant term $u$ has been omitted because it will cancel out on taking difference. The error term $e$ has also been omitted because its sum vanishes in the block total and treatment total. To further simplify the matter, we may concentrate our attention on treatment (1), since the situation with any other treatment is the same. Consequently only the entries relevant to $Q_1$ are explicitly shown in Table 27.5.

*Table 27.5* **Linear model of a balanced incomplete block design; $t = b = 7$, $k = r = 3$. (The general mean $u$ and error $e$ have been omitted.)**

| Blocks | Treatments | | | | | | | Sum of $k$ units Block total $B_i$ |
|---|---|---|---|---|---|---|---|---|
| | (1) | (2) | (3) | (4) | (5) | (6) | (7) | |
| 1 | $b_1 + t_1$ | $b_1 + t_2$ | | $b_1 + t_4$ | | | | $3b_1 + t_1 + t_2 + t_4$ |
| 2 | | $\cdots$ | $\cdots$ | $\cdots$ | | | | $\cdots$ |
| 3 | | | $\cdots$ | $\cdots$ | | $\cdots$ | | $\cdots$ |
| 4 | | | | $\cdots$ | $\cdots$ | | $\cdots$ | $\cdots$ |
| 5 | $b_5 + t_1$ | | | | $b_5 + t_5$ | $b_5 + t_6$ | | $3b_5 + t_5 + t_6 + t_1$ |
| 6 | | $\cdots$ | | | | $\cdots$ | $\cdots$ | $\cdots$ |
| 7 | $b_7 + t_1$ | | $b_7 + t_3$ | | | | $b_7 + t_7$ | $3b_7 + t_7 + t_1 + t_3$ |
| Sum of $r$ units, $T_j$ | $T_1$ | $T_2$ | $T_3$ | $T_4$ | $T_5$ | $T_6$ | $T_7$ | $B_1* = B_1 + B_5 + B_7$ |

Although in our particular example $k = r$ and $t = b$, there will be no difficulty in arriving at a more general expression if we remember that each block total $B_i$ is the sum of $k$ units and each treatment total $T_j$ is the sum of $r$ units. The total of treatment (1) is

$$T_1 = 3t_1 + b_1 + b_5 + b_7$$

The block totals are obtained the same way; they are shown in the table. The sum of the block totals in which treatment (1) occurs is

$$B_1^* = B_1 + B_5 + B_7 = 3(b_1 + b_5 + b_7) + 2t_1$$

the other $t$'s being dropped out because

$$t_1 + t_2 + t_3 + t_4 + t_5 + t_6 + t_7 = 0$$

Now it is clear that

$$Q_1 = 3T_1 - B_1^* = 9t_1 - 2t_1 = 7t_1$$

which is a multiple of $t_1$ only and does not contain other treatment effects or any block effect. More generally,

$$T_1 = rt_1 + b_1 + \cdots + b_r \qquad [r \text{ } b\text{'s associated with (1)}]$$
$$kT_1 = krt_1 + k(b_1 + \cdots + b_r)$$
$$B_1^* = k(b_1 + \cdots + b_r) + (r - \lambda)t_1$$

where $\lambda$ is the number of times that $t_1$ occurs with each of the other $t$'s. Then

$$Q_1 = kT_1 - B_1^* = (kr - r + \lambda)t_1 = (t\lambda)t_1$$

because of the relationship $r(k - 1) = \lambda(t - 1)$ established in Sec. 2. This expression also explains the procedure of calculating treatment effects $t_j$ and the adjusted treatment means in Sec. 5.

## 7. Intrablock differences

It is instructive to obtain the $Q$ values in an entirely different manner; for it gives us more insight to their meaning. According to the linear model shown in Table 27.5, the three observations in the same block all contain the same block effect $b_i$. It follows that the difference between any two observations of the same block does not contain the block effect.

$$\text{Treatments } (1) - (2) = 10 - 15 = -5$$
$$(1) - (3) = 5 - 14 = -9, \ldots$$

and

$$\text{Treatments } (1) - (7) = 5 - 21 = -16$$

If we do this for every treatment, calculating treatments $(2) - (1) = 5$, etc., we find that the differences for each treatment are as given in Table 27.6. We see that the sum of these intrablock differences is identical with the $Q$ values obtained before (Table 27.2). Algebraically, the differences for treatment (1), for instance, are

$$t_1 - t_2 + e_{11} - e_{12}$$
$$t_1 - t_3 + e_{71} - e_{73}$$
$$t_1 - t_4 + e_{11} - e_{14}$$
$$t_1 - t_5 + e_{51} - e_{55}$$
$$t_1 - t_6 + e_{51} - e_{56}$$
$$t_1 - t_7 + e_{71} - e_{77}$$

The sum of the $e$'s is zero because $\Sigma e_{ij} = 0$ whether summation is with respect to block or with respect to treatment. To obtain the sum of the $t$'s, we add $0 = t_1 - t_1$ and obtain $7t_1 = Q_1$. Thus the intrablock differences give us the same $Q$ values.

*Table 27.6* **Differences between treatments in the same block**

| Treatments | Treatments | | | | | | |
|---|---|---|---|---|---|---|---|
| | (1) | (2) | (3) | (4) | (5) | (6) | (7) |
| $-(1)$ | $\cdots$ | $+5$ | $+9$ | $+1$ | $+2$ | $+9$ | $+16$ |
| $-(2)$ | $-5$ | $\cdots$ | $+8$ | $-4$ | $+11$ | $+5$ | $+6$ |
| $-(3)$ | $-9$ | $-8$ | $\cdots$ | $+9$ | $+3$ | $+5$ | $+7$ |
| $-(4)$ | $-1$ | $+4$ | $-9$ | $\cdots$ | $+5$ | $-4$ | $+5$ |
| $-(5)$ | $-2$ | $-11$ | $-3$ | $-5$ | $\cdots$ | $+7$ | $0$ |
| $-(6)$ | $-9$ | $-5$ | $-5$ | $+4$ | $-7$ | $\cdots$ | $+1$ |
| $-(7)$ | $-16$ | $-6$ | $-7$ | $-5$ | $0$ | $-1$ | $\cdots$ |
| Sum $= Q$ | $-42$ | $-21$ | $-7$ | $0$ | $14$ | $21$ | $35$ |

## 8. Adjusted treatment *ssq* and joint *ssq*

For orthogonal experiments the treatment *ssq* is simply $r\Sigma t_j^2$, where $t_j$ is the effect of treatment $(j)$ and $r$ is the number of replications of these treatments. For incomplete blocks, the adjusted treatment *ssq* is always smaller than $r\Sigma t_j^2$. The reason for this can be easily seen from the block totals shown in Table 27.5. The square of the first block total is

$$B_1^2 = 9b_1^2 + 6b_1(t_1 + t_2 + t_4) + (t_1 + t_2 + t_4)^2$$

Squaring each block total, summing the totals together, and dividing the sum by $k$, we obtain the raw block *ssq*. This expression may be greatly simplified by the fact that $\Sigma t_j = 0$ leads to $(\Sigma t_j)^2 = \Sigma t_j^2 + 2\Sigma t_j t_{j'} = 0$, where $j' > j$. If the reader carries out this process, he will find that the raw block *ssq* contains not only the sum of squares $\Sigma b_i^2$ and the sum of products $\Sigma b_i t_j$ reflecting the nonorthogonality but also $\left(\dfrac{r - \lambda}{k}\right)\Sigma t_j^2$. This much $\Sigma t_j^2$ has been included in the raw block *ssq*. Consequently, the adjusted treatment *ssq* is

$$\left(r - \frac{r - \lambda}{k}\right)\sum t_j^2 - \frac{t\lambda}{k}\sum t_j^2 = \frac{\Sigma Q_j^2}{kt\lambda}$$

which is the formula adopted for numerical calculation in Sec. 4.

As a verification of these relationships, let us first calculate the quantity (see Sec. 5)

$$\Sigma t_j^2 = (-6)^2 + (-3)^2 + (-1)^2 + 0^2 + 2^2 + 3^2 + 5^2 = 84$$

This value would be multiplied by $r = 3$ for complete blocks, but here it should be multiplied by

$$r - \frac{r - \lambda}{k} = 3 - \frac{3 - 1}{3} = \frac{7}{3} = \frac{t\lambda}{k}$$

This adjusted treatment *ssq* is then $\frac{7}{3}(84) = 196$, in agreement with our previous result.

The concept of efficiency is also based on the loss of some $\Sigma t_j^2$ to the raw block *ssq*. The proportion of $\Sigma t_j^2$ that is retained in the adjusted treatment *ssq*, relative to complete block designs, is

$$\frac{r - (r - \lambda)/k}{r} = \frac{t\lambda}{rk} = \frac{t(k - 1)}{k(t - 1)} = \frac{7}{9}$$

which is the efficiency factor defined previously. And this is why, in comparing two adjusted means (really two $t$'s), the error variance should be enlarged by multiplying it by the reciprocal of the efficiency factor.

We may also mention in passing, without giving details, that the *ssq* due to blocks and treatments remains the same whether we adjust treatments and leave the blocks alone or adjust blocks and leave the treatments alone. In our particular example, $t = b$ and the relationship is symmetrical, and the adjusted block *ssq* may be found by a procedure similar to that employed for adjusted treatment *ssq*. In the more general case in which there are more blocks than treatments, a modified formula is required. Whether the more general formula or the present method is used, the results are the same for our example.

$$
\begin{array}{ll}
\text{Raw block } ssq = 132 & \text{Adjusted block } ssq = 112 \\
\text{Adjusted treatment } ssq = 196 & \text{Raw treatment } ssq = \underline{216} \\
\text{Joint } ssq = \overline{328} & \text{Joint } ssq = \overline{328}
\end{array}
$$

In a large and more complicated experiment, the adjusted block *ssq* is needed but there is no simple direct method of obtaining it. For numerical calculation, it is then obtained through subtraction by using the relationship above. Thus,

$$
\begin{array}{c}
\text{Adjusted} \\
\text{block } ssq
\end{array}
=
\begin{array}{c}
\text{raw} \\
\text{block } ssq
\end{array}
+
\begin{array}{c}
\text{adjusted} \\
\text{treatment } ssq
\end{array}
-
\begin{array}{c}
\text{raw} \\
\text{treatment } ssq
\end{array}
$$

For detailed methods of dealing with large experiments, the reader should consult a more comprehensive textbook.

### 9. Balanced designs

There are many different, interesting, intriguing, and elegant methods of constructing balanced incomplete block designs. Some of these methods are based upon certain branches of advanced mathematics which we cannot possibly describe here. The pioneer designs for simple cases are due to Yates (1936).[†] The application of formal and systematic methods for more complicated cases is to be found in the book-length classical paper of Bose (1939).[‡] These findings were quickly followed by a number of other statisticians (Cox, Fisher, etc.). Fortunately, for most practical purposes, these designs have been tabulated in Fisher and Yates' statistical tables (1963) and also given as "plans" at the end of chapter 11 of Cochran and Cox (1957), so it is not necessary for the experimenter to know the mathematical methods of construction. He is advised to choose one of the published designs that suits his situation best.

The "size" of an incomplete block (that is, $k$, the number of units in a block) ranges from 2 to $t - 1$, where $t$ is the number of treatments to

[†] F. Yates, Incomplete randomized blocks, *Ann. Eugenics London*, 7:121–140 (1936).
[‡] R. C. Bose, On the construction of balanced incomplete block designs, *Ann. Eugenics London*, 9:353–400 (1939).

be studied. The case of $k = 2$ will be dealt with in the next chapter. When $k = t - 1$, the balanced design is accomplished by omitting one treatment from each block: for example, omitting treatment 1 from block 1, omitting treatment 2 from block 2, and so on. No special methods are needed for such cases. It is the intermediate cases that require ingenuity to construct.

For $t = 4$, the value of $k$ can only be 2 or 3. When $t = 5$, $k = 2, 3$, or 4. Once the design for $k = 2$ is obtained, the complementary part of that design is that for $k = 3$. The design for $k = 4$ takes four blocks, each treatment being omitted from a different block. Hence the problem of construction really begins with $t = 6$ and $k = 3$. We shall work out an example of this design in the exercises.

In closing this section, it is appropriate to mention the particular method by which the design in our example ($t = b = 7$ and $k = r = 3$) has been constructed. There are several methods available for this case, but by far the simplest is that of *cyclic substitution*. For this method to work, it is necessary to choose an appropriate *initial* block. Once this is chosen, the other blocks may be obtained successively by changing treatment 1 to treatment 2, treatment 2 to treatment 3, . . . , treatment 7 back to treatment 1 in each block. The initial block is (1, 2, 4). Then the second block is (2, 3, 5). The third block is (3, 4, 6), and so on. The cyclic nature of this method is shown in Fig. 27.1. Since it constitutes a cyclic formation, it is immaterial which one of the seven blocks is taken as the initial one. For convenience, we use (1, 2, 4). At this stage, the

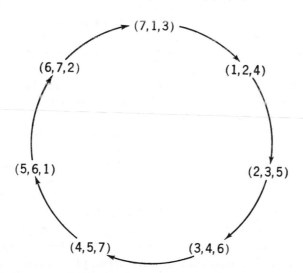

**Fig. 27.1** Balanced incomplete blocks generated by cyclic substitutions.

reader should check Fig. 27.1 with Table 27.1 and be satisfied that this actually is the design adopted, with the order of treatments within a block randomized.

In some books the same design has been represented as consisting of blocks (1, 2, 3), (2, 4, 5), etc., and they do not have the cyclic property. The reader may think that is entirely a different solution. But it must be realized that the numbering of the treatments is arbitrary. If we write 3 for 4 and write 4 for 3 in our cyclic solution, we obtain the design (1, 2, 3), (2, 4, 5) etc. It is really the same solution in spite of the very different appearances.

The solution presented in Fig. 27.1 also provides a solution for $t = b = 7$ and $k = r = 4$. Corresponding to the block of three treatments (1, 2, 4), there is a complementary block of four treatments (3, 5, 6, 7), and so on. These seven complementary blocks form a balanced design for seven treatments with $k = 4$. The design is also cyclic.

### 10. Youden rectangle

This section ties together the incomplete blocks of this chapter and the incomplete Latin square of the preceding chapter. The purpose of blocking is to form homogeneous blocks to eliminate the heterogeneity among the experimental units. In a Latin square the blocking procedure is done in two directions simultaneously so that we have a two-way elimination of the heterogeneity among the experimental units. Similarly, under certain circumstances, the incomplete blocks may also be arranged into an incomplete Latin square (of which that with one row missing is a particular case) so that two-way control of heterogeneity may be achieved. Once more, we shall use the example with seven treatments to illustrate the situation. The seven blocks, each with three units, may be so arranged that each column consists of all seven treatments, as shown in the left half of Table 27.7. Such a *rectangular* arrangement is usually known in the literature as the *Youden square*. We shall call a rectangle a rectangle.

The analysis of a Youden rectangle design is simple. Since the columns are orthogonal to treatments and to blocks, the *ssq* due to columns (replications) is calculated in the usual way. The rest of the analysis is the same as that outlined for balanced incomplete blocks in the first few sections of this chapter.

In the right half of Table 27.7 is shown the complementary Youden rectangle for seven treatments in blocks of four units each. These two Youden rectangles together make a complete Latin square. Thus, a Youden rectangle is a particular kind of incomplete Latin square. Not all Latin squares have this property. Table 27.8 gives another pair of

*Table 27.7* Youden rectangles for $t = b = 7$

| $k = r = 3$ | | | | $k = r = 4$ | | | |
|---|---|---|---|---|---|---|---|
| 1 | 2 | 4 | | 3 | 5 | 6 | 7 |
| 2 | 3 | 5 | | 4 | 6 | 7 | 1 |
| 3 | 4 | 6 | | 5 | 7 | 1 | 2 |
| 4 | 5 | 7 | | 6 | 1 | 2 | 3 |
| 5 | 6 | 1 | | 7 | 2 | 3 | 4 |
| 6 | 7 | 2 | | 1 | 3 | 4 | 5 |
| 7 | 1 | 3 | | 2 | 4 | 5 | 6 |

*Table 27.8* Youden rectangles for $t = b = 11$

| $k = r = 5$ | | | | | $k = r = 6$ | | | | | |
|---|---|---|---|---|---|---|---|---|---|---|
| 1 | 2 | 3 | 5 | 8 | 4 | 6 | 7 | 9 | 10 | 11 |
| 2 | 3 | 4 | 6 | 9 | 5 | 7 | 8 | 10 | 11 | 1 |
| 3 | 4 | 5 | 7 | 10 | 6 | 8 | 9 | 11 | 1 | 2 |
| 4 | 5 | 6 | 8 | 11 | 7 | 9 | 10 | 1 | 2 | 3 |
| 5 | 6 | 7 | 9 | 1 | 8 | 10 | 11 | 2 | 3 | 4 |
| 6 | 7 | 8 | 10 | 2 | 9 | 11 | 1 | 3 | 4 | 5 |
| 7 | 8 | 9 | 11 | 3 | 10 | 1 | 2 | 4 | 5 | 6 |
| 8 | 9 | 10 | 1 | 4 | 11 | 2 | 3 | 5 | 6 | 7 |
| 9 | 10 | 11 | 2 | 5 | 1 | 3 | 4 | 6 | 7 | 8 |
| 10 | 11 | 1 | 3 | 6 | 2 | 4 | 5 | 7 | 8 | 9 |
| 11 | 1 | 2 | 4 | 7 | 3 | 5 | 6 | 8 | 9 | 10 |

Youden rectangles for $t = 11$.  Instead of using a complete $11 \times 11$ Latin square, we may use either the left rectangle with $k = 5$ or the right rectangle with $k = 6$.  Note that each of these may be derived by cyclic substitution, starting from the initial block (1, 2, 3, 5, 8) or its complementary block.

These tables are given for a double purpose.  If we ignore the restrictions on its columns, the Youden rectangle becomes simply a design for balanced incomplete blocks.

## EXERCISES

**27.1  Design for $t = 6$ treatments in blocks of $k = 3$ units each**
First, form all possible combinations of three numbers out of six.  There are 20 such possible combinations; that is, we have at most

$$b = \binom{t}{k} = \binom{6}{3} = 20 \text{ blocks}$$

These 20 blocks are listed in Table 27.9.

*Table 27.9*  Balanced incomplete blocks with $t = 6$ treatments in blocks of $k = 3$ units each

| Block | Treatments | Block | Treatments |
|---|---|---|---|
| 1 | (1, 2, 3) | 11 | (4, 5, 6) |
| 2 | (1, 2, 4) | 12 | (3, 5, 6) |
| 3 | (1, 3, 5) | 13 | (2, 4, 6) |
| 4 | (1, 4, 6) | 14 | (2, 3, 5) |
| 5 | (1, 5, 6) | 15 | (2, 3, 4) |
| 6 | (2, 3, 6) | 16 | (1, 4, 5) |
| 7 | (2, 4, 5) | 17 | (1, 3, 6) |
| 8 | (2, 5, 6) | 18 | (1, 3, 4) |
| 9 | (3, 4, 5) | 19 | (1, 2, 6) |
| 10 | (3, 4, 6) | 20 | (1, 2, 5) |

$b = 10, \quad r = 5, \quad \lambda = 2$ | $b = 20, \quad r = 10, \quad \lambda = 4$

Second, we observe that each block has a complementary block.  Thus, the complement of (1, 2, 3) is (4, 5, 6).  These two blocks together form a complete replication of the six treatments.  This is possible only when $t$ is a multiple of $k$.  The blocks in the table are so arranged that two complementary blocks are in the same row.  Thus, block 3 and block 13 are complements.

Third, and this is crucial, the blocks are so divided into two groups that each group of 10 blocks constitutes a balanced incomplete block design by itself.  Thus, blocks

1 to 10 form a good design by themselves, and so do blocks 11 to 20. The reader should check carefully that in the first 10 blocks, each of the 15 possible pairs of treatments appears exactly twice. The same is also true in the second 10 blocks.

Because of these observations, we see that Table 27.9 gives us three designs. Either the first 10 or the second 10 blocks may be used as a separate design, each with $r = 5$ and $\lambda = 2$. If we use all 20 blocks, we have $b = 20$, $r = 10$, and $\lambda = 4$. In the following numerical example, the second 10 blocks (11 to 20) have been used.

**27.2** The data of an experiment with $t = 6$, $k = 3$, $r = 5$, $b = 10$, and $\lambda = 2$ are collected and arranged for calculation purpose in the accompanying table. All the quantities shown are self-explanatory except $B_j^*$. It is the sum of the

| Block | Treatments | | | | | | Block total $B_i$ |
|---|---|---|---|---|---|---|---|
| | (1) | (2) | (3) | (4) | (5) | (6) | |
| 1 | 11 | 13 | | | 12 | | 36 |
| 2 | 18 | 22 | | | | 23 | 63 |
| 3 | 19 | | 21 | 23 | | | 63 |
| 4 | 19 | | 14 | | | 15 | 48 |
| 5 | 3 | | | 9 | 3 | | 15 |
| 6 | | 23 | 21 | 16 | | | 60 |
| 7 | | 28 | 13 | | 19 | | 60 |
| 8 | | 19 | | 13 | | 25 | 57 |
| 9 | | | 19 | | 15 | 26 | 60 |
| 10 | | | | 5 | 6 | 7 | 18 |
| $T_j$ | 70 | 105 | 88 | 66 | 55 | 96 | 480 |
| $kT_j$ | 210 | 315 | 264 | 198 | 165 | 288 | 1,440 |
| $B_j^*$ | 225 | 276 | 291 | 213 | 189 | 246 | 1,440 |
| $kT_j - B_j^* = Q_j$ | $-15$ | $+39$ | $-27$ | $-15$ | $-24$ | $+42$ | 0 |
| $Q_j/t\lambda = t_j$ | $-1.25$ | $+3.25$ | $-2.25$ | $-1.25$ | $-2.00$ | $+3.50$ | 0 |

totals of the blocks in which the treatment under consideration appears. For example, treatment (6) appears in blocks 2, 4, 8, 9, 10. The sum of these five block totals is

$$B_6^* = 63 + 48 + 57 + 60 + 18 = 246$$

For analysis of variance we need the following four quantities:

$$A = 11^2 + \cdots = 9,050 \qquad B = \frac{\Sigma B_i^2}{k} = 8,712$$

$$C = \frac{480^2}{30} = 7,680 \qquad \Sigma Q_j^2 = 5,040$$

Hence, we have the analysis of variance table.

| Source | df | ssq | msq | |
|--------|-----|-----|-----|-----|
| Blocks, raw | $b - 1 = 9$ | $B - C = 1{,}032$ | | |
| Treatments, adj. | $t - 1 = 5$ | $\Sigma Q^2/kt\lambda = \quad 140$ | 28.0 | $F = 2.12$ |
| Error | $rt - b - r + 1 = 15$ | (subt) $= \quad 198$ | 13.2 | |
| Total | $rt - 1 = 29$ | $A - C = 1{,}370$ | | |

**27.3**  Find the adjusted treatment means $\bar{y} + t_j$ of the experiment in Exercise 27.2 and calculate the variance and standard error of the difference between two adjusted treatment means.

**27.4**  When $t$ is a perfect square (for example, 4, 9, 16, 25, etc.), the construction is particularly simple by using orthogonal Latin squares of corresponding size. To illustrate the method, let us consider the case $t = 9$ and $k = 3$.

At the right we write out a $3 \times 3$ Graeco-Latin square, and at the left we write out the nine treatments in the form of a square. We then collect the three treatments into blocks by rows, by columns, by Greek letters, and finally by Latin letters. The results are as tabulated. For this design, $t \times r = 9 \times 4 = 36$, $k \times b = 3 \times 12 = 36$, and $\lambda = 1$. Furthermore, the twelve blocks fall into four replications. Designs of this nature are called *balanced lattices*. For this particular small lattice, we may use the method described in this chapter. For larger lattices a more complicated method is available. It is clear, however, that lattice designs in general are for a very large number of treatments and are seldomly used in human and medical experiments.

$$
\begin{array}{ccc}
\alpha\ A & \beta\ B & \gamma\ C \\
\gamma\ B & \alpha\ C & \beta\ A \\
\beta\ C & \gamma\ A & \alpha\ B
\end{array}
\qquad
\begin{array}{ccc}
1 & 2 & 3 \\
4 & 5 & 6 \\
7 & 8 & 9
\end{array}
$$

| By rows | By columns | By Greek | By Latin |
|---------|------------|----------|----------|
| (1, 2, 3) | (1, 4, 7) | (1, 5, 9) | (1, 6, 8) |
| (4, 5, 6) | (2, 5, 8) | (2, 6, 7) | (2, 4, 9) |
| (7, 8, 9) | (3, 6, 9) | (3, 4, 8) | (3, 5, 7) |

**27.5**  It will be recalled that at the end of the preceding chapter we have used an incomplete $3 \times 3$ Latin square to exhibit the full algebraic details. By the same token, the reader may use an extremely simple balanced incomplete block design (that still retains the general properties) to gain an understanding of the mathematical descriptions given in this chapter. Suppose that there are four treatments ($t = 4$) to be tested in blocks of three units each ($k = 3$). Then we need four blocks ($b = 4$) to achieve balance. The four blocks are: I (2, 3, 4), II (1, 3, 4), III (1, 2, 4), IV (1, 2, 3).

Omitting the general constant term and the error term of the linear model, we may write out the remaining terms of the model for each observation as in the accompanying table, remembering that $\Sigma b_i = 0$ and $\Sigma t_j = 0$. Verify all the mathematical relationships described in this chapter, particularly that for adjusted treatment sum of squares. Note that $\lambda = 2$ in this design.

| Blocks | Treatments | | | | Block total |
| | (1) | (2) | (3) | (4) | $B_i$ |
|---|---|---|---|---|---|
| I | $\cdots$ | $t_2 + b_1$ | $t_3 + b_1$ | $t_4 + b_1$ | $3b_1 - t_1$ |
| II | $t_1 + b_2$ | $\cdots$ | $t_3 + b_2$ | $t_4 + b_2$ | $3b_2 - t_2$ |
| III | $t_1 + b_3$ | $t_2 + b_3$ | $\cdots$ | $t_4 + b_3$ | $3b_3 - t_3$ |
| IV | $t_1 + b_4$ | $t_2 + b_4$ | $t_3 + b_4$ | $\cdots$ | $3b_4 - t_4$ |
| $T_j$ | $3t_1 - b_1$ | $3t_2 - b_2$ | $3t_3 - b_3$ | $3t_4 - b_4$ | 0 |

**27.6 Missing value.** Referring to Table 27.2, suppose that the observation on treatment (1) in the first block is missing, so that $B_1' = B_1 + y = 26 + y$ and $T_1' = T_1 + y = 13 + y$. The symbol with prime includes the missing $y$ and that without prime denotes the actual (incomplete) numerical value. Then

$$Q_1' = k(T_1 + y) - (B_1 + y) - B_5 - B_7 = Q_1 + (k - 1)y$$

where $Q_1 = kT_1 - B_1 - B_5 - B_7$ is the value calculated without $y$. The missing value also affects $Q_2'$ and $Q_4'$ because treatments (2) and (4) are also in the first block. Thus,

$$Q_2' = kT_2 - (B_1 + y) - B_2 - B_6 = Q_2 - y$$

Similarly, $Q_4' = Q_4 - y$. In the analysis of variance (Table 27.3), the sum of squares for error is obtained by subtraction,

$$ssq_E = y^2 - \frac{(B_1 + y)^2}{k} - \frac{[Q_1 + (k - 1)y]^2 + (Q_2 - y)^2 + (Q_4 - Y)^2}{kt\lambda} + \text{numbers}$$

Setting the derivative of $ssq_E$ with respect to $y$ to zero, we obtain the equation

$$y - \frac{B_1 + y}{k} - \frac{(k - 1)Q_1 + (k - 1)^2y - (Q_2 - y) - (Q_4 - y)}{kt\lambda} = 0$$

In general, the terms involving $y$ in the numerator of the last factor is $(k - 1)^2y + (k - 1)y = k(k - 1)y$. The rest is $kQ_1 - (Q_1 + Q_2 + Q_4)$. Solving (and dropping the subscript for generality) we obtain

$$y = \frac{t\lambda B + kQ - \Sigma Q}{(t\lambda - k)(k - 1)}$$

where $\Sigma Q$ means the sum of the $Q$'s for all treatments in the block under consideration, and $Q$ is the particular $Q$ for the missing treatment. Note the similarity in form of this expression to that on page 229 for ordinary randomized blocks.

# 28

# Symmetrical pairs

The smallest possible block is that of two experimental units only. This is a special case of the balanced incomplete blocks designs dealt with in the preceding chapter. Blocks of "size" 2 arise in nature frequently, especially in human experiments. For instance, in testing insect repellents, the most uniform block is the two arms of the same individual. (TV commercials for detergents and hand lotions adopt this principle, but the difference between the two hands of the attractive lady shown on the screen is not necessarily due more to the product under consideration than to some other "treatments.") Identical twins, of human or of other animals, form a natural homogeneous block of experimental units. For certain small-scale experiments in which the heterogeneity from individual to individual is great, it is most desirable to find pairs of comparatively homogeneous units.

## 1. All possible pairs of treatments

Let there be $t$ treatments to be studied, and let the number of units per block be $k = 2$. In this special case, the number of blocks $b$ and the number of replications $r$ needed for balance may be expressed in terms of $t$. Obviously,

$$r = t - 1 \qquad b = \binom{t}{2} = \frac{t(t-1)}{2}$$

so that the total number of units needed for all possible pairs of treatments to appear in one block is

$$N = bk = tr = t(t-1) \qquad \text{and} \qquad \lambda = \frac{r(k-1)}{t-1} = 1$$

All the above refer to one basic balanced design. Of course, the entire design may be replicated several times when needed.

The upper portion of Table 28.1 enumerates all possible pairs of treat-

*Table* 28.1   Possible pairs of treatments and number of blocks.
$k = 2, r = t - 1, \lambda = 1$ (upper portion of table is cumulative)

| $t$ | Pairs of treatments | | | | | | | | $b$ |
|---|---|---|---|---|---|---|---|---|---|
| 2 | (1, 2) | | | | | | | | 1 |
| 3 | (1, 3) | (2, 3) | | | | | | | 3 |
| 4 | (1, 4) | (2, 4) | (3, 4) | | | | | | 6 |
| 5 | (1, 5) | (2, 5) | (3, 5) | (4, 5) | | | | | 10 |
| 6 | (1, 6) | (2, 6) | (3, 6) | (4, 6) | (5, 6) | | | | 15 |
| 7 | (1, 7) | (2, 7) | (3, 7) | (4, 7) | (5, 7) | (6, 7) | | | 21 |
| 8 | (1, 8) | (2, 8) | (3, 8) | (4, 8) | (5, 8) | (6, 8) | (7, 8) | | 28 |
| 9 | (1, 9) | (2, 9) | (3, 9) | (4, 9) | (5, 9) | (6, 9) | (7, 9) | (8, 9) | 36 |
| · · · | · · · · · · · · · · · · · · · · · · · · · · · · | | | | | | | | · · · |

Pairs collected into replications

| | I | II | III | IV | V | | | $b$ |
|---|---|---|---|---|---|---|---|---|
| | (1, 2) | (1, 3) | (1, 4) | (1, 5) | (1, 6) | | | |
| $t = 6$ | (3, 4) | (2, 5) | (2, 6) | (2, 4) | (2, 3) | | | 15 |
| | (5, 6) | (4, 6) | (3, 5) | (3, 6) | (4, 5) | | | |

| | I | II | III | IV | V | VI | VII | $b$ |
|---|---|---|---|---|---|---|---|---|
| $t = 4$ | (1, 2) | (1, 3) | (1, 4) | (1, 5) | (1, 6) | (1, 7) | (1, 8) | 6 |
| | (3, 4) | (2, 4) | (2, 3) | (2, 6) | (2, 5) | (2, 8) | (2, 7) | |
| $t = 8$ | (5, 6) | (5, 7) | (5, 8) | (3, 7) | (3, 8) | (3, 5) | (3, 6) | 28 |
| | (7, 8) | (6, 8) | (6, 7) | (4, 8) | (4, 7) | (4, 6) | (4, 5) | |

ments and the number of blocks up to $t = 9$. The pattern of enumeration of possible pairs is so obvious that the reader can extend the table to cases $t = 10, 11, 12$, etc. The design that includes all possible pairs of treatments in blocks of 2 units is called symmetrical pairs (Yates, 1936). When $t$ is large, the symmetrical pair design ceases to be practical because it requires too many replications. For instance, when $t = 16$, we need 120 pairs (240 single observations) and each treatment must be replicated 15 times. In such cases we must use larger blocks. For $t = 16$, it is possible to have a balanced design with $r = k = 6$ and only 96 observations. Thus we see that symmetrical pair design is essentially for small experiments.

When the number of treatments is even ($t = 4, 6, 8$, etc.), it is possible to collect the pairs into complete replications in which each treatment appears once, just as in the general case wherein blocks of $k$ units may be collected into replications when $t$ is a multiple of $k$. Thus, the 15 pairs (or blocks) for 6 treatments may be collected into 5 replications as shown in the lower portion of Table 28.1. Similarly, for $t = 4$, the 6 blocks may be grouped into 3 replications. When $t = 8$, the 28 blocks may be arranged into 7 replications. When the number of treatments is odd, the grouping into replications is obviously impossible.

## 2. Procedure of analysis

The example chosen for illustration is the case for five treatments with ten blocks (pairs). In practice, the two treatments are assigned to the units of a pair at random. Then the data are collected into the form of Table 28.2 for purpose of calculation. The total number of observations is $N = t(t - 1) = 5 \times 4 = 20$.

The method of analysis is the same as that prescribed in the preceding chapter, except that $k = 2$ in the present case. First, we obtain the block totals $B_i$ and treatment totals $T_j$ and $2T_j$. The quantity $B_j^*$ is the sum of the block totals in which the treatment $j$ appears. Thus, for treatment (1), which appears in blocks 1, 2, 4, 7,

$$B_1^* = B_1 + B_2 + B_4 + B_7 = 36 + 34 + 42 + 40 = 152$$

The quantities in the last row are $Q_j = 2T_j - B_j^*$. Thus,

$$Q_1 = 120 - 152 = -32$$

The sum of these $Q$'s is zero, which provides an arithmetic check. This completes the preliminary calculations in Table 28.2, ignoring the column $D_i$ for the time being.

*Table* 28.2 Data of a symmetrical pair experiment with five treatments and ten blocks

| Block | Treatments | | | | | Block total $B_i$ | Pair difference $D_i$ |
|---|---|---|---|---|---|---|---|
| | (1) | (2) | (3) | (4) | (5) | | |
| 1 | 17 | 19 | | | | 36 | $-2$ |
| 2 | 11 | | 23 | | | 34 | $-12$ |
| 3 | | 17 | 25 | | | 42 | $-8$ |
| 4 | 13 | | | 29 | | 42 | $-16$ |
| 5 | | 22 | | 28 | | 50 | $-6$ |
| 6 | | | 21 | 25 | | 46 | $-4$ |
| 7 | 19 | | | | 21 | 40 | $-2$ |
| 8 | | 18 | | | 18 | 36 | 0 |
| 9 | | | 20 | | 18 | 38 | $+2$ |
| 10 | | | | 20 | 16 | 36 | $+4$ |
| $T_j$ | 60 | 76 | 89 | 102 | 73 | 400 | |
| $2T_j$ | 120 | 152 | 178 | 204 | 146 | $\Sigma B_i^2 = 16,232$ | $\Sigma D_i^2 = 544$ |
| $B_j^*$ | 152 | 164 | 160 | 174 | 150 | | |
| $Q_j$ | $-32$ | $-12$ | 18 | 30 | $-4$ | $\Sigma Q_j = 0$ | $\Sigma Q_j^2 = 2,408$ |

For analysis of variance the following three basic quantities (areas) are calculated the usual way,

$$A = 17^2 + 19^2 + \cdots + 16^2 = 8,388$$
$$B = \sum \frac{B_i^2}{2} = \frac{16,232}{2} = 8,116$$
$$C = \frac{Y^2}{N} = \frac{400^2}{20} = 8,000$$

so that total $ssq = A - C - 388$ and block $ssq = B - C = 116$, unadjusted. The

$$\text{Adjusted treatment } ssq = \frac{\Sigma Q^2}{kt\lambda} = \frac{2,408}{2 \times 5 \times 1} = 240.8$$

*Table* 28.3 Analysis of variance for data in Table 28.2

| Variation | df | ssq | msq |
|---|---|---|---|
| Blocks, raw | $b - 1 = 9$ | $B - C = 116.0$ | |
| Treatments, adj. | $t - 1 = 4$ | $\Sigma Q_j^2/2t = 240.8$ | 60.2 |
| Intrablock error | $rt - b - t + 1 = 6$ | (subt) 31.2 | $5.2 = s^2$ |
| Total | $rt - 1 = 19$ | $A - C = 388.0$ | |

The error *ssq* is then obtained by subtraction.  Note that the error *df* is

$$rt - b - t + 1 = t(t - 1) - \tfrac{1}{2}t(t - 1) - t + 1 = \tfrac{1}{2}(t - 1)(t - 2)$$

The analysis of variance is given in Table 28.3.  The *F* value is $60.2/5.2 = 11.58$, being highly significant.

### 3. Adjusted treatment means

The treatment effect $t_j$ is equal to $Q_j/t\lambda = Q/t = Q/5$.  The adjusted treatment means are $\bar{y} + t_j$, where $\bar{y}$ is the general mean of the entire

| Treatment | (1) | (2) | (3) | (4) | (5) | Mean |
|-----------|-----|-----|-----|-----|-----|------|
| $Q_j$ | −32 | −12 | +18 | +30 | −4 | 0 |
| $t_j$ | −6.4 | −2.4 | +3.6 | +6.0 | −0.8 | 0 |
| Adj. mean | 13.6 | 17.6 | 23.6 | 26.0 | 19.2 | 20.0 |

experiment.   In our example, $\bar{y} = {}^{400}\!/_{20} = 20$.   The variance of the difference between two adjusted treatment means (or two *t*'s) is

$$\frac{2s^2}{r}\left[\frac{k(t - 1)}{t(k - 1)}\right] = \frac{4s^2}{t} = \frac{4(5.2)}{5} = 4.16$$

and the standard error is $\sqrt{4.16} = 2.04$.   Treatment (4) is significantly higher than treatments (1), (2), and (5).   Also see Chap. 31.

### 4. Method of pair differences

It was pointed out in the preceding chapter that the difference between two observations in the same block does not contain the block effects. Hence, when a block contains only two units, a special method is available for eliminating the block effects.   Instead of taking the block totals $B_i$, we may take the difference $D_i$ between the two observations in each block, as shown in the last column of Table 28.2.   This procedure is quite analogous to the Student pairing method for two groups (Chap. 10), which is equivalent to the usual method of two-way analysis of variance.

After taking the differences, we now have, instead of the original $N = t(t - 1)$ single observations, $\tfrac{1}{2}t(t - 1)$ differences with as many degrees of freedom.   The analysis is then entirely based on these differences without referring to blocks anymore.   In our example the total *ssq* for the 10 differences is $\Sigma D_i^2 = 544$ with 10 *df*.   No correction term is needed.   One should, however, remember that this *ssq* is in units of $D$ and no longer in units of the original observation *y*.   From Table 28.3

we observe that the total *ssq* for treatments (adjusted) and error is

$$ssq \text{ (treatments and error)} = 240.8 + 31.2 = 272$$

and here

$$ssq_D = \Sigma D_i^2 = 544 = 272 \times 2$$

It is then clear that the *ssq* of the $D$'s contain only treatment and error but no block *ssq*. That it is twice as large is due to the fact that $D$ is the difference of two single values. The algebra is similar to Exercise 10.3.

To obtain the treatment effects from the $D$ values, we arrange the $D$ values into a square, the first column of which contains the differences $(1) - (2)$, $(1) - (3)$, $(1) - (4)$, $(1) - (5)$. The second column contains differences $(2) - (1)$, $(2) - (3)$, etc. The arrangement is given in Table 28.4. The sums of these differences are denoted by $Q_j$ because they are identical with those shown in Table 28.2. In units of $D$, the

$$\text{Adjusted treatment } ssq = \frac{\Sigma Q_j^2}{t\lambda} = \frac{2408}{5 \times 1} = 481.6$$

The error *ssq* may then be obtained by subtraction. The analysis of variance is given in Table 28.5. This is a direct analysis of treatment and

**Table 28.4  Differences between treatments in the same block**

|  | Treatments | | | | |
|---|---|---|---|---|---|
|  | (1) | (2) | (3) | (4) | (5) |
| $-(1)$ | $\cdots$ | $+2$ | $12$ | $16$ | $2$ |
| $-(2)$ | $-2$ | $\cdots$ | $8$ | $6$ | $0$ |
| $-(3)$ | $-12$ | $-8$ | $\cdots$ | $4$ | $-2$ |
| $-(4)$ | $-16$ | $-6$ | $-4$ | $\cdots$ | $-4$ |
| $-(5)$ | $-2$ | $0$ | $2$ | $4$ | $\cdots$ |
| $Q_j$ | $-32$ | $-12$ | $18$ | $30$ | $-4$ |

**Table 28.5  Analysis of variance of pair differences (same data)**

| Variation | df | ssq | msq |
|---|---|---|---|
| Treatment, adj. | $t - 1 = 4$ | $\Sigma Q_j^2/t = 481.6$ | $120.4$ |
| Intrablock error | $\frac{1}{2}(t-1)(t-2) = 6$ | (subt) $\quad 62.4$ | $10.4 = s_D^2$ |
| Treatment and error | $\frac{1}{2}t(t-1) = 10$ | $\Sigma D_i^2 = 544.0$ | |

error only, without calculating the block *ssq*.  Since the *ssq* entries are all twice those in Table 28.3, the error variance is also twice as large; $s_D{}^2 = 10.4 = 2 \times 5.2$.  It is the estimate of the variance of the *D* values, not of single observations.  The *F* value remains the same as before, and so does the variance of the difference between two adjusted means except that $4s^2$ may be written $2s_D{}^2$.

## 5. Repetitions of basic design

Suppose that there are four treatments to be tested in blocks of two units each.  There are only 6 possible pairs (Table 28.1) in the balanced design, with 12 single observations.  Of the total 11 *df*, there will be 5 for blocks, 3 for treatments, and only 3 left for the estimation of error.  An experiment of such a size is usually regarded as too small to give reliable conclusions.  Under this circumstance the entire set of six pairs may be replicated as many times as the investigator sees fit.

Repetition of the entire design is sometimes particularly desirable when individuals are employed as blocks rather than as units of experimentation.  Consider four repellents once more.  Each individual (block) receives two of the four treatments, one on his left arm and one on his right, to be assigned at random.  The basic design needs six individuals (six pairs of arms).  However, if 12 individuals are available for experimentation, the results will be much more reliable, and in this particular case, we may reverse the left or right assignment of the two treatments on the arms in the second individual.

Similarly, if there are four drugs to be tested and each has only temporary effect, so that an individual may be used as a block of two units, the order of application of the drugs may be reversed in the second set of pairs.  This practice of using individuals as blocks not only makes maximum use of the number of patients available for experimentation but also eliminates the heterogeneity from individual to individual, which is usually very considerable.

It has been mentioned that when the number of treatments is even, the blocks may be collected into replications of treatments, but in the following example it is not necessary to so collect them.  The variation among the replications will be included in the block *ssq*.

Table 28.6 gives the data for four treatments in blocks of two units.  The first six blocks constitute a balanced design, and the second six blocks is another balanced design with the order of treatments reversed.  Individual 1's left arm receives repellent (1), individual 7's left arm receives repellent (2), and so on.

The procedure of analysis is exactly the same as before except that now the values of *r*, *b*, and *λ* have all been doubled.  Thus, for $t = 4$,

*Table 28.6* **Data and analysis of four treatments tested in twelve pairs**

| Block (individual) | Left (first) | Right (second) | Treatments (1) | (2) | (3) | (4) | Block total $B_i$ | Difference $D_i$ (see Exercise 28.1) |
|---|---|---|---|---|---|---|---|---|
| 1 | (1) | (2) | 21 | 11 | | | 32 | |
| 2 | (1) | (3) | 33 | | 23 | | 56 | |
| 3 | (2) | (3) | | 8 | 24 | | 32 | |
| 4 | (1) | (4) | 31 | | | 13 | 44 | |
| 5 | (2) | (4) | | 10 | | 14 | 24 | |
| 6 | (3) | (4) | | | 27 | 17 | 44 | |
| 7 | (2) | (1) | 28 | 12 | | | 40 | |
| 8 | (3) | (1) | 28 | | 26 | | 54 | |
| 9 | (3) | (2) | | 14 | 20 | | 34 | |
| 10 | (4) | (1) | 26 | | | 8 | 34 | |
| 11 | (4) | (2) | | 13 | | 11 | 24 | |
| 12 | (4) | (3) | | | 20 | 18 | 38 | |
| Treatment total $T_j$ | | | 167 | 68 | 140 | 81 | 456; $\Sigma B_i^2 = 18{,}480$ | |
| $2T_j$ | | | 334 | 136 | 280 | 162 | $912 = 2(456)$ | |
| $B_j^*$ | | | 260 | 186 | 258 | 208 | $912$; $kt\lambda = 2 \times 4 \times 2$ | |
| $2T_j - B_j^* = Q_j$ | | | 74 | $-50$ | 22 | $-46$ | $0$  $\Sigma Q_j^2 = 10576$ | |
| $Q_j/t\lambda = Q_j/8 = t_j$ | | | 9.25 | $-6.25$ | 2.75 | $-5.75$ | $0$ | |
| Adj. mean $\bar{y} + t_j$ | | | 28.25 | 12.75 | 21.75 | 13.25 | $19.00 = \bar{y}$ | |

Usual quantities:

$A = \Sigma y^2 = 10{,}002$    $B = \Sigma B_i^2/2 = 9{,}240$    $C = 456^2/24 = 8{,}664$

Analysis of variance

| Variation | df | ssq | msq |
|---|---|---|---|
| Blocks, raw | 11 | $B - C =$ 576 | |
| Treatments, adj. | 3 | $\Sigma Q^2/kt\lambda =$ 661 | 220.33  $F = 19.6$ |
| Error | 9 | (subt) 101 | 11.22 |
| Total | 23 | $A - C = 1{,}338$ | |

$r = 2(t - 1) = 6$, $b = 6(t - 1) = 12$, and $\lambda = 2$. The arithmetic details and analysis of variance are all given in the lower part of Table 28.6 and no special explanation is needed. The conclusion of the analysis is that treatments (1) and (3) are significantly better than (2) and (4).

The variance of the difference between two adjusted treatment means is

$$\frac{2s^2}{r}\left[\frac{k(t - 1)}{t(k - 1)}\right] = \frac{2(11.22)}{6}\left(\frac{2 \times 3}{4}\right) = 5.61$$

and the standard error is $\sqrt{5.61} = 2.37$.   Student's $t$ test for treatments
(1) and (2) yields $= (9.25 - 2.75)/2.37 = 2.74$ with 9 $df$, being significant
at the 5 per cent level.

This completes the three chapters on incomplete blocks.   The reader
should be reminded that we have barely touched upon the very wide
variety of possible designs with incomplete blocks.   However, the more
complicated designs are aimed at testing large numbers of crop varieties
or treatments (for example, $t = 125 = 5^3$) and are more suitable for agricultural field experiments than for medical research with human units.
The simple designs outlined here are generally applicable to almost any
type of laboratory experiments.


### EXERCISES

**28.1**   Analyze the data of Table 28.6 by the method of pair differences.   That is,
instead of taking the block totals $B_i$, we obtain the difference $D_i$ between the two
observations in each block.   These differences, each used twice with opposite signs,
may be arranged to yield treatment effects free of block influences.
*Partial Ans.:* $\Sigma D_i^2 = 1{,}524 = 2 \times (661 + 101)$

**28.2**   There are six treatments tested in fifteen pairs (blocks of two units each).
The observations are as follows:

| (1) (2) | (1) (3) | (2) (3) | (1) (4) | (2) (4) |
|---|---|---|---|---|
| 13  11 | 22  12 | 18   6 | 23  13 | 19  23 |

| (1) (5) | (2) (5) | (3) (5) | (4) (5) | (3) (4) |
|---|---|---|---|---|
| 28  14 | 19  21 | 15  13 | 5  21 | 3   9 |

| (1) (6) | (2) (6) | (3) (6) | (4) (6) | (5) (6) |
|---|---|---|---|---|
| 19  23 | 3  15 | 19  25 | 16  26 | 19   7 |

Collect the data into the form of Table 28.2 to facilitate calculation.   This step is
not necessary for experienced workers, but it does make the arithmetic more systematic and minimize the risk of error.   Do an analysis of variance and test the significance of treatment effects.

*Partial Ans.:*

|  | (1) | (2) | (3) | (4) | (5) | (6) |
|---|---|---|---|---|---|---|
| $Q$: | $+32$ | $-8$ | $-32$ | $-26$ | $+14$ | $+20$ |

| Source | $df$ | $ssq$ |
|---|---|---|
| Blocks, raw | 14 | 720 |
| Treatments, adj. | 5 | 282 |
| Error | 10 | 368 |

# 29

# Residual effects

In subject matter this chapter is a continuation of Chap. 18, in which change-over sequences of treatment on the same experimental animal are considered. The assumption there is made that the rest period between successive treatments is long enough for the effects of the preceding treatment to wear off. A long rest period between two treatments is, however, not always practical (in nutrition tests, for example), and may even introduce other complicating factors due to change in age or in environmental conditions. One compromise is to allow a reasonable rest period whenever possible and make allowance for residual effects in the statistical model and analysis. The introduction of designs to cope with residual effects is postponed to this chapter because the analysis bears certain resemblance to that for incomplete Latin squares and incomplete block designs. There are numerous possibilities and variations in designs for estimating residual effects. Again, we can mention only the simplest for a small number of treatments with a small number of experimental animals. These designs are potentially useful in clinical trials with various drugs on a limited number of patients.

## 1. Balanced sequences

All of the balanced designs for analyzing both direct and residual effects of treatments are based on complete or incomplete Latin squares. Suppose there are three ($t = 3$) treatments, designated as $a$, $b$, $c$. These may be arranged into two $3 \times 3$ Latin squares, each with three experimental animals treated in three successive periods, as shown in Table 29.1. With three letters ($a$, $b$, $c$), there are $3! = 3 \times 2 \times 1 = 6$ possible sequences, and these are the six columns shown in Table 29.1. When $t$ is large, it is

*Table* 29.1   Balanced sequences for three treatments

| Period | Animals | | | Animals | | |
|---|---|---|---|---|---|---|
|  | (1) | (2) | (3) | (4) | (5) | (6) |
| I | $a$ | $b$ | $c$ | $a$ | $b$ | $c$ |
| II | $b$ | $c$ | $a$ | $c$ | $a$ | $b$ |
| III | $c$ | $a$ | $b$ | $b$ | $c$ | $a$ |

unnecessary to have all $t!$ possible sequences. The criterion for balanced sequences is that every treatment is preceded or followed by all the other treatments the same number of times. Thus, in the design above, each of the two-period sequences $ab$, $ba$, $ac$, $ca$, $bc$, $cb$ occurs twice.

The reason that we concentrate on two-period sequences is that we shall assume that the residual effects of a treatment are felt only in the immediately following period and not after that. Let $t_a$ be the direct effect of treatment $a$ and $r_a$ be $a$'s residual effect in the immediately following period. Then for the three animals in the first Latin square, the treatment effects are as given in Table 29.2. In practice the two squares should be repeated a certain number of times. Twelve animals are needed for two complete replications.

Designs of this type are simple enough, but, unfortunately, the analysis is long and tedious owing to the nonorthogonality between the direct and residual effects. If, however, we prolong the experiment for an additional period with the same treatment as in the last, direct and residual effects will become orthogonal and the analysis then becomes much simpler. This is the design we shall illustrate with a numerical example in the next section.

*Table 29.2* Model for direct and residual effects

| Period | Animals | | |
|---|---|---|---|
| | (1) | (2) | (3) |
| I | $t_a$ | $t_b$ | $t_c$ |
| II | $t_b + r_a$ | $t_c + r_b$ | $t_a + r_c$ |
| III | $t_c + r_b$ | $t_a + t_c$ | $t_b + r_a$ |

## 2. Extra-period balanced sequences

The treatments in the first three periods of Table 29.3 are the same as those indicated in the preceding section. The distinctive feature is that an extra period (IV) is being added with the same treatments as in period III. Now, every treatment is preceded or followed by every other treatment *including itself* the same number of times, so that the $t$'s and $r$'s are orthogonal. A practical experiment will involve at least two complete replications with 12 animals, but we shall illustrate the procedure of analysis with 6 animals and 24 observations.

Some new terminology and symbols are needed. We have already used the word "animal" instead of the usual "unit," because several observations are made on the same individual (cow, patient, etc.). Each group of 3 animals $\times$ 4 periods will be referred to as a "square" (although it actually constitutes a rectangle) instead of the usual "block" to avoid possible misunderstanding.

Let $p$ be the number of periods for an ordinary balanced sequence design, so that $p + 1$ is the total number of periods in the corresponding extra-period design. Let $t$ be the number of treatments, $k$ be the number of animals in a square, and $q$ be the number of squares. The total number of animals is then $kq$, and the total number of observations is $(p + 1)kq$. In the simple example of Table 29.3, where complete Latin squares form

*Table 29.3* Balanced extra-period design with three treatments $(a, b, c)$

| Period | Animals | | | Row | Animals | | | Row | Overall period |
|---|---|---|---|---|---|---|---|---|---|
| | (1) | (2) | (3) | | (4) | (5) | (6) | | |
| I | a 16 | b 16 | c 18 | 50 | a 22 | b 10 | c 8 | 40 | 90 = I |
| II | b 17 | c 23 | a 28 | 68 | c 23 | a 27 | b 20 | 70 | 138 |
| III | c 16 | a 25 | b 16 | 57 | b 16 | c 19 | a 16 | 51 | 108 |
| (Extra) IV | c 24 | a 28 | b 19 | 71 | b 18 | c 25 | a 30 | 73 | 144 |
| Animal total | | 73 | 92 | 81 | 246 | 79 | 81 | 74 | 234 | 480 = G |

the basis of design, we have $t = k = p = 3$ and $q = 2$, so that there are $kq = 6$ animals and $4 \times 6 = 24$ observations. We mention these parameters because, in general, the values of $t$, $k$, $p$ are not necessarily equal, especially when $t$ is large. The case $t = k = p$ is the simplest type of the design, which is useful only when there are a few treatments to be tested.

## 3. Preliminary calculations

The first step of analysis is to obtain the marginal totals shown in Table 29.3. The second step is to calculate the direct and residual treatment totals, designated as $T$ and $R$, respectively, in Table 29.4 The $T$'s are the ordinary treatment totals. Thus, $T_a = 192$ is the sum of the eight observations under treatment $a$ in Table 29.3. The $R$ values are new; $R_a$ is the sum of observations immediately following treatment $a$ (regardless of the treatments for the observations themselves). For instance, proceeding from left to right of Table 29.3,

$$R_a = 17 + 28 + 16 + 23 + 19 + 30 = 133$$

The values of $R_b$ and $R_c$ are obtained the same way. The observations in period I are not included in the $R$'s because the initial observations have no residual effects. An arithmetic check is that

$$\Sigma R = G - I = 480 - 90 = 390$$

Another new quantity, $U$, is also needed. It is the sum of columns (animals) in which this treatment occurs in the final (extra) period. Thus, $U_a = 92 + 74 = 166$. The sum of the $U$'s is $G = 480$.

*Table 29.4*   Calculation of direct and residual treatment totals

| Treatment | $T$ | $R$ | $U$ | $U + G$ |
|-----------|-----|-----|-----|---------|
| $a$ | 192 | 133 | 166 | 646 |
| $b$ | 132 | 119 | 160 | 640 |
| $c$ | 156 | 138 | 154 | 634 |
| Total | 480 | 390 | 480 | 1920 |
| Check | $G$ | $G - I$ | $G$ | $4G$ |

| Treatment | $(p + 1)T$ $= 4T$ | $pR$ $= 3R$ | $\hat{T} =$ $4T - (U + G)$ | $\hat{R} =$ $3R - (G - I)$ |
|-----------|-----|-----|-----|-----|
| $a$ | 768 | 399 | $+122$ | $+9$ |
| $b$ | 528 | 357 | $-112$ | $-33$ |
| $c$ | 624 | 414 | $-10$ | $+24$ |
| Total | 1,920 | 1,170 | 0 | 0 |

The direct effects $(t_a, t_b, t_c)$ and residual effects $(r_a, r_b, r_c)$ of treatments are separated by forming the following expressions:

$$\hat{T} = (p + 1)T - U - G = 4T - (U + G)$$
$$\hat{R} = pR - (G - I) = 3R - (G - I)$$

These values are shown in the lower portion of Table 29.4.  Note that the sum of these $\hat{T}$'s and $\hat{R}$'s is zero, providing an arithmetic check.

These are all we need for the simplest extra-period balanced sequence designs based on complete Latin squares.  For balanced designs without the extra period or when $t > k$, several other quantities are needed.

## 4. Analysis of variance

The analysis of variance is set out in Table 29.5.  The sum of squares due to the six experimental animals is calculated the usual way:

$$\tfrac{1}{4}(73^2 + 92^2 + \cdots + 74^2) - \tfrac{1}{24}(480^2) = 9,658 - 9,600 = 58$$

and the *ssq* due to periods within each "square" is also obtained the usual way:

$$\tfrac{1}{3}(50^2 + 68^2 + \cdots + 51^2 + 73^2) - \tfrac{1}{12}(246^2 + 234^2)$$
$$= 9,948 - 9,606 = 342$$

The *ssq*'s due to direct and residual effects of treatments are respectively

$$\frac{\Sigma \hat{T}^2}{qt(t + 1)(t + 2)} = \frac{122^2 + 112^2 + 10^2}{2 \times 3 \times 4 \times 5} = \frac{27,528}{120} = 229.40$$

$$\frac{\Sigma \hat{R}^2}{qt^3} = \frac{9^2 + 33^2 + 24^2}{2 \times 3 \times 3 \times 3} = \frac{1,746}{54} = 32.33$$

The total *ssq* for the 24 observations is $A - C = 10,324 - 9,600 = 724$, and the error *ssq* is obtained by subtraction.  Mean square and variance ratio are calculated the usual way.  In this particular example, the direct

**Table 29.5** Analysis of variance of an extra-period design experiment (based on data of Table 29.3)

| Source of variation | df | ssq | msq | F |
|---|---|---|---|---|
| All animals | 5 | 58.00 | | |
| Periods within square | 6 | 342.00 | | |
| Direct effects | 2 | 229.40 | 114.70 | 14.74 |
| Residual effects | 2 | 32.33 | 16.17 | 2.08 |
| Error | 8 | 62.27 | 7.78 | |
| Total | 23 | 724.00 | | |

effects are significant at the 1 per cent level and the residual effects are not significant at the 5 per cent level.

## 5. Treatment means and variances

The direct and residual effects on a per observation basis are

$$\hat{t} = \frac{\hat{T}}{qt(t+2)} \qquad \hat{r} = \frac{\hat{R}}{qt^2}$$

where $q$ = number of squares and $t$ = number of treatments for designs based on complete Latin squares with an extra period. In our example, $q = 2$ and $t = p = 3$, so that $qt(t+2) = qp(p+2) = 30$, and $qt^2 = 18$. The estimated effects (from Table 29.4) are given in Table 29.6. The observed treatment means are simply the ordinary means. For treatment $a$, the observed mean is $192/8 = 24.00$ and the general mean is $\bar{y} = 20.00$. The treatment mean adjusted for residual effects is $\bar{y} + \hat{t}$. Since the residual effects are small in our example, the adjusted and crude means do not differ very much.

The variance of $\hat{t}$ and $\hat{r}$ for the simple design under consideration is

$$V(\hat{t}) = \frac{(p+1)s^2}{qp(p+2)} = \frac{4(7.78)}{2 \times 3 \times 5} = 1.037$$

$$V(\hat{r}) = \frac{(p+1)s^2}{qp(p+1)} = \frac{s^2}{qp} = \frac{7.78}{6} = 1.297$$

The latter is larger than the former, but the two are much closer than they would be without the extra period. The difference between two adjusted treatment means is simply the difference between two $\hat{t}$'s with variance $2V(\hat{t})$. The variance of the difference between two observed (unadjusted) treatment means, containing both direct and residual effects, is

$$2[V(\hat{t}) + V(\hat{r})] = \frac{2(p+1)}{qp}\left(\frac{1}{p+2} + \frac{1}{p+1}\right)s^2$$

Table 29.6  Treatment effects and adjusted means

| Treatment | Observed mean | Direct effect $\hat{t}$ | Adjusted mean $\bar{y} + \hat{t}$ | Residual effect $\hat{r}$ |
|---|---|---|---|---|
| $a$ | 24.00 | +4.07 | 24.07 | +0.50 |
| $b$ | 16.50 | −3.73 | 16.27 | −1.83 |
| $c$ | 19.50 | −0.33 | 19.67 | +1.33 |
| Mean | 20.00 | 0 | 20.00 | 0 |

In using the formulas, it should always be remembered that they refer only to the simple case $t = k = p$ based on complete Latin squares and that $p + 1$ is the number of periods including the extra period.

### 6. Pairs of experimental animals

When there are only two ($k = 2$) experimental animals in each square, the balanced sequences with extra period for three treatments ($a$, $b$, $c$) take the following form:

| Period | Animals | | | | | |
|---|---|---|---|---|---|---|
| | (1) | (2) | (3) | (4) | (5) | (6) |
| I | $a$ | $b$ | $a$ | $c$ | $b$ | $c$ |
| II | $b$ | $a$ | $c$ | $a$ | $c$ | $b$ |
| (Extra) III | $b$ | $a$ | $c$ | $a$ | $c$ | $b$ |

Note that each pair of treatments forms a complete $2 \times 2$ Latin square in the first two periods. This type of design is a combination of symmetrical pairs (Chap. 28) and sequential treatments. Table 29.7 gives an example for four treatments ($a$, $b$, $c$, $d$):

$$t = 4 \qquad \text{number of treatments}$$
$$q = \binom{t}{2} = 6 \qquad \text{number of squares}$$
$$q \times k = 12 \qquad \text{total number of animals}$$
$$p = k = 2 \qquad \text{animals per square}$$
$$p + 1 = 3 \qquad \text{total number of periods}$$
$$(p + 1)qk = 36 \qquad \text{total number of observations}$$

In practice there should be at least two complete replications of the design with 24 animals to provide more than 10 degrees of freedom for error variance. Here we are merely concerned with illustrating the numerical procedure of analysis.

The first step is to obtain the animal and row (period) totals for each "square" as shown in Table 29.7. The grand total of all observations in the first period is $I = 234$, and the grand total for all observations in the entire experiment is $G = 720$. The calculation of $T$, $R$, $U$ is exactly the same as before, $T$ being the treatment total, $R$ the total of observations immediately following the treatment, and $U$ the sum of

*Table 29.7* **Extra-period balanced sequences with** $t = 4$ **treatments and** $k = 2$ **animals in each square (block of sequence)**

| Period | Animals | | | | | | | | | | | | Total |
|---|---|---|---|---|---|---|---|---|---|---|---|---|---|
| | (1) | (2) | (3) | (4) | (5) | (6) | (7) | (8) | (9) | (10) | (11) | (12) | |
| I | $a$ 12 | $b$ 8 | $a$ 13 | $c$ 11 | $a$ 26 | $d$ 14 | $b$ 28 | $c$ 13 | $b$ 21 | $d$ 24 | $c$ 33 | $d$ 31 | 234 |
| II | $b$ 26 | $a$ 8 | $c$ 18 | $a$ 11 | $d$ 16 | $a$ 28 | $c$ 22 | $b$ 20 | $d$ 30 | $b$ 27 | $d$ 24 | $c$ 28 | 258 |
| III | $b$ 24 | $a$ 19 | $c$ 10 | $a$ 17 | $d$ 20 | $a$ 24 | $c$ 14 | $b$ 18 | $d$ 18 | $b$ 24 | $d$ 17 | $c$ 23 | 228 |
| Animal total | 62 | 35 | 41 | 39 | 62 | 66 | 64 | 51 | 69 | 75 | 74 | 82 | 720 |

Row (period) totals

| | | | | | | | Total |
|---|---|---|---|---|---|---|---|
| I | 20 | 24 | 40 | 41 | 45 | 64 | 234 |
| II | 34 | 29 | 44 | 42 | 57 | 52 | 258 |
| III | 43 | 27 | 44 | 32 | 42 | 40 | 228 |
| Square total | 97 | 80 | 128 | 115 | 144 | 156 | 720 |

Calculation of direct and residual effects

| Treatment | $T$ | $R$ | $U$ | $V$ | $P_1$ | $V - P_1$ |
|---|---|---|---|---|---|---|
| $a$ | 158 | 120 | 140 | 305 | 84 | 221 |
| $b$ | 196 | 126 | 188 | 356 | 106 | 250 |
| $c$ | 172 | 102 | 187 | 351 | 129 | 222 |
| $d$ | 194 | 138 | 205 | 428 | 149 | 279 |
| Total | 720 | 486 | 720 | 1440 | 468 | 972 |
| Check | $G$ | $G - I$ | $G$ | $kG$ | $kI$ | $k(G - I)$ |

| Treatment | $(p+1)T$ $= 3T$ | $(p+1)R$ $= 3R$ | $\hat{T} =$ $3T - U - V$ | $\hat{R} =$ $3R - \frac{3}{2}(V - P_1)$ |
|---|---|---|---|---|
| $a$ | 474 | 360 | $+29$ | $+28.5$ |
| $b$ | 588 | 378 | $+44$ | $+3.0$ |
| $c$ | 516 | 306 | $-22$ | $-27.0$ |
| $d$ | 582 | 414 | $-51$ | $-4.5$ |
| Total | 2,160 | 1,458 | 0 | 0 |
| Sum of squares | | | $\Sigma\hat{T}^2 = 5,862$ | $\Sigma\hat{R}^2 = 1,570.50$ |

animal (column) totals in which the treatment appears in the final period. The following are examples:

$$R_a = 26 + 19 + 18 + 17 + 16 + 24 = 120$$
$$U_a = 35 + 39 + 66 = 140$$

Since a "square" does not contain all treatments (it contains only $k$ out of $t$), two new quantities are to be calculated: $V$ is the sum of row totals that include the treatment and $P_1$ is the sum of all first rows that include the treatment. For instance,

$$V_a = (20 + 34 + 43) + (24 + 29 + 27) + (40 + 44 + 44)$$
$$= 97 + 80 + 128 = 305$$

For this particular type of balanced design, $V$ is simply the sum of square totals that include the treatment. In the example of Table 29.3, in which every row of every square contains all the treatments, $V = G$ for every treatment and need not be calculated. $P_1$ is a similar quantity, only limited to the first row. Thus, for treatment $a$,

$$P_1 = 20 + 24 + 40 = 84$$

Again, in the example of Table 29.3, in which each row contains all the treatments, $P_1 = I$, grand total of period I, for every treatment and need not be calculated. Thus, we see that $V$ and $P_1$ for squares not containing all treatments are certain parts of $G$ and $I$. In calculating the residual effects, we now use $V - P_1$ instead of $G - I$ as in the case of complete Latin-square designs. The values of $\hat{T}$ and $\hat{R}$ are then calculated according to the following formulas:

$$\hat{T} = (p + 1)T - U - V$$
$$\hat{R} = (p + 1)\left(R - \frac{V - P_1}{k}\right)$$

These values are shown in the bottom portion of Table 29.7, in which $(p + 1)/k = \frac{3}{2}$. The $\hat{T}$'s and $\hat{R}$'s all add up to zero. This completes the preliminary calculations.

## 7. Analysis of variance

As before, the sums of squares due to all animals and periods (within squares) are calculated the usual way. Thus, for the 12 animals (11 $df$),

$$\tfrac{1}{3}(62^2 + 35^2 + \cdots + 82^2) - \tfrac{1}{36}(720^2)$$
$$= 15,251.33 - 14,400.00 = 851.33$$

and for periods within squares ($2 \times 6 = 12\ df$),

$$\tfrac{1}{2}(20^2 + 34^2 + \cdots + 40^2) - \tfrac{1}{6}(97^2 + \cdots + 156^2)$$
$$= 15{,}465.00 - 15{,}081.67 = 383.33$$

To facilitate the calculation of the *ssq* due to direct and residual effects, it is desirable to give some general expressions which are applicable to all extra-period balanced sequence designs.   Let

*General*        *In example*

$$\beta = \frac{(p+1)(t-1)}{qk \cdot (p-1)(p+2)} = \frac{3 \times 3}{12 \times 1 \times 4} = \frac{3}{16}$$

$$\beta' = \frac{(p+1)(t-1)}{qp(kp-1)} = \frac{3 \times 3}{6 \times 2 \times 3} = \frac{1}{4}$$

where $t$ is the number of treatments, $k$ is the number of animals per square, $q$ is the number of squares, $p$ is the number of periods *without* the extra period.   In our example, $t = 4$, $q = 6$, and $p = k = 2$.   The *ssq* due to direct and residual effects are respectively

$$\frac{\beta \Sigma \hat{T}^2}{(p+1)^2} = \frac{3 \times 5{,}862}{16 \times 3 \times 3} = 122.125$$

$$\frac{\beta' \Sigma \hat{R}^2}{(p+1)^2} = \frac{1{,}570.50}{4 \times 3 \times 3} = 43.625$$

each with $t - 1 = 3$ degrees of freedom.   These values are entered in Table 29.8.   The total *ssq* is $A - C = 16{,}028 - 14{,}400 = 1{,}628$, and the error *ssq* is obtained by subtraction.   It is noticed that when $k = p = 2$, most of the degrees of freedom have been taken up by animals and periods within squares and there are only six left for estimating the error variance.   That is why we need at least two complete replications of the sequences to provide more than 10 *df* for error.

*Table 29.8*  **Analysis of variance (data of Table 29.7)**

| Variation | df | ssq | msq | F |
|---|---|---|---|---|
| All animals | 11 | 851.333 | | |
| Periods within square | 12 | 383.333 | | |
| Direct effects | 3 | 122.125 | 40.71 | 1.07 |
| Residual effects | 3 | 43.625 | 14.54 | |
| Error | 6 | 227.584 | $37.93 = s^2$ | |
| Total | 35 | 1,628.000 | | |

## 8. Treatment means and variances

The estimates of the direct and residual effects of treatments are

$$\hat{t} = \frac{\beta \hat{T}}{p+1} \qquad \hat{r} = \frac{\beta' \hat{R}}{p+1}$$

with variance

$$V(\hat{t}) = \beta s^2 \qquad V(\hat{r}) = \beta' s^2$$

The adjusted mean of a treatment is $\bar{y} + \hat{t}$. The variance of the difference between two adjusted means is $2V(\hat{t})$, and that between two observed (unadjusted) means is

$$V(\hat{t}_u + \hat{r}_a - \hat{t}_b - \hat{r}_b) = 2[V(\hat{t}) + V(\hat{r})] = 2(\beta + \beta')s^2$$

We leave the numerical calculations as an exercise.

In this chapter we have dealt only with the extra-period balanced sequence design for its simplicity. For balanced sequences without the extra period and for partially balanced designs the reader may consult the comprehensive review by Patterson and Lucas [Change-over Designs, *N. Carolina State Coll. Agr. Expt. Sta. Tech. Bull.* 147 (1962)]. If the residual effects are small, we may use either the simple designs of Chap. 18 or the general balanced sequences without the extra period.

### EXERCISES

**29.1** The components for animals and periods (within squares) in the analysis of variance (Table 29.5) are given in brief form, which is sufficient for most practical purposes. A more detailed subdivision of the ssq's for animals and periods may be obtained in a straightforward manner by the usual method. Since some of the basic quantities have already been given in Sec. 4, it should be easy to verify the subdivisions in the accompanying table. Perform a similar subdivision for the data of Table 29.7 and the components shown in Table 29.8.

| Subdivision | df | ssq | Subdivision | df | ssq |
|---|---|---|---|---|---|
| Between squares | 1 | 6 | Overall periods | 3 | 324 |
| Animals within square | 4 | 52 | Periods × square | 3 | 18 |
| (Table 29.5) All animals | 5 | 58 | Periods within square | 6 | 342 |

**29.2** The values of $\hat{T}$ and $\hat{R}$ given in Sec. 3 are multiples of $\hat{t}$ and $\hat{r}$. Different multiples may be used. For instance, instead of $\hat{R} = pR - (G - I)$, we may calculate

$$\hat{R} = (p+1)\left(R - \frac{G-I}{k}\right) = 4\left(R - \frac{390}{3}\right)$$

which are $4(+3, -11, +8) = +12, -44, +32$, respectively, for treatments $a$, $b$, $c$. The ratio of these values to those in Table 29.4 is $4:3$.   To obtain the $t$'s, the denominator should now be $q \cdot p(p + 1) = 2 \times 3 \times 4 = 24$ instead of the former $q \cdot p^2 = 18$. The $ssq$ due to residual effects is then

$$\frac{\Sigma \hat{\hat{R}}^2}{q \cdot p(p + 1)^2} = \frac{12^2 + 44^2 + 32^2}{2 \times 3 \times 4 \times 4} = \frac{3,104}{96} = 32.33$$

which is identical with that found in Sec. 4.

**29.3**   Calculate the numerical values of $\hat{t}$, $\hat{r}$, and their variances (Sec. 8) for the data of Table 29.7.   Also calculate the adjusted treatment means.

**29.4**   In order to have an opportunity to get familiar with the basic arithmetic procedure, consider the data in the accompanying table.   Make a complete analysis of the data, following the procedures of Secs. 2 to 5.

| Period | Animals (1) | (2) | (3) | Row | Animals (4) | (5) | (6) | Row | Overall period |
|--------|------|------|------|-----|------|------|------|-----|---------|
| I    | c 9  | b 11 | a 23 |   | c 19 | b 17 | a 17 |   | = I |
| II   | b 21 | a 28 | c 24 |   | a 29 | c 24 | b 18 |   |   |
| III  | a 17 | c 20 | b 17 |   | b 17 | a 26 | c 17 |   |   |
| IV   | a 31 | c 26 | b 19 |   | b 20 | a 29 | c 25 |   |   |
| Total |     |      |      |   |      |      |      |   | = G |

# Some related topics

# 30

# Multiple measurements

Throughout this book we have been analyzing one variable ($y$) only and seeing if the mean values of various treatment groups differ more than is allowable by random sampling error.  In the chapters on regression and concomitant observations, another variable $x$ has been brought into consideration, but in those cases the primary purpose has been to eliminate the influence of the initial variation of $x$ on $y$ before testing treatment effects.  We are still interested only in the differences in $y$, not in $x$.  The $x$ variable represents the heterogeneity of experimental units and is not the result of treatment.  The so-called "analysis of covariance" (that is, regression followed by analysis of variance) is not to be confused with the subject of the present chapter—the analysis of multiple measurements simultaneously.

In order to avoid complicated notation with multiple subscripts, we shall use different letters to denote different variables.  Consider for example, the effects of various diets.  A diet may affect the weight of an individual; it may also affect the individual's height and some other characteristics.    In measuring the effect of a diet, therefore, the investigator may measure both the weight $x$ and the height $y$ of the same individual (or experimental unit).    In such a case both $x$ and $y$ are results of treatments: they measure different aspects of the treatment effects.    In other words, the treatments have effects on more than one variable.    Then, $x$ and $y$ are called multiple measurements on the same individual.    The reader should now see that here the variable $x$ has the same status as $y$ and is not the concomitant variable previously denoted by the same symbol.

Examples of multiple effects and hence multiple measurements are plentiful in biology and medicine.    In this chapter we shall describe a method of analyzing multiple measurements ($x$ and $y$) simultaneously. This also serves as an introduction to the difficult field known as multivariate analysis.    The method described in this chapter is limited to two (treatment) groups; this simple case can be mastered and used by biologists and medical research workers profitably.

### 1. Single variable; a reorientation

In Table 30.1 are exhibited the data on two characteristics of each individual.    Group I consists of seven individuals ($n_1 = 7$), and group II consists of five individuals ($n_2 = 5$).    The usual term "observation" is not quite adequate when there are multiple measurements; instead we use the term *individual*, meaning experimental unit in general.    To avoid abstractness, the reader may think of the two characteristics as weight $x$ and height $y$, as blood pressure $x$ and cholesterol level $y$, or, better still, as two variables with which he happens to be working.

Before proceeding with the method of analyzing $x$ and $y$ simultaneously, we shall review the ordinary case of a single variable ($x$ or $y$) and introduce the concept of *distance* between two samples, so that the method of multivariate analysis will not strike the reader as something wholly unconnected with what we have been studying.

Let $d$ be the difference between two sample means.    For example, if we are considering the variable $x$, then $d_1 = \bar{x}_1 - \bar{x}_2$, where $\bar{x}_1$ and $\bar{x}_2$ are the mean values of groups I and II respectively.    As before, let $s^2$ be the pooled estimate of variance from the two groups; that is,

$$s^2 = \frac{ssq_W}{n_1 + n_2 - 2}$$

where $ssq_W$ is the total within-group sum of squares.    Then the distance

*Table* 30.1   Measurements of two variables on each individual

| | Group I | | Group II | | Group sizes | |
|---|---|---|---|---|---|---|
| | $x_1$ | $y_1$ | $x_2$ | $y_2$ | | |
| | 19 | 9 | 16 | 12 | | |
| | 17 | 8 | 19 | 15 | $n_1 = 7$ | $n_2 = 5$ |
| | 18 | 8 | | | $n_1 + n_2 - 2 = 10$ | |
| | 20 | 12 | 15 | 9 | | |
| | 15 | 6 | 16 | 13 | Difference between | |
| | 18 | 7 | | | group means | |
| | 19 | 13 | 14 | 11 | | |
| Total | 126 | 63 | 80 | 60 | $d_1 = \bar{x}_1 - \bar{x}_2 = 2$ | |
| Mean | 18 | 9 | 16 | 12 | $d_2 = \bar{y}_1 - \bar{y}_2 = -3$ | |
| | Sum of squares | | Sum of squares | | Pooled within groups | |
| (area) $A$ | 2,284 | 607 | 1,294 | 740 | | |
| (area) $C$ | 2,268 | 567 | 1,280 | 720 | $(xx) = 16 + 14 = 30$ | |
| $ssq$ | 16 | 40 | 14 | 20 | | |
| | Sum of products | | Sum of products | | $(yy) = 40 + 20 = 60$ | |
| (area) $A$ | 1,154 | | 974 | | | |
| (area) $C$ | 1,134 | | 960 | | | |
| $spl$ | 20 | | 14 | | $(xy) = 20 + 14 = 34$ | |

between the two samples is defined as

$$D = \frac{d}{s} = \frac{\bar{x}_1 - \bar{x}_2}{s_x}$$

for variable $x$. It is the distance between the two sample means measured in units of standard deviation. It is, so to speak, a standardized distance. The square of the distance is then

$$D^2 = \frac{d^2}{s^2} = \frac{(\bar{x}_1 - \bar{x}_2)^2}{s_x^2}$$

With the distance so defined, we see that its relationship with Student's $t$ is

$$t^2 = \frac{D^2}{1/n_1 + 1/n_2} = \frac{n_1 n_2}{n_1 + n_2} D^2$$

Let $p$ be the number of variables under study. For a single variable, $p = 1$. With $n_1 + n_2 = N$, the total number of individuals, the relationship between $t^2$ and the variance ratio $F$ is generally

$$F = \frac{N - p - 1}{(N - 2)p} t^2$$

It should be noted that when $p = 1$, the formula above reduces to $F = t^2$. The sums of squares within groups are given in Table 30.1 in the columns of the respective variables. For example, for the column under $x_1$, the basic quantities are $A = \Sigma x_1^2 = 19^2 + 17^2 + \cdots = 2,284$ and $C = 126^2/7 = 2,268$, so that $ssq = A - C = 16$ for group I. The other $ssq$'s are calculated in a similar way. The pooled within-group $ssq$ for $x$ is designated by $(xx)$ and that for $y$ by $(yy)$. These values have been given in the right-hand margin of the table. The pooled estimate of variance of $x$ is $s_x^2 = (xx)/(N - 2) = {}^{30}\!/_{10} = 3$, and that of $y$ is $s_y^2 = (yy)/(N - 2) = {}^{60}\!/_{10} = 6$. Testing the significance of the difference between the two sample means for $x$ and for $y$ separately, we have

$$F = t^2 = \frac{n_1 n_2}{N} \cdot \frac{d^2}{s^2} = \frac{7 \times 5}{12} \cdot \frac{4}{3} = 3.889 \qquad \text{for } x$$

$$F = t^2 = \frac{n_1 n_2}{N} \cdot \frac{d^2}{s^2} = \frac{7 \times 5}{12} \cdot \frac{9}{6} = 4.375 \qquad \text{for } y$$

each with $\nu_1 = 1$ and $\nu_2 = 10$ $df$. The tabulated value of $F$ at the 5 per cent significance level is 4.96. The conclusion is that neither $x$ nor $y$ has attained the significance level.

Now we are ready to embark on the joint analysis of two variables. For purpose of generalization, the squared distance between two samples may be rewritten as

$$D^2 = (d)(s^2)^{-1}(d)$$

which is, of course, $d^2/s^2$. From the theory of multivariate normal population, however, the generalization takes the form $(d)(s^2)^{-1}(d)$.

## 2. Joint analysis

The purpose of joint analysis is to see if the two groups differ if we take into consideration both $x$ and $y$ at the same time. To do this, we need the sum of products $(spt)$ of the two variables within the groups. These are calculated the usual way. Thus, for group I,

$$A = \Sigma x_1 y_1 = (19 \times 9) + (17 \times 8) + \cdots = 1,154$$

and $C = (126)(63)/7 = 1,134$, so that $spt = A - C = 20$. The total within-group $spt$ is $(xy) = 34$, so that the estimate of covariance is

$(xy)/(n_1 + n_2 - 2) = {}^{34}\!/_{10} = 3.4$. We shall use the following notations to denote the various quantities:

$$d_1 = \bar{x}_1 - \bar{x}_2 = 2 \qquad\qquad d_2 = \bar{y}_1 - \bar{y}_2 = -3$$

$$s_1{}^2 = s_x{}^2 = \frac{(xx)}{N-2} = 3.0 \qquad\qquad s_2{}^2 = s_y{}^2 = \frac{(yy)}{N-2} = 6.0$$

$$s_{12} = s_{xy} = \frac{(xy)}{N-2} = 3.4$$

The squared distance (Mahalanobis's $D^2$) between the two samples considering both variables is defined as

$$D^2 = (d_1 \quad d_2) \begin{pmatrix} s_1{}^2 & s_{12} \\ s_{12} & s_2{}^2 \end{pmatrix}^{-1} \begin{pmatrix} d_1 \\ d_2 \end{pmatrix}$$

It is not practical to explain matrix algebra in this book, but even those without knowledge of it should be able to see that the expression above is a direct generalization of that for one variable: $D^2 = (d)(s^2)^{-1}(d)$, whereas a single $d$ is replaced by two $d$'s and a single $s^2$ is replaced by the variance-covariance matrix. Fortunately, there exists a very simple method of calculating $D^2$ that does not require any knowledge of matrix manipulation, and it is the procedure now to be described. For those who do know matrix algebra, the next section explains that the arithmetic procedure adopted here is identical with the matrix expression given above. Write out two linear equations as follows:

$$(xx)a_1 + (xy)a_2 = (N-2)d_1$$
$$(xy)a_1 + (yy)a_2 = (N-2)d_2$$

Substituting the numerical values in our example,

$$30a_1 + 34a_2 = 10 \times 2 = 20$$
$$34a_1 + 60a_2 = 10(-3) = -30$$

Solving for $a_1$ and $a_2$, we obtain

$$a_1 = \frac{555}{161} = 3.447205 \qquad\qquad a_2 = \frac{-395}{161} = -2.453416$$

Then the squared distance between the two groups is

$$D^2 = a_1 d_1 + a_2 d_2 = 6.89441 + 7.36025 = 14.25466$$

Once $D^2$ is known, the rest of the calculation follows the familiar procedure. Thus, analogously to the one variable case, Hotelling's $T^2$, a generalization of Student's $t^2$, is

$$T^2 = \frac{n_1 n_2}{n_1 + n_2} D^2 = \frac{7 \times 5}{12} (14.25466) = 41.576$$

and, with $p = 2 =$ the number of variables taken into consideration,

$$F = \frac{N - p - 1}{(N - 2)p} T^2 = \frac{9}{10 \times 2} (41.576) = 18.71$$

with $\nu_1 = p = 2$, and $\nu_2 = N - p - 1 = 9$ degrees of freedom. The tabulated value of $F$ at 1 per cent significance level is 8.02, so the difference between the two groups is highly significant. The reader may recall that neither $x$ alone nor $y$ alone has attained the 5 per cent significance level, but considering both variables at the same time, the difference between the two groups are quite striking. We say that multivariate analysis has more discriminating power than single-variable analysis. This completes the numerical analysis of the data.

### 3. Mathematical notes on $D^2$

This section may well be omitted by those lacking background in matrix algebra. However, those with some knowledge of the elements of matrix theory may follow quite easily. Let the determinant of the variance-covariance matrix be $\Delta$; that is,

$$\begin{vmatrix} s_1^2 & s_{12} \\ s_{12} & s_2 \end{vmatrix} = s_1^2 s_2^2 - (s_{12})^2 = \Delta$$

First, consider the expression for $D^2$ in the matrix form. The inverse of the variance-covariance matrix is then

$$\begin{pmatrix} s_1^2 & s_{12} \\ s_{12} & s_2^2 \end{pmatrix}^{-1} = \begin{pmatrix} s_2^2 & -s_{12} \\ -s_{12} & s_1^2 \end{pmatrix} \frac{1}{\Delta}$$

Hence

$$D^2 = (d_1 \ \ d_2) \begin{pmatrix} s_2^2 & -s_{12} \\ -s_{12} & s_1^2 \end{pmatrix} \begin{pmatrix} d_1 \\ d_2 \end{pmatrix} \cdot \frac{1}{\Delta}$$

$$= \frac{1}{\Delta} (s_2^2 d_1^2 - 2 s_{12} d_1 d_2 + s_1^2 d_2^2)$$

On the other hand, if we set up the following equations (which are equivalent to those we adopted for numerical calculation)

$$s_1^2 a_1 + s_{12} a_2 = d_1$$
$$s_{12} a_1 + s_2^2 a_2 = d_2$$

we obtain the solutions

$$a_1 = \frac{1}{\Delta} \begin{vmatrix} d_1 & s_{12} \\ d_2 & s_2^2 \end{vmatrix} = \frac{s_2^2 d_1 - s_{12} d_2}{\Delta}$$

$$a_2 = \frac{1}{\Delta} \begin{vmatrix} s_1^2 & d_1 \\ s_{12} & d_2 \end{vmatrix} = \frac{s_1^2 d_2 - s_{12} d_1}{\Delta}$$

Then

$$D^2 = a_1d_1 + a_2d_2 = \frac{1}{\Delta} (s_2{}^2d_1{}^2 - 2s_{12}d_1d_2 + s_1{}^2d_2{}^2)$$

which is identical with the expression in matrix form.

In our numerical example, $\Delta = 3 \times 6 - (3.4)^2 = 6.44$ and

$$D^2 = \frac{1}{6.44} [6(2)^2 - 2(3.4)(2)(-3) + 3(-3)^2] = \frac{91.80}{6.44} = 14.254658$$

in agreement with that obtained before.

### 4. Discriminant function

From the analysis of the data in Table 30.1 we have concluded that neither $x$ alone nor $y$ alone attains significance; yet, when both $x$ and $y$ are taken into consideration, the two groups differ quite strikingly. In order to get a visual presentation of the situation, the scatter diagram of the seven points in group I and the five points in group II on the $xy$ plane is given in Fig. 30.1. When we consider $x$ values alone, that is, when we look at the projections of these points on the $x$ axis, we see that the two groups of points overlap to a very large extent. Similarly, if we look at their $y$ values alone (that is, projections on the $y$ axis), they also overlap considerably. But the points plotted on the $xy$ plane (that is, according

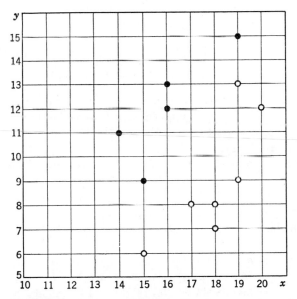

**Fig. 30.1** Scatter diagram for two groups of points $(x,y)$ (data from Table 30.1).

to their $x$ and $y$ values) clearly fall into two groups.   In this particular example, a straight line can be drawn such that all points of group I are on one side and all points of group II are on the other side of it.   Hence, the very high value of $F = 18.71$.   In less striking cases, however, the points may not fall into two separate groups; they intermingle and the $F$ value may or may not be significant, depending on the extent to which the two groups of points are intermingled.

These considerations lead to the attempt to construct a so-called discriminant function for the measured variables.   The problem is essentially this: how can we best combine the measurements $x$ and $y$ so that the two groups of individuals exhibit the maximum difference?   Restricting ourselves to linear combinations of $x$ and $y$, the question reduces to finding a function of the form $L = a_1x + a_2y$ so that the $L$ values of the two groups of individuals will achieve a maximum difference.   In so doing, the two separate variables $x$ and $y$ are replaced by the single compounded variable $L$.   Such a function is called a discriminant function because it has the maximum discriminating power for the two groups of individuals.

The formulation of the discriminant problem given above may sound profound and difficult, but the solution is pleasingly simple.   It turns out that the values of $a_1$ and $a_2$ obtained in the preceding section for the purpose of calculating $D^2$ and $T^2$ are also the required coefficients of the $L$ function.   Thus, the discriminant function for the data in Table 30.1 is simply

$$L = a_1x + a_2y = 3.4472x - 2.4534y$$

For accurate numerical calculations we may first calculate the quantity

$$161L = 555x - 395y$$

and then divide it by 161.   The $L$ value of each individual is given in Table 30.2.   In practice it is unnecessary to give all of the values; they are given here to show that the $L$ values of the two groups do indeed differ a great deal more than the original $x$ and $y$.   In group I the $L$ varies from 33.6 to 44.9; in group II it varies from 21.3 to 29.6.   The difference is quite clear-cut.

The compounded variable $L$ is closely connected with the (squared) distance between the two groups.   In fact, the (squared) distance as calculated in the preceding two sections applies to the variable $L$ directly, because

$$\begin{aligned} D^2 &= a_1d_1 + a_2d_2 = a_1(\bar{x}_1 - \bar{x}_2) + a_2(\bar{y}_1 - \bar{y}_2) \\ &= (a_1\bar{x}_1 + a_2\bar{y}_1) - (a_1\bar{x}_2 + a_2\bar{y}_2) = \bar{L}_1 - \bar{L}_2 \\ &= 39.968944 - 25.714286 = 14.254658 \end{aligned}$$

which is precisely the value obtained before.   Thus, we see that $D^2$ is merely the difference between the mean $L$ values of the two groups.

Table **30.2** Values of compounded variable
$L = a_1x + a_2y$, where $a_1 = 3.44720497$,
$a_2 = -2.45341615$

| | Group I<br>$L_1 = a_1x_1 + a_2y_1$ | Group II<br>$L_2 = a_1x_2 + a_2y_2$ |
|---|---|---|
| | 43.416149 | |
| | | 25.714286 |
| | 38.975155 | |
| | | 28.695652 |
| | 42.422360 | |
| | | 29.627330 |
| | 39.503105 | |
| | | 23.260870 |
| | 36.987578 | |
| | | 21.273292 |
| | 44.875776 | |
| | 33.602484 | |
| Total | 279.782607 | 128.571430 |
| Mean | 39.968944 | 25.714286 |
| | $\bar{L}_1 = a_1\bar{x}_1 + a_2\bar{y}_1$ | $\bar{L}_2 = a_1\bar{x}_2 + a_2\bar{y}_2$ |
| $\Sigma L^2$ | 11,275.2198 | 3,356.0647 |
| Corr | 11,182.6154 | 3,306.1225 |
| $ssq_L$ | 92.6044 | 49.9422 |

Furthermore, the statistic $T^2$ defined previously for $x$ and $y$ is nothing more than the ordinary $t^2$ applied to the single compounded variable $L$. To see this, let us recall that the squared distance between two groups for one variable is

$$(\text{Distance})^2 = \frac{(\bar{L}_1 - \bar{L}_2)^2}{s_L^2}$$

where $s_L^2$ is the pooled estimate of the variance of $L$. The sum of squares for $L$ within each group has been shown at the bottom of Table 30.2. The total within-group $ssq_L = 92.6044 + 49.9422 = 142.5466$. Dividing this by $n_1 + n_2 - 2 = 10$, we obtain $s_L^2 = 14.25466$, which is equal to $D^2 = \bar{L}_1 - \bar{L}_2$. Therefore the expression $(\bar{L}_1 - \bar{L}_2)^2/s_L^2$ is the same $D^2$ as defined and calculated in the preceding two sections.

The only remaining point is to demonstrate that $s_L^2$ is in general equal to $D^2$, although this has been numerically verified above. To prove it, let us use once more the two linear equations:

$$a_1s_1^2 + a_2s_{12} = d_1$$
$$a_2s_{12} + a_2s_2^2 = d_2$$

Multiplying the first equation by $a_1$ and the second by $a_2$ and adding the results, we obtain

$$a_1{}^2 s_1{}^2 + 2a_1 a_2 s_{12} + a_2{}^2 s_2{}^2 = a_1 d_1 + a_2 d_2 = D^2$$

But

$$s_L{}^2 = s^2_{(a_1 x + a_2 y)} = a_1{}^2 s_1{}^2 + 2a_1 a_2 s_{12} + a_2{}^2 s_2{}^2$$

also. Hence,

$$s_L{}^2 = D^2$$

and

$$T^2 \text{ (for } x \text{ and } y) = \frac{n_1 n_2}{n_1 + n_2} D^2 = \frac{n_1 n_2}{n_1 + n_2} \frac{(\bar{L}_1 - \bar{L}_2)^2}{s_L{}^2} = t^2 \qquad \text{for } L$$

The discriminant function has many practical applications in as varied fields as taxonomy and animal and plant breeding. For instance, the construction of a selection index in breeding work is essentially the problem of finding a suitable discriminant function. As to classification problems, suppose that we observe an individual whose measurements are $x = 15$ and $y = 10$. By inspecting the $x$ and $y$ values of Table 30.1, it is not very clear to which group this individual is likely to belong. The discriminant value of this individual, however, is

$$L = a_1 x + a_2 y = 51.7 - 24.5 = 27.2$$

Referring to the $L$ values of Table 30.2, particularly the two mean $L$ values, we see that this individual is far more likely to belong to group II than to group I. We cannot go into any of these specific topics in this general text. The reader must consult special books for special subjects. The purpose of including a section on the discriminant function here is to show that the compounding of variables and the generalized $T^2$ test for several variables are closely related—both being based on the same coefficients $a_1$ and $a_2$.

## 5. Uncorrelated variables

The variables $x$ and $y$ are in general correlated. In our numerical example the total within-group sum of products is $(xy) = 34$, so that the estimated covariance is $(xy)/(N - 2) = {}^{34}\!/_{10} = 3.4$ and the estimated correlation coefficient between $x$ and $y$ is $(xy)/\sqrt{(xx)(yy)} = 34/\sqrt{30 \times 60} = 0.80$ approximately. If the variables are uncorrelated within groups, $(xy) = 0$. Substituting in our preceding formulas, we obtain

$$a_1 = \frac{(N - 2)d_1}{(xx)} = \frac{d_1}{s_1{}^2}$$

Similarly, $a_2 = d_2/s_2{}^2$. Hence

$$D^2 = a_1d_1 + a_2d_2 = \frac{d_1{}^2}{s_1{}^2} + \frac{d_2{}^2}{s_2{}^2}$$

In other words, the total squared distance is simply the sum of the separate squared distances for the single variables. It follows that

$$T^2 = t_1{}^2 + t_2{}^2$$

and $F = F_1 + F_2$, where the subscripts 1 and 2 represent $x$ and $y$, respectively. If the separate analysis of each variable yields approximately the same value of $F$, the joint analysis will double the $F$ value.

## 6. Three variables: preliminary calculations

The method of analyzing two variables ($x$ and $y$) measured on each individual may be extended to three variables ($x$, $y$, $z$) or more. In general, the more variables we are trying to utilize, the larger the number of individuals we should measure. Each additional variable adds 1 $df$ to the numerator of $F$ and decreases the $df$ by 1 for the denominator, the error variance.

Since the arithmetic is fairly long for three variables, we shall use the $x$ and $y$ values of Table 30.1 to which a third variable $z$ is added, as shown in Table 30.3. Since we have learned from previous analysis that the two groups differ significantly with respect to $x$ and $y$ jointly, it goes without saying that they will also differ with respect to three variables. Nevertheless, the arithmetic procedure of analyzing three simultaneous measurements is given here in full without regard to our previous knowledge. This section may be read or consulted independently by those who happen to have three variables to analyze without knowing the details of the preceding sections.

Suppose that there are two (treatment) groups with $n_1$ and $n_2$ individuals, the total number of individuals studied being $n_1 + n_2 = N$. Measurements of three variables are taken on each individual. The reader may think of the variables as blood pressure $x$, cholesterol level $y$, and daily fat intake $z$. The data are recorded in Table 30.3, units of measurements being immaterial. The problem is to test the significance of the difference between the two groups with respect to the three variables taken as a whole.

First, we find the mean values of the three variables for each group and the differences between the group means. Thus,

$$d_1 = \bar{x}_1 - \bar{x}_2 = 18 - 16 = 2$$
$$d_2 = \bar{y}_1 - \bar{y}_2 = 9 - 12 = -3$$
$$d_3 = \bar{z}_1 - \bar{z}_2 = 11 - 10 = 1$$

*Table* 30.3 **Measurements of three variables on each individual**

| | Group I | | | Group II | | | Group sizes | |
|---|---|---|---|---|---|---|---|---|
| | $x_1$ | $y_1$ | $z_1$ | $x_2$ | $y_2$ | $z_2$ | | |
| | 19 | 9 | 10 | 16 | 12 | 9 | $n_1 = 7$ | $n_2 = 5$ |
| | 17 | 8 | 14 | 19 | 15 | 7 | $n_1 + n_2 - 2 = 10$ | |
| | 18 | 8 | 11 | | | | | |
| | 20 | 12 | 8 | 15 | 9 | 12 | | |
| | 15 | 6 | 14 | 16 | 13 | 10 | Difference between | |
| | 18 | 7 | 11 | | | | group means | |
| | 19 | 13 | 9 | 14 | 11 | 12 | | |
| | | | | | | | $d_1 = \bar{x}_1 - \bar{x}_2 = 2$ | |
| Total | 126 | 63 | 77 | 80 | 60 | 50 | $d_2 = \bar{y}_1 - \bar{y}_2 = -3$ | |
| Mean | 18 | 9 | 11 | 16 | 12 | 10 | $d_3 = \bar{z}_1 - \bar{z}_2 = 1$ | |

| | Sum of squares | | | Sum of squares | | | Within groups | |
|---|---|---|---|---|---|---|---|---|
| $A$ | 2,284 | 607 | 879 | 1,294 | 740 | 518 | $(xx) = 30$ | |
| $C$ | 2,268 | 567 | 847 | 1,280 | 720 | 500 | $(yy) = 60$ | |
| $ssq$ | 16 | 40 | 32 | 14 | 20 | 18 | $(zz) = 50$ | |

| | Sum of products | | | Sum of products | | | | |
|---|---|---|---|---|---|---|---|---|
| | $xy$ | $xz$ | $yz$ | $xy$ | $xz$ | $yz$ | | |
| $A$ | 1,154 | 1,365 | 664 | 974 | 785 | 583 | $(xy) =$ | 34 |
| $C$ | 1,134 | 1,386 | 693 | 960 | 800 | 600 | $(xz) =$ | $-36$ |
| $spt$ | 20 | $-21$ | $-29$ | 14 | $-15$ | $-17$ | $(yz) =$ | $-46$ |

These values are shown in the upper part of Table 30.3. The second step is to calculate the sum of squares of each of the three variables for each group; that is, the $ssq$ within groups. As we have always done, the basic quantity (area) $A$, for the variable $y$ in group I, for example, is

$$A = 9^2 + 8^2 + \cdots + 13^2 = 607$$

(area) $C = 63^2/7 = 567$, so that $ssq = A - C = 40$. This is done for each of the six columns of numbers. Then the total within-group $ssq$ is obtained:

$$(xx) = 16 + 14 = 30$$
$$(yy) = 40 + 20 = 60$$
$$(zz) = 32 + 18 = 50$$

These values are given in the middle portion of Table 30.3. The final step is to calculate the sum of products for all possible pairs of variables.

For the pair $y$ and $z$ in group I for example, the basic quantity

$$\text{(area)} \; A = (9 \times 10) + (8 \times 14) + \cdots + (13 \times 9) = 664$$

and $C = 63 \times 77/7 = 693$, so that $spt = A - C = -29$. The total within-group sums of products are

$$
\begin{aligned}
(xy) &= 20 + 14 = 34 \\
(yz) &= -21 - 15 = -36 \\
(yz) &= -29 - 17 = -46
\end{aligned}
$$

This completes the preliminary calculations shown in Table 30.3.

## 7. Three variables: significance test

The second stage of the analysis begins with the writing down of three linear equations as follows:

$$
\begin{aligned}
(xx)a_1 + (xy)a_2 + (xz)a_3 &= (N-2)d_1 \\
(xy)a_1 + (yy)a_2 + (yz)a_3 &= (N-2)d_2 \\
(xz)a_1 + (yz)a_2 + (zz)a_3 &= (N-2)d_3
\end{aligned}
$$

Substituting the numerical values and remembering that $N - 2 = 10$,

$$
\begin{aligned}
30a_1 + 34a_2 - 36a_3 &= 20 \\
34a_1 + 60a_2 - 46a_3 &= -30 \\
-36a_1 - 46a_2 + 50a_3 &= 10
\end{aligned}
$$

Solving for the $a$'s by any method at the disposal of the reader,

$$a_1 = 6.9955 \qquad a_2 = -1.52466 \qquad a_3 = 3.8341$$

The quantity known as the squared distance between the two groups is

$$D^2 = a_1 d_1 + a_2 d_2 + a_3 d_3 = 22.399$$

and the generalized (Hotelling's) $T^2$ is

$$T^2 = \frac{n_1 n_2}{n_1 + n_2} D^2 = \frac{7 \times 5}{12} (22.399) = 65.33$$

Let $p = 3 =$ number of variables considered. The variance ratio is

$$F = \frac{N - p - 1}{(N-2)p} T^2 = \frac{8}{10 \times 3} (65.33) = 17.42$$

with $v_1 = p = 3$ and $v_2 = N - p - 1 = 8$ degrees of freedom. The tabulated value of $F$ at the 1 per cent level is 7.59, so the difference between the two groups is highly significant. All of the discussions given for the two-variable case apply equally well for the three variable case. The procedure outlined here may be extended to four or more variables.

## EXERCISES

**30.1**   Ignore the variable $x$ from Table 30.3 and assume that the measurements have been made only on $y$ and $z$.   The relevant data would be as shown in the accom-

| | Group I | | Group II | | Difference or pooled quantities |
|---|---|---|---|---|---|
| | $y_1$ | $z_1$ | $y_2$ | $z_2$ | |
| Mean | 9 | 11 | 12 | 10 | $d_1 = 9 - 12 = -3$ |
| | | | | | $d_2 = 11 - 10 = 1$ |
| ssq | 40 | 32 | 20 | 18 | $(yy) = 40 + 20 = 60$ |
| | | | | | $(zz) = 32 + 18 = 50$ |
| spt | | $-29$ | | $-17$ | $(yz) = -29 - 17 = -46$ |

panying table.   The two linear equations are then

$$(yy)a_1 + (yz)a_2 = (N - 2)d_1$$
$$(yz)a_1 + (zz)a_2 = (N - 2)d_2$$

Calculate the values of Mahalanobis's $D^2$ and Hotelling's $T^2$ and make an $F$ test of the significance of the difference between the two groups with respect to $y$ and $z$ as a whole.   Draw a scatter diagram for the values $y$ and $z$.   Does the conclusion of nonsignificance make sense?

*Partial Ans.:*  $a_1 = -1.17647$;  $a_2 = -0.88235$;  $D^2 = 2.647$;  $T^2 = 7.72$;  $F = 3.47$.

**30.2**   Three variables have been measured on each individual, and there are seven individuals in each group.   The data are given in the accompanying table.   Make an $F$ test to see if the two groups differ significantly.

| | Group I | | | Group II | | |
|---|---|---|---|---|---|---|
| | $x_1$ | $y_1$ | $z_1$ | $x_2$ | $y_2$ | $z_2$ |
| | 19 | 13 | 9 | 14 | 11 | 12 |
| | 18 | 7 | 11 | 16 | 13 | 10 |
| | 15 | 6 | 14 | 15 | 9 | 12 |
| | 20 | 12 | 8 | 19 | 15 | 7 |
| | 18 | 8 | 11 | 16 | 12 | 9 |
| | 17 | 8 | 14 | 17 | 11 | 9 |
| | 19 | 9 | 10 | 15 | 13 | 11 |
| Total | 126 | 63 | 77 | 112 | 84 | 70 |

*Partial Ans.:*  $a_1 = 7.70925$,  $a_2 = -0.9956$,  $a_3 = 5.022$,  $D^2 = 23.427$.

The discriminant function may be taken as approximately $L = 7.7x - y + 5z$.

**30.3**  If you like algebra, show that

$$D^2 = (d_1 \quad d_2 \quad d_3) \begin{pmatrix} s_1{}^2 & s_{12} & s_{13} \\ s_{12} & s_2{}^2 & s_{23} \\ s_{13} & s_{23} & s_3{}^2 \end{pmatrix}^{-1} \begin{pmatrix} d_1 \\ d_2 \\ d_3 \end{pmatrix}$$

$$= a_1 d_1 + a_2 d_2 + a_3 d_3$$

where the $a$'s are given by the linear equations in Sec. 7.

*Mathematical notes:* Write $s_{11}$ for $s_1{}^2$, etc., in the interest of generality. Then let

$$\begin{pmatrix} a_1 \\ a_2 \\ a_3 \end{pmatrix} = \begin{pmatrix} s_{11} & s_{12} & s_{13} \\ s_{12} & s_{22} & s_{23} \\ s_{13} & s_{23} & s_{33} \end{pmatrix}^{-1} \begin{pmatrix} d_1 \\ d_2 \\ d_3 \end{pmatrix}$$

Multiplying both sides by the variance-covariance matrix, we obtain

$$\begin{pmatrix} s_{11} & s_{12} & s_{13} \\ s_{12} & s_{22} & s_{23} \\ s_{13} & s_{23} & s_{33} \end{pmatrix} \begin{pmatrix} a_1 \\ a_2 \\ a_3 \end{pmatrix} = \begin{pmatrix} d_1 \\ d_2 \\ d_3 \end{pmatrix}$$

This is the set of equations given in Sec. 7.  Also note that

$$D^2 = (d_1 \quad d_2 \quad d_3) \begin{pmatrix} a_1 \\ a_2 \\ a_3 \end{pmatrix} = (a_1 \quad a_2 \quad a_3) \begin{pmatrix} d_1 \\ d_2 \\ d_3 \end{pmatrix} = (a_1 \quad a_2 \quad a_3) \begin{pmatrix} s_{11} & s_{12} & s_{13} \\ s_{12} & s_{22} & s_{23} \\ s_{13} & s_{23} & s_{33} \end{pmatrix} \begin{pmatrix} a_1 \\ a_2 \\ a_3 \end{pmatrix}$$

Thus we see that there are two complementary forms for the quantity $D^2$: one in terms of the $d$'s and $(s_{ij})^{-1}$ and one in terms of the $a$'s and $(s_{ij})$.  For numerical calculation, however, the most convenient is the product form, $D^2 = a_1 d_1 + a_2 d_2 + a_3 d_3$.

**30.4**  The test procedure of converting $T^2$ into $F$ is for convenience.  If you wish to test $T^2$ as such, its critical value may be obtained from the $F$ table first and then compared with the observed value.  For example, from the data of Table 30.3 we obtain $T^2 = 65.33$.  The critical value of $F$ at 1 per cent significance level is 7.59 with $p = 3$ and $N - p - 1 = 8$ degrees of freedom.  The corresponding critical value of $T^2$ is then

$$\text{Critical } T^2 = \frac{(N-2)p}{N-p-1} F = \frac{10 \times 3}{8} (7.59) = 28.46$$

Since observed $T^2 = 65.33 >$ critical $T^2 = 28.40$, the difference between the two groups is highly significant.

# 31

# Multiple comparisons

Throughout this book we have been testing the significance of differences (or contrasts) which are orthogonal to each other; that is, in nonmathematical language, one difference has had no necessary implications for the other differences. [Hartley, *Commun. Pure Appl. Math.*, **8**:47–72 (1955) shows that orthogonality is not a condition for tabulated probability levels to be valid.] Such tests are simple and clear-cut by the use of the standard $t$ and $F$ tables. Furthermore, all of the differences to be tested are supposed to be formulated prior to actual execution of the experiment. The probability levels indicated in the $t$ or $F$ tables are thus valid.

When several comparisons are not independent of each other, we cannot use the standard $t$ or $F$ to judge their significance one by one separately. To obtain a joint probability statement concerning a set of nonorthogonal comparisons is a difficult mathematical problem. It also frequently happens that after the examination of the data, the experimenter wishes to test certain particular contrasts suggested by the data themselves. In such a case, the ordinary $t$ and $F$ tests are not valid either. Yet, these occasions arise very frequently. After all, the experimenter argues, we want the data themselves to make suggestions or to provide leads as to the nature of the treatments. No one can deny that this type of "data snooping" activity may lead to discoveries and new understandings. But the statistical treatment of such a posteriori comparisons is very involved and can hardly be disposed of in a few pages. In this chapter, we shall give only three comparatively simple methods of dealing with such situations.

## 1. One control and several treatments

Probably the most frequently encountered situation is the comparison of one control with each of several treatments. In medicine as well as in industry, we may wish to try several new drugs or processes and compare each of them with a standard drug or process. This comparison is not limited to any specific type of experimental design; a numerical example may be chosen from almost any chapter. All we need for illustration is a set of observed treatment means and the estimated error variance $s^2$, assuming that the general conditions for the analysis of variance are satisfied. Let us consider the data in Table 31.1. (These data are taken from set I, Table 16.3, in which we have found $F = 10.2$, significant. All the relevant information are given here; it is helpful but unnecessary for the reader to look back at that table.) Let us assume that treatment $a$ is the control, designated by (0). The four treatment vs. control com-

*Table* 31.1  One control and four treatments

|        | Treatments |       |       |       |       |
|--------|------------|-------|-------|-------|-------|
|        | $a$        | $b$   | $c$   | $d$   | $e$   |
|        | (0)        | (1)   | (2)   | (3)   | (4)   |
|        | .          | .     | .     | .     | .     |
| Blocks | .          | .     | .     | .     | .     |
|        | .          | .     | .     | .     | .     |
| Total  | 68         | 84    | 64    | 88    | 96    |
| Mean   | 17         | 21    | 16    | 22    | 24    |

Estimated error variance is $s^2 = 4.5$ with 12 $df$

parisons are

$$d_1 = \bar{y}_1 - \bar{y}_0 = 21 - 17 = 4$$
$$d_2 = \bar{y}_2 - \bar{y}_0 = 16 - 17 = -1$$
$$d_3 = \bar{y}_3 - \bar{y}_0 = 22 - 17 = 5$$
$$d_4 = \bar{y}_4 - \bar{y}_0 = 24 - 17 = 7$$

Since the variance of a single observation is $s^2$, the variance of a mean is $s^2/n$, where $n = 4$, the number of replications. The variance of the difference $d$ between two treatment means is $2s^2/n$, and the standard error of $d$ is

$$s_d = \sqrt{\frac{2s^2}{n}} = \sqrt{\frac{2 \times 4.5}{4}} = 1.50$$

If there were only two means to compare, the ordinary Student's $t$ would be $d/s_d$ and the confidence interval for $d$ would be $d \pm ts_d$, where $t$ is tabulated by its $df$ and probability level. So far, we have merely reviewed the general procedure we learned before.

Now the four comparisons $(\bar{y}_i - \bar{y}_0)$ are not independent of each other (as judged by the criteria of orthogonal contrasts, Chap. 12). To obtain a joint confidence statement (or significance tests) in regard to the simultaneous comparisons, Dunnett [*J. Am. Statist. Assoc.*, **50**:1096–1121 (1955) and *Biometrics*, **20,** in press (1964)] has constructed "$t$" tables corresponding to the number of treatments to be compared with the control, two of which are reproduced here (Tables 31.2 and 31.3). The larger the number of treatments, the greater the value of Dunnett's $t$. Obviously, when there is only one treatment to be compared with the control (that is, two groups altogether), Dunnett's $t$ is identical with the ordinary Student's $t$, as the reader may see in the first column of the tables.

The application is simple. Instead of using the ordinary $t$, we use Dunnett's $t$ according to the number of treatments to be compared with the fixed control. In our example, we calculate the four "$t$" values for the four differences:

$$\frac{d_1}{s_d} = \frac{4.0}{1.5} = 2.67 \qquad \text{nonsignificant}$$

$$\frac{d_2}{s_d} = \frac{-1.0}{1.5} = -0.67 \qquad \text{nonsignificant}$$

$$\frac{d_3}{s_d} = \frac{5.0}{1.5} = 3.33 \qquad \text{significant}$$

$$\frac{d_4}{s_d} = \frac{7.0}{1.5} = 4.67 \qquad \text{significant}$$

For an ordinary two-tail test, we look up Table 31.2 in column "1 & 4" (one control and four treatments) and the row with 12 $df$ and find that

*Table* 31.2   Dunnett's $t$ at 5 per cent level for two-tail comparisons between several treatments and a control

| df | One control and number of treatment | | | | | | | | |
|---|---|---|---|---|---|---|---|---|---|
| | 1 & 1 | 1 & 2 | 1 & 3 | 1 & 4 | 1 & 5 | 1 & 6 | 1 & 7 | 1 & 8 | 1 & 9 |
| 5 | 2.57 | 3.03 | 3.29 | 3.48 | 3.62 | 3.73 | 3.82 | 3.90 | 3.97 |
| 6 | 2.45 | 2.86 | 3.10 | 3.26 | 3.39 | 3.49 | 3.57 | 3.64 | 3.71 |
| 7 | 2.36 | 2.75 | 2.97 | 3.12 | 3.24 | 3.33 | 3.41 | 3.47 | 3.53 |
| 8 | 2.31 | 2.67 | 2.88 | 3.02 | 3.13 | 3.22 | 3.29 | 3.35 | 3.41 |
| 9 | 2.26 | 2.61 | 2.81 | 2.95 | 3.05 | 3.14 | 3.20 | 3.26 | 3.32 |
| 10 | 2.23 | 2.57 | 2.76 | 2.89 | 2.99 | 3.07 | 3.14 | 3.19 | 3.24 |
| 11 | 2.20 | 2.53 | 2.72 | 2.84 | 2.94 | 3.02 | 3.08 | 3.14 | 3.19 |
| 12 | 2.18 | 2.50 | 2.68 | 2.81 | 2.90 | 2.98 | 3.04 | 3.09 | 3.14 |
| 13 | 2.16 | 2.48 | 2.65 | 2.78 | 2.87 | 2.94 | 3.00 | 3.06 | 3.10 |
| 14 | 2.14 | 2.46 | 2.63 | 2.75 | 2.84 | 2.91 | 2.97 | 3.02 | 3.07 |
| 15 | 2.13 | 2.44 | 2.61 | 2.73 | 2.82 | 2.89 | 2.95 | 3.00 | 3.04 |
| 16 | 2.12 | 2.42 | 2.59 | 2.71 | 2.80 | 2.87 | 2.92 | 2.97 | 3.02 |
| 17 | 2.11 | 2.41 | 2.58 | 2.69 | 2.78 | 2.85 | 2.90 | 2.95 | 3.00 |
| 18 | 2.10 | 2.40 | 2.56 | 2.68 | 2.76 | 2.83 | 2.89 | 2.94 | 2.98 |
| 19 | 2.09 | 2.39 | 2.55 | 2.66 | 2.75 | 2.81 | 2.87 | 2.92 | 2.96 |
| 20 | 2.09 | 2.38 | 2.54 | 2.65 | 2.73 | 2.80 | 2.86 | 2.90 | 2.95 |
| 24 | 2.06 | 2.35 | 2.51 | 2.61 | 2.70 | 2.76 | 2.81 | 2.86 | 2.90 |
| 30 | 2.04 | 2.32 | 2.47 | 2.58 | 2.66 | 2.72 | 2.77 | 2.82 | 2.86 |
| 40 | 2.02 | 2.29 | 2.44 | 2.54 | 2.62 | 2.68 | 2.73 | 2.77 | 2.81 |
| 60 | 2.00 | 2.27 | 2.41 | 2.51 | 2.58 | 2.64 | 2.69 | 2.73 | 2.77 |
| 120 | 1.98 | 2.24 | 2.38 | 2.47 | 2.55 | 2.60 | 2.65 | 2.69 | 2.73 |
| ∞ | 1.96 | 2.21 | 2.35 | 2.44 | 2.51 | 2.57 | 2.61 | 2.65 | 2.69 |

Courtesy of Dr. C. W. Dunnett, *Biometrics*, **20**, in press (1964).

the tabulated $t = 2.81$ at the 5 per cent significance level. The conclusion is that treatments (1) and (2) are not significantly different from the control, but treatments (3) and (4) are. The probability for the joint conclusions to be correct is 0.95.

As usual, the 95 per cent confidence intervals for the four differences may be constructed. They are $d \pm (dut)s_d$, where $dut$ is the tabulated Dunnett's $t$ at the required probability level. In our example, $dut = 2.81$. The quantity

$$(dut)s_d = (dut)\sqrt{\frac{2s^2}{n}} = 2.81 \times 1.50 = 4.22$$

*Table* 31.3   Dunnett's *t* at 5 per cent level for one-tail
comparisons between several treatments and a control

| df | \multicolumn{9}{c}{One control and number of treatment} |
| --- | --- | --- | --- | --- | --- | --- | --- | --- | --- |
|  | 1 & 1 | 1 & 2 | 1 & 3 | 1 & 4 | 1 & 5 | 1 & 6 | 1 & 7 | 1 & 8 | 1 & 9 |
| 5 | 2.02 | 2.44 | 2.68 | 2.85 | 2.98 | 3.08 | 3.16 | 3.24 | 3.30 |
| 6 | 1.94 | 2.34 | 2.56 | 2.71 | 2.83 | 2.92 | 3.00 | 3.07 | 3.12 |
| 7 | 1.89 | 2.27 | 2.48 | 2.62 | 2.73 | 2.82 | 2.89 | 2.95 | 3.01 |
| 8 | 1.86 | 2.22 | 2.42 | 2.55 | 2.66 | 2.74 | 2.81 | 2.87 | 2.92 |
| 9 | 1.83 | 2.18 | 2.37 | 2.50 | 2.60 | 2.68 | 2.75 | 2.81 | 2.86 |
| 10 | 1.81 | 2.15 | 2.34 | 2.47 | 2.56 | 2.64 | 2.70 | 2.76 | 2.81 |
| 11 | 1.80 | 2.13 | 2.31 | 2.44 | 2.53 | 2.60 | 2.67 | 2.72 | 2.77 |
| 12 | 1.78 | 2.11 | 2.29 | 2.41 | 2.50 | 2.58 | 2.64 | 2.69 | 2.74 |
| 13 | 1.77 | 2.09 | 2.27 | 2.39 | 2.48 | 2.55 | 2.61 | 2.66 | 2.71 |
| 14 | 1.76 | 2.08 | 2.25 | 2.37 | 2.46 | 2.53 | 2.59 | 2.64 | 2.69 |
| 15 | 1.75 | 2.07 | 2.24 | 2.36 | 2.44 | 2.51 | 2.57 | 2.62 | 2.67 |
| 16 | 1.75 | 2.06 | 2.23 | 2.34 | 2.43 | 2.50 | 2.56 | 2.61 | 2.65 |
| 17 | 1.74 | 2.05 | 2.22 | 2.33 | 2.42 | 2.49 | 2.54 | 2.59 | 2.64 |
| 18 | 1.73 | 2.04 | 2.21 | 2.32 | 2.41 | 2.48 | 2.53 | 2.58 | 2.62 |
| 19 | 1.73 | 2.03 | 2.20 | 2.31 | 2.40 | 2.47 | 2.52 | 2.57 | 2.61 |
| 20 | 1.72 | 2.03 | 2.19 | 2.30 | 2.39 | 2.46 | 2.51 | 2.56 | 2.60 |
| 24 | 1.71 | 2.01 | 2.17 | 2.28 | 2.36 | 2.43 | 2.48 | 2.53 | 2.57 |
| 30 | 1.70 | 1.99 | 2.15 | 2.25 | 2.33 | 2.40 | 2.45 | 2.50 | 2.54 |
| 40 | 1.68 | 1.97 | 2.13 | 2.23 | 2.31 | 2.37 | 2.42 | 2.47 | 2.51 |
| 60 | 1.67 | 1.95 | 2.10 | 2.21 | 2.28 | 2.35 | 2.39 | 2.44 | 2.48 |
| 120 | 1.66 | 1.93 | 2.08 | 2.18 | 2.26 | 2.32 | 2.37 | 2.41 | 2.45 |
| ∞ | 1.64 | 1.92 | 2.06 | 2.16 | 2.23 | 2.29 | 2.34 | 2.38 | 2.42 |

By permission of JASA and Dr. C. W. Dunnett.

is known as the *allowance*, which is half the width of the confidence interval.   Thus,

$$d_1 \pm \text{allowance} = \quad 4.00 \pm 4.22 \qquad \text{that is, } -0.22 \text{ to } \ 8.22$$
$$d_2 \pm \text{allowance} = -1.00 \pm 4.22 \qquad \text{that is, } -5.22 \text{ to } \ 3.22$$
$$d_3 \pm \text{allowance} = \quad 5.00 \pm 4.22 \qquad \text{that is, } \ \ 0.78 \text{ to } \ 9.22$$
$$d_4 \pm \text{allowance} = \quad 7.00 \pm 4.22 \qquad \text{that is, } \ \ 2.78 \text{ to } 11.22$$

The probability for the joint intervals to be true is 0.95.   Hence we can
say with 95 per cent confidence that treatment (4) exceeds the control by
at least 2.78, etc.

If the number of replications are not the same for all treatments, the factor $2/n$ in the variance formula should be replaced by $1/n_0 + 1/n_i$, where $n_0$ is the number of replications for the control and $n_i$ is that for the $i$th treatment. This situation usually arises in a one-way classification experiment. For more complicated experiments, the number of replications for each group should be the same.

## 2. One-tail test

In comparing several treatments with a control, the investigator is often interested only in whether a treatment is better than the control. Such tests are called one-tail or one-sided tests (Chap. 8). Dunnett's $t$'s for 95 per cent confidence intervals for one-sided comparisons are given in Table 31.3. Note that the first column (1 control and 1 treatment) is the ordinary Student's $t$ at the significance level $\alpha = 0.10$. We are now only using the upper (right-hand tail) 0.05 of the distribution. The $t$ values for the one-tail 95 per cent confidence (Table 31.3) are, of course, smaller than these for two-tail 95 per cent confidence (Table 31.2).

Let us use the same numerical example ($s^2 = 4.5$ with 12 $df$) to illustrate the difference in calculations and conclusions. For one-sided comparisons between one control and four treatments, Dunnett's tabulated $t = 2.41$ at the 5 per cent level. Treatment (1) will be now considered significantly better than the control, because $2.67 > 2.41$. The "allowance" is now

$$(dut)s_d = (dut) \sqrt{\frac{2s^2}{n}} = 2.41 \times 1.50 = 3.62$$

The lower limits of the four differences are then

$$
\begin{aligned}
d_1 - \text{allowance} &= \phantom{-}4.00 - 3.62 = \phantom{-}0.38 \\
d_2 - \text{allowance} &= -1.00 - 3.62 = -4.62 \\
d_3 - \text{allowance} &= \phantom{-}5.00 - 3.62 = \phantom{-}1.38 \\
d_4 - \text{allowance} &= \phantom{-}7.00 - 3.62 = \phantom{-}3.38
\end{aligned}
$$

The conclusion is that treatment (4) exceeds the control by at least 3.38, etc. The probability for the joint limit statements to be correct is 0.95.

For one-sided comparisons, the number of replications for each group must be equal. If not, the replacement of the factor $2/n$ in the variance formula by $1/n_0 + 1/n_i$ will yield only an approximate answer.

## 3. Comparing all pairs of treatments

In an endeavor to find where the "significance" lies, another common type of comparison made by investigators is to take the differences between all possible pairs of treatment means. For $t$ treatments, there are $\frac{1}{2}t(t-1)$ pairs to compare, although there are only a total of $t-1$ degrees of freedom for treatments. Clearly, not all comparisons are independent or orthogonal to each other. Furthermore, one of the pair comparisons is bound to be the largest treatment mean vs. the smallest treatment mean. Even if the general $F$ test for the $t$ treatments as a whole is nonsignificant, the extreme difference (largest minus smallest) may attain the conventional significance level. It should be carefully noted that this extreme difference may happen to be $\bar{y}_1 - \bar{y}_2$, $\bar{y}_1 - \bar{y}_3$, $\bar{y}_2 - \bar{y}_3$, or any other of the possible pairs. Therefore, it is not a predetermined particular hypothesis, and the conventional significance level is not valid for comparison of the two extremes.

At this stage we may digress briefly to mention another statistic. In any sample, there will be in general a largest value ($y_{max}$) and a smallest value ($y_{min}$). The extreme difference $Q = y_{max} - y_{min}$ is called the *range* of the sample. Under random sampling the range tends to be longer as the sample size increases. This is common experience. The range is in some physical unit. Let $s$ be the sample estimate of standard deviation of $y$. When the original range is expressed in units of $s$,

$$q = \frac{Q}{s} = \frac{y_{max} - y_{min}}{s}$$

is appropriately called the "studentized range," without a physical unit. The sampling distribution of $q$ from normal populations has been known for many years and tabulated by various workers. A recent tabulation is that of Wright Air Development Center [*Tech. Rept.* **2**:58–484 (1959), by Harter, Glemm, and Guthrie]. The distribution of $q$ varies with the sample size and the $df$ of $s$.

Returning to our problem of comparing the extreme difference between two treatment means, the reader will see immediately that our problem is closely connected with the distribution of studentized range. In fact, Tukey's method of testing all possible pairs of treatment means, to be described below, is based entirely on the $q$ table. [J. W. Tukey, "The Problem of Multiple Comparisons," Princeton University Press, Princeton, N.J. (1953) 396 pp., Dittoed.] When all comparisons are of the type $\bar{y}_i - \bar{y}_j$, we may convert the published studentized range table into Tukey's $t$ table (which we shall call *tuk* for brevity) by the formula

$$tuk = \frac{q}{\sqrt{2}} = q \times 0.7071$$

Tukey's $t$ values are shown in Table 31.4. Again, when there are only two groups to compare, Tukey's $t$ is identical with Student's $t$, as shown in the first column of the table.

*Table* 31.4  Tukey's $t$ at 5 per cent level, adjusted for making all pair-comparisons

| $df$ | Number of treatment means | | | | | | | | |
|---|---|---|---|---|---|---|---|---|---|
| | 2 | 3 | 4 | 5 | 6 | 7 | 8 | 9 | 10 |
| 5 | 2.57 | 3.25 | 3.69 | 4.01 | 4.26 | 4.48 | 4.65 | 4.81 | 4.94 |
| 6 | 2.45 | 3.07 | 3.46 | 3.75 | 3.98 | 4.16 | 4.33 | 4.47 | 4.59 |
| 7 | 2.36 | 2.94 | 3.32 | 3.58 | 3.79 | 3.97 | 4.12 | 4.24 | 4.36 |
| 8 | 2.31 | 2.86 | 3.20 | 3.46 | 3.66 | 3.82 | 3.96 | 4.08 | 4.19 |
| 9 | 2.26 | 2.79 | 3.13 | 3.37 | 3.55 | 3.71 | 3.84 | 3.96 | 4.06 |
| 10 | 2.23 | 2.74 | 3.06 | 3.29 | 3.47 | 3.62 | 3.75 | 3.86 | 3.95 |
| 11 | 2.20 | 2.70 | 3.01 | 3.23 | 3.41 | 3.56 | 3.68 | 3.78 | 3.88 |
| 12 | 2.18 | 2.67 | 2.97 | 3.19 | 3.36 | 3.50 | 3.62 | 3.73 | 3.82 |
| 13 | 2.16 | 2.64 | 2.93 | 3.15 | 3.32 | 3.45 | 3.57 | 3.67 | 3.76 |
| 14 | 2.14 | 2.62 | 2.91 | 3.12 | 3.28 | 3.42 | 3.53 | 3.63 | 3.71 |
| 15 | 2.13 | 2.60 | 2.88 | 3.09 | 3.25 | 3.38 | 3.49 | 3.59 | 3.68 |
| 16 | 2.12 | 2.58 | 2.86 | 3.06 | 3.22 | 3.35 | 3.46 | 3.56 | 3.64 |
| 17 | 2.11 | 2.57 | 2.84 | 3.04 | 3.20 | 3.33 | 3.44 | 3.53 | 3.61 |
| 18 | 2.10 | 2.55 | 2.83 | 3.03 | 3.17 | 3.30 | 3.41 | 3.51 | 3.59 |
| 19 | 2.09 | 2.54 | 2.81 | 3.01 | 3.16 | 3.29 | 3.39 | 3.48 | 3.56 |
| 20 | 2.09 | 2.53 | 2.80 | 2.99 | 3.15 | 3.27 | 3.37 | 3.46 | 3.54 |
| 24 | 2.06 | 2.50 | 2.76 | 2.95 | 3.09 | 3.21 | 3.31 | 3.40 | 3.48 |
| 30 | 2.04 | 2.47 | 2.72 | 2.90 | 3.04 | 3.15 | 3.25 | 3.34 | 3.42 |
| 40 | 2.02 | 2.43 | 2.68 | 2.86 | 2.99 | 3.10 | 3.20 | 3.27 | 3.35 |
| 60 | 2.00 | 2.40 | 2.64 | 2.81 | 2.94 | 3.05 | 3.14 | 3.22 | 3.29 |
| 120 | 1.98 | 2.38 | 2.61 | 2.77 | 2.90 | 3.00 | 3.08 | 3.17 | 3.22 |
| ∞ | 1.96 | 2.34 | 2.57 | 2.73 | 2.85 | 2.95 | 3.03 | 3.10 | 3.16 |

For an illustration of the *tuk* test, let us rearrange the five treatment means of our numerical example in decreasing order of magnitude. The $\frac{1}{2}(5 \times 4) = 10$ possible differences $d$ between pairs are given in the accompanying table. The standard error of the difference between two

| | | | (1) −24 | (2) −22 | (3) −21 | (4) −17 | (5) +16 |
|---|---|---|---|---|---|---|---|
| $e$ | (1) | 24 | 0 | 2 | 3 | 7 | 8 |
| $d$ | (2) | 22 | | 0 | 1 | 5 | 6 |
| $b$ | (3) | 21 | | | 0 | 4 | 5 |
| $a$ | (4) | 17 | | | | 0 | 1 |

means is $s_d = \sqrt{2s^2/n} = 1.50$, as we have obtained before. Then, the observed $tuk = d/s_d$. For five treatments and 12 $df$ for $s$, Tukey's $t$ table (Table 31.4) gives $tuk = 3.19$ at the 5 per cent significance level. Hence the significant difference must be at least as large as

$$tuk \times s_d = 3.19 \times 1.50 = 4.78$$

Using this as the criterion, we see that treatments (1) and (2) are considered significantly different from (4) and (5), and (3) is also better than (5). All the other differences are nonsignificant.

Confidence intervals may also be constructed in the usual way by using the "allowance" $tuk \times s_d = 4.78$ for each difference. The probability for the joint statements to be true is 0.95.

Tukey's method may be extended to more complicated comparisons by using the values $const \times q$, where the constant depends on the type of comparison, but it is not very sensitive. For all other types of comparison, the following general procedure may be resorted to.

## 4. All types of contrasts

For a set of orthogonal contrasts preconceived as a set of hypotheses to be tested, the ordinary $t^2 = F$ test is appropriate for each contrast, as we have used it in many preceding chapters. But, analogously to comparing all pairs of treatments, the investigator may wish to test a number of contrasts, not necessarily all independent of each other, after examination of the data. This of course includes the comparison between the largest and the smallest treatment means. To cope with the general snooping type of contrasts, Scheffe [A Method for Judging All Contrasts in the Analysis of Variance, *Biometrika*, **40**:87–104 (1953)] developed a test method based on the ordinary $F$ table. For $k$ treatments and $\nu$ degrees of freedom for $s^2$, his statistic is

$$chit = \sqrt{(k-1)F_{(k-1),\nu}}$$

where $F_{(k-1),\nu}$ is the value of $F$ with $k-1$ degrees of freedom in the numerator and $\nu$ degrees of freedom in the denominator. Again, when there are only two groups ($k = 2$ and $k - 1 = 1$), Scheffe's $chit$ is $\sqrt{F}$, identical with Student's $t$. Also, when $\nu$ is very large, $(k-1)F$ approaches the value of chi square, and thus $\sqrt{(k-1)F}$ is approximately the value of chi. Hence we call Scheffe's statistic $chit$ (instead of the usual notation $S$). The reader may easily construct a table of $chit$ from published tables of $F$ if he has frequent occasion to use it.

As a numerical example of Scheffe's test method, let us consider the five treatment means once more, viz., $\bar{y}_1 = 24$, $\bar{y}_2 = 22$, $\bar{y}_3 = 21$, $\bar{y}_4 = 17$, $\bar{y}_5 = 16$, for $n = 4$ replications with $s^2 = 4.50$ having 12 $df$ and suppose that we wish to test the contrast of (1), (2), (3) vs. (4), (5). The calculations proceed as follows:

$$\theta = 2\bar{y}_1 + 2\bar{y}_2 + 2\bar{y}_3 - 3\bar{y}_4 - 3\bar{y}_5 = 134 - 99 = 35$$

$$V(\theta) = (2^2 + 2^2 + 2^2 + 3^2 + 3^2)V(\bar{y}) = 30\left(\frac{s^2}{n}\right) = 33.75$$

$$\frac{\theta}{\sqrt{V(\theta)}} = \frac{35}{\sqrt{33.75}} = \frac{35}{5.81} = 6.02$$

while from the $F$ table,

$$chit = \sqrt{4 \times 3.26} = \sqrt{13.04} = 3.61$$

where 4 is $k - 1$ and 3.26 is the tabulated value of $F$ with 4 and 12 degrees of freedom at the 5 per cent significance level. Since 6.02 is greater than 3.61, the contrast is significant; that is, $\theta$ is significantly different from zero. The 95 per cent confidence interval for the true value of $\theta$ may be constructed the usual way, the allowance being

$$chit \sqrt{V(\theta)} = 3.61 \times 5.81 = 20.97 \doteq 21$$

The method outlined above may be applied to any number of contrasts suggested by the data.

## 5. Summary

In order to appreciate the relative merits of the various methods of multiple comparisons, let us apply Scheffe's method to the simple paired comparison:

$$\theta = \bar{y}_3 - \bar{y}_5 = 21 - 16 = 5$$

$$V(\theta) = (1^2 + 1^2)V(\bar{y}) = \frac{2 \times 4.5}{1} = 2.25$$

$$\frac{\theta}{\sqrt{V(\theta)}} = \frac{5}{\sqrt{2.25}} = \frac{5}{1.50} = 3.33$$

$$chit = \sqrt{4 \times 3.26} = \sqrt{13.04} = 3.61$$

The observed value 3.33 being smaller than $chit = 3.61$, the conclusion here is that treatment (3) is not significantly different from treatment (5), while Tukey's method gives a significant result in Sec. 3. We thus see that for paired comparisons, Tukey's method is more sensitive and Scheffe's method should not be used. On the other hand, for more complicated contrasts, it has been shown that Scheffe's method is more sensitive than Tukey's method based on the studentized range.

Both as a review of the numerical calculations and as a summary of the three methods of multiple comparisons, we set the following rules for practical application.

1. For independent comparisons, use Student's test ($t = 2.18$).
2. For comparisons with a control, use Dunnett's method ($dut = 2.81$).
3. For all paired comparisons, use Tukey's method ($tuk = 3.19$).
4. For other types of contrasts, use Scheffe's method ($chit = 3.61$).

Note that for a given set of data,

$$t < dut < tuk < chit$$

This reflects the general fact of life:

> The more general the testing method,
> the larger the required value of statistic,
> the wider the confidence intervals,
> the less the sensitivity of the test.

The comparatively low sensitivity of Scheffe's method is the price we pay for its general ability to test any and all types of contrasts.

## EXERCISES

**31.1**   A small portion of the studentized range ($q$) table at the 0.05 level is reproduced in the accompanying table. Convert it into Tukey's $t$ table by multiplying its entries by $\sqrt{1/2} = 0.7071$ and check your answers with Table 31.4.

| $df$ of $s^2$ | Value of $q$ | | | | Value of $tuk$ | | | |
|---|---|---|---|---|---|---|---|---|
| | $k$ | | | | $k$ | | | |
| | 3 | 4 | 5 | 6 | 3 | 4 | 5 | 6 |
| 12 | 3.77 | 4.20 | 4.51 | 4.75 | | | 3.19 | |
| 15 | 3.68 | 4.07 | 4.37 | 4.60 | | | | |
| 20 | 3.58 | 3.96 | 4.23 | 4.45 | | | | |

**31.2**   A small portion of the $F$ table at the 0.05 level is reproduced in the accompanying table. Convert it into Scheffe's $t$ table by the relation $chit = \sqrt{(k-1)F}$ and enter your results in the blanks in the table for possible future use.

| $df$ of $s^2$ | Value of $F$ | | | | Value of $chit$ | | | |
|---|---|---|---|---|---|---|---|---|
| | $k - 1$ | | | | $k - 1$ | | | |
| | 2 | 3 | 4 | 5 | 2 | 3 | 4 | 5 |
| 12 | 3.88 | 3.49 | 3.26 | 3.11 | | | 3.61 | |
| 15 | 3.68 | 3.29 | 3.06 | 2.90 | | | | |
| 20 | 3.49 | 3.10 | 2.87 | 2.71 | | | | |

**31.3** In Exercise 16.2 (page 186) the following data on six treatments were obtained from four ($r = 4$) replications:

| Treatment | a | b | c | d | e | f |
|-----------|----|----|----|----|----|----|
| Mean | 22 | 26 | 17 | 16 | 21 | 18 |

It was further found that $s^2 = 8.0$ with 15 degrees of freedom. Before doing any test, calculate the standard error of the difference $d$ between two treatment means:

$$s_d = \sqrt{\frac{2s^2}{r}} = \sqrt{\frac{2 \times 8}{4}} = 2.00$$

**a** Treatment $c$ is the control. Compare each of the other treatments with the control by Dunnett's $t$ method. What treatments are significantly different (two-tail test) from the control? What treatments are significantly better (one-tail test) than the control? Significance level is $\alpha = 0.05$.

**b** Make comparisons between all possible pairs by Tukey's $t$ method. Which pairs are considered to be significantly different?

**c** After examining the data, we wish to test the contrast

$$\theta = 3c - d - e - f$$

by Scheffe's method. Is this comparison significant at the 0.05 level?

# 32

# Unequal group variances

Throughout this book it has been assumed that the various (treatment) groups have the same variance. It is one of the mathematical conditions under which the $F$ and $t$ tests are valid. However, there is plenty of empirical evidence indicating that when the group variances are not too widely different, the $F$ test we have been using is still valid for practical purposes. The mathematically exact test for unequal group variances is very troublesome, and as of now I do not think that mathematical statisticians have settled on an acceptable method so that we, the users of statistical methods, may have a standard procedure to follow. In this chapter a comparatively simple approximate method will be described. In the course of presenting this method, the properties, limitations, and difficulties in practical application will also be discussed.

## 1. Review of the $t$ test

Before we introduce the new method it is best to review briefly the ordinary $t$ test for purposes of comparison. Consider the two samples (or

*Table* **32.1** Two independent samples and estimates of mean
and variance

| | Single values $y_{i\alpha}$ | | | | | $n_i$ | $df_i$ | $\bar{y}_i$ | $ssq_i$ | $s_i^2$ | $s_i^2/n_i$ |
|---|---|---|---|---|---|---|---|---|---|---|---|
| (1) | 18 13 12 14 19 | | | | | 10 | 9 | 15 | 72 | 8.0 | 0.80 |
| | 17 11 15 18 13 | | | | | | | | | | |
| (2) | 12 8 15 9 | | | | | 4 | 3 | 11 | 30 | 10.0 | 2.50 |
| Pooled | | | | | | | 12 | | 102 | 8.5 | 3.30 |

two treatment groups) in Table 32.1. The first sample has 10 observations with $ssq_1 = 72$, so that the estimate of variance is

$$s_1^2 = \frac{ssq_1}{df_1} = \frac{72}{9} = 8.0$$

The second sample has four observations with $ssq_2 = 30$, so that
$s_2^2 = {}^{30}\!/_3 = 10.0$. These two estimates of variance are not too unequal.
In fact, we may perform a variance-ratio test to see if they are signifi-
cantly different. $F = {}^{10}\!/_8 = 1.25$ with three and nine degrees of freedom.
The tabulated value of $F$ is 3.86 at the 5 per cent significance level.
Hence $s_1^2$ and $s_2^2$ are not significantly different and the ordinary $t$ test
is valid.

In performing the ordinary $t$ test, we pool the sum of squares and
degrees of freedom for the two samples and obtain a single estimate of the
variance.

$$s^2 = \frac{ssq_1 + ssq_2}{df_1 + df_2} = \frac{72 + 30}{9 + 3} = \frac{102}{12} = 8.5$$

as shown at the bottom line of Table 32.1. Then

$$t = \frac{\bar{y}_1 - \bar{y}_2}{\sqrt{s^2(1/n_1 + 1/n_2)}} = \frac{4.0}{\sqrt{8.5({}^1\!/_{10} + {}^1\!/_4)}} = \frac{4.0}{\sqrt{2.975}} = 2.32$$

with $n_1 + n_2 - 2 = df_1 + df_2 = 12$ degrees of freedom. The tabulated
$t$ value is 2.18 at the 5 per cent significance level, so that the difference
between the two sample means must be judged significant.

## 2. The *t*-like statistic

Now let us suppose that the two group variances are different. The
variance of sample mean is $V(\bar{y}_1) = s_1^2/n_1$ for the first sample and
$V(\bar{y}_2) = s_2^2/n_2$ for the second sample, so that the variance of the differ-

ence between two sample means is

$$V(\bar{y}_1 - \bar{y}_2) = \frac{s_1{}^2}{n_1} + \frac{s_2{}^2}{n_2}$$

Then a *t*-like statistic is defined as

$$t^* = \frac{\bar{y}_1 - \bar{y}_2}{\sqrt{s_1{}^2/n_1 + s_2{}^2/n_2}} = \frac{4.0}{\sqrt{\frac{8}{10} + 10\frac{1}{4}}} = \frac{4.0}{\sqrt{3.30}} = 2.20$$

($t^* = 2.2019$ more exactly.)    It is called a *t-like statistic* because it does *not* have the exact *t* distribution.    The denominator of the Student's *t* is one variable, whereas the denominator of $t^*$ is the sum of two variables. The exact sampling distribution of $t^*$ is unknown.    (It must involve parameters $\nu_1 = 1$ for two groups and two additional ones $df_1$ and $df_2$ for the denominator.)

In order to be able to make use of the ordinary *t* table, a certain number of degrees of freedom ($df^*$) is to be assigned to $t^*$ so that it is equivalent to *t* with that number of degrees of freedom.    This method is due to Welch (1951; see Brownlee, 1960, pp. 235–239, 265–268).    The formula for the number of degrees of freedom is an awkwardly balanced three-story high expression:

$$df^* = \frac{(s_1{}^2/n_1 + s_2{}^2/n_2)^2}{\dfrac{(s_1{}^2/n_1)^2}{n_1 - 1} + \dfrac{(s_2{}^2/n_1)^2}{n_2 - 1}}$$

It may be somewhat simplified in appearance.    The numerical value of $df^*$, however, may be readily calculated from the values listed in Table 32.1.    Thus, in our example,

$$df^* = \frac{(3.30)^2}{(0.80)^2/9 + (2.50)^2/3} = \frac{10.89}{2.1544} = 5.055$$

For approximately five degrees of freedom, the tabulated value of *t* is 2.57 at the 5 per cent significance level.    Since $t^* = 2.20$, we conclude that the difference between the two sample means is not significant.

Before proceeding further we may summarize the results of the two methods as applied to the given set of data shown in Table 32.1.    These

| *Ordinary t* | *t-like* |
|---|---|
| $t = \dfrac{\bar{y}_1 - \bar{y}_2}{\sqrt{s^2(1/n_1 + 1/n_2)}} = 2.32$ | $t^* = \dfrac{\bar{y}_1 - \bar{y}_2}{\sqrt{s_1{}^2/n_1 + s_2{}^2/n_2}} = 2.20$ |
| $df = 12 \qquad t_{0.05} \qquad = 2.18$ | $df^* = 5$ approx. $\qquad t_{0.05} = 2.57$ |
| Difference significant | difference nonsignificant |

two methods make us to draw two different conclusions with respect to the same set of data. This constitutes a practical difficulty that users of this method should be aware of. Usually a more general mathematical method would reduce to the simpler case when certain conditions are satisfied. This is not the case here. Even if $s_1^2 = s_2^2$ (as they are nearly so in our example), these two methods still give two different results. Apparently, the $t$-like is not a "generalization" of $t$, but is a different method based on different assumptions. Hence, the two could give different results. (The $t$-like is not like $t$.) In this particular example and similar cases, I would use the ordinary $t$ or $F$ test.

## 3. The value of $df^*$

From the summary results above it is seen that $t = 2.32$ and $t^* = 2.20$ differ only moderately, but $df = 12$ and $df^* = 5$ are widely different. Let us investigate the expression for $df^*$ in more detail. It may be shown by differential calculus that the maximum value of $df^*$ is $n_1 + n_2 - 2$ when (Brownlee, 1960, p. 237)

$$\frac{s_1^2}{n_1(n_1 - 1)} = \frac{s_2^2}{n_2(n_2 - 1)}$$

but not when $s_1^2 = s_2^2$. The minimum value of $df^*$ is $n_1 - 1$ or $n_2 - 1$, whichever is smaller. This occurs when $s_2^2/n_2$ is much larger than $s_1^2/n_1$, or vice versa, as can be seen directly from the full expression for $df^*$. In our example, $s_2^2/n_2 = 2.50$ is more than three times as large as $s_1^2/n_1 = 0.80$, and $df^* = 5$, being much closer to the smaller size of the sample than to $n_1 + n_2 - 2$. It also follows that when $s_1^2 = s_2^2$ the value of $df^*$ is entirely determined by the sizes of the two samples. For instance, when $n_1$ is much larger than $n_2$, the value of $df^*$ approaches its minimum value of $n_2 - 1$. This makes sense, because when $n_1$ is much larger than $n_2$, the distribution of $t$-like will be essentially determined by $s_2^2$ with $n_2 - 1$ degrees of freedom. In other words, $\bar{y}_1$ and $s_1^2$ based on a very large $n_1$ will behave like constants, and thus $t^*$ behaves like $t$ with $n_2 - 1$ degrees of freedom.

## 4. Two groups of equal size

When two samples are of the same size, $t^*$ and $t$ have the same value because, writing $n_1 = n_2 = n$,

$$\frac{s_1^2}{n} + \frac{s_2^2}{n} = \frac{ssq_W}{n(n - 1)} = s^2 \left( \frac{1}{n} + \frac{1}{n} \right)$$

where $ssq_w = ssq_1 + ssq_2$ is the total *ssq* within groups. The expression for $df^*$ also simplifies considerably. Substituting $n_1 = n_2 = n$ in the general expression for $df^*$, we obtain

$$df^* = \frac{(n-1)(s_1^2 + s_2^2)^2}{(s_1^2)^2 + (s_2^2)^2}$$

Table 32.2 gives an example of two groups of the same size $(n = 10)$. The pooled estimate of variance is $406/18 = 22.56 = s^2$, and

$$22.55\left(\frac{1}{10} + \frac{1}{10}\right) = \frac{35.11}{10} + \frac{10.00}{10} = \frac{406}{10 \times 9} = 4.511$$

Hence, noting that $\bar{y}_2 - \bar{y}_1 = 23 - 17 = 6$,

$$t^* = t = \frac{6}{\sqrt{5.411}} = \frac{6}{2.124} = 2.825$$

For the ordinary $t$, $df = 20 - 2 = 18$; for $t$-like, however,

$$df^* = \frac{9(45.11)^2}{(35.11)^2 + (10.00)^2} = 13.7$$

Since $df^*$ is usually not an integer, the critical value for $t^*$ is to be obtained by interpolation from the ordinary $t$ table. Thus, at 5 per cent significance level,

For $df = 13$:     $t = 2.160$
For $df = 14$:     $t = 2.145$     for $df = 13.7$     $t = 2.15$

In our particular example (Table 32.2), $t^* = t = 2.825$, being significant by either method. However, should the observed $t^* = t = 2.13$, it would be judged nonsignificant by the $t^*$ method but significant by the ordinary $t$ test with $df = 18$, whereas $t_{0.05} = 2.10$. This is the same practical problem mentioned before.

When two groups are of equal size, there is one pleasing fact, viz., the value of $df^*$ does reduce to $n_1 + n_2 - 2 = 2(n - 1)$ when $s_1^2 = s_2^2$. This shows that when group variances are unequal, it is even more desir-

*Table* **32.2**  Two samples of equal size and estimates of variance

|  | Single values $y_{i\alpha}$ | | | | | | | | | | $Y_i$ | $ssq_i$ | $s_i^2$ | $s_i^2/n$ |
|---|---|---|---|---|---|---|---|---|---|---|---|---|---|---|
| (1) | 11 | 18 | 15 | 26 | 25 | 22 | 11 | 13 | 19 | 10 | 170 | 316 | 35.1i | 3.511 |
| (2) | 26 | 21 | 19 | 22 | 19 | 27 | 28 | 22 | 24 | 22 | 230 | 90 | 10.00 | 1.000 |
|  | | | | $n_1 = n_2 = n = 10$ | | | | | | | | 406 | 22.5ṡ | 4.511 |

able to have groups of equal size. In an experimental setup, the investigator should by all means strive for equal groups, which minimize the disturbances caused by unequal variances.

## 5. The $F$-like statistic

Let us first introduce the *F-like* ($F^*$) *statistic* to the comparison between only two groups so that its value may be compared with that of $t$-like ($t^*$) obtained previously, using the same numerical example. The weight of a group mean is defined as the reciprocal of its variance, viz.,

$$w_i = \frac{n_i}{s_i^2}$$

for the $i$th group, so that the weighted general mean of the group means is

$$\bar{y}^* = \frac{\Sigma w_i \bar{y}_i}{\Sigma w_i} = \frac{\Sigma w_i \bar{y}_i}{W}$$

where $W = \Sigma w_i$. It should be noted that $\bar{y}^*$ is in general not equal to the ordinary general mean $\bar{y} = \Sigma n_i \bar{y}_i / \Sigma n_i$ derived on the assumption that all group variances are equal. With the weights and the weighted general mean so calculated, the $F$-like statistic for two groups ($k = 2$) is defined as

$$F^* = \Sigma w_i(\bar{y}_i - \bar{y}^*)^2 = w_1(\bar{y}_1 - \bar{y}^*)^2 + w_2(\bar{y}_2 - \bar{y}^*)^2$$

with degrees of freedom $\nu_1 = k - 1 = 1$, and, writing $f_i = n_i - 1$ for brevity,

$$\frac{1}{\nu_2} = \Lambda = \Sigma \frac{1}{f_i}\left(1 - \frac{w_i}{W}\right)^2$$

For illustration of the method, Table 32.3 is given in two parts: the upper portion lists the arithmetic details for calculating $F^*$ and the lower portion lists the details for calculating the number of degrees of freedom for the same set of data as shown in Table 32.1. It will be noted that

$$F^* = 4.848 = (2.2019)^2 = t^{*2}$$

and that the number of degrees of freedom is

$$\nu_2 = \frac{1}{0.19783} = 5.055 = df^*$$

Thus, the $F$-like is to $t$-like just as the ordinary $F$ is to the ordinary $t$ for two groups. When there are only two groups, it is arithmetically much simpler to use $t^*$. The comparatively long calculations shown in Table 32.3, however, pave the way to generalization for $k$ groups.

*Table* 32.3   Calculation of $F$-like and degrees of freedom for two groups (data from Table 32.1)

| $n_i$ | $s_i{}^2$ | $w_i = n_i/s_i{}^2$ | $\bar{y}_i$ | $w_i\bar{y}_i$ | $\bar{y}_i - y^*$ | $w_i(\bar{y}_i - y^*)^2$ |
|---|---|---|---|---|---|---|
| 10 | 8.0 | 1.25 | 15 | 18.75 | $+0.97$ | 1.176 |
| 4 | 10.0 | 0.40 | 11 | 4.40 | $-3.03$ | 3.672 |
| | | $W = 1.65$ | | $\Sigma w_i\bar{y}_i = 23.15$ | | 4.848 |
| | | | | $\bar{y}^* = 14.03$ | | $\Sigma w_i(\bar{y}_i - \bar{y}^*)^2 = F^*$ |

| $f_i = n_i - 1$ | $\dfrac{w_i}{W}$ | $1 - \dfrac{w_i}{W}$ | $\left(1 - \dfrac{w_i}{W}\right)^2$ | $\dfrac{1}{f_i}\left(1 - \dfrac{w_i}{W}\right)^2$ |
|---|---|---|---|---|
| 9 | 0.7576 | 0.2424 | 0.05877 | 0.00653 |
| 3 | 0.2424 | 0.7576 | 0.57392 | 0.19131 |
| | 1.0000 | | | $\Lambda = 0.19783 = 1/\nu_2$ |

## 6. The general $F$-like

When there are more than two groups, the expressions for $F^*$ and its degrees of freedom are quite formidable, although the arithmetic procedure is essentially the same as that outlined in Table 32.3. The weights $w_i$ and the weighted general mean $\bar{y}^*$ are defined the same way. Let $W = \Sigma w_i$, $f_i = n_i - 1$, and

$$\Lambda = \sum \frac{1}{f_i}\left(1 - \frac{w_i}{W}\right)^2$$

Then

$$F^* = \frac{\Sigma w_i(\bar{y}_i - \bar{y}^*)^2/(k - 1)}{1 + \dfrac{2(k - 2)}{k^2 - 1}\Lambda}$$

with

$$\nu_1 = k - 1 \qquad \text{and} \qquad \nu_2^* = \frac{k^2 - 1}{3\Lambda}$$

It is to be observed that when $k = 2$, the denominator of $F^*$ becomes unity, and so does the factor $(k^2 - 1)/3$, and the expressions reduce to those of the preceding section.

Since the arithmetic procedure is long, we choose to use an extremely simple numerical example for illustration. The data are taken from Table 5.1 and reproduced here for convenience of reference.

Group 1:    6  9  3               $Y_1 = 18$    $ssq_1 = 18$

Group 2:  10  4  9  12  8  11    $Y_2 = 54$    $ssq_2 = 40$

Group 3:    2  6  4              $Y_3 = 12$    $ssq_3 = \phantom{0}8$

The rest of the arithmetic is listed systematically in Table 32.4. Substituting $k = 3$, $\Sigma w_i(\bar{y}_i - \bar{y}^*)^2 = 9.4432$, and $\Lambda = 0.57913$ in the formulas above, we have

$$F^* = \frac{9.4432/2}{1 + \frac{2}{8}(0.57913)} = \frac{4.7216}{1.1448} = 4.124$$

with

$$\nu_1 = 2 \qquad \text{and} \qquad \nu_2^* = \frac{8}{3(0.57913)} = 4.6$$

The critical value of $F^*$ is to be found by interpolation from the ordinary $F$ table. Thus, for $\nu_1 = 2$ at the 5 per cent significance level,

$$\begin{aligned} \nu_2 &= 4 & F &= 6.94 \\ \nu_2 &= 5 & F &= 5.79 \end{aligned} \qquad \nu_2^* = 4.6 \qquad F^* = 6.25$$

In this particular example the interpolation is not necessary, since the observed $F^* = 4.12$ is smaller than the tabulated 5.79. The difference among the three groups must then be judged nonsignificant.

When the same data are analyzed in the ordinary way, we have $\nu_2 = 9$ and $F = 3.68$, as shown in Table 8.1. The tabulated $F$ is 4.26 at the 5 per cent significance level. The conclusion is also nonsignificant.

*Table 32.4* Calculations for $F$-like and its degrees of freedom (data from Table 5.1)

| $n_i$ | $s_i^2$ | $w_i$ | $\bar{y}_i$ | $w_i\bar{y}_i$ | $\bar{y}_i - \bar{y}^*$ | $w_i(\bar{y}_i - \bar{y}^*)^2$ |
|---|---|---|---|---|---|---|
| 3 | 9.0 | 0.333 | 6 | 2.00 | −0.4091 | 0.0558 |
| 6 | 8.0 | 0.750 | 9 | 6.75 | 2.5909 | 5.0346 |
| 3 | 4.0 | 0.750 | 4 | 3.00 | −2.4091 | 4.3528 |
| | | 1.833 | | $\Sigma w_i\bar{y}_i = 11.75$ | | 9.4432 |
| | | | | $\bar{y}^* = 6.4091$ | | $\Sigma w_i(\bar{y}_i - \bar{y}^*)^2$ |

| $f_i = n_i - 1$ | $\dfrac{w_i}{W}$ | $1 - \dfrac{w_i}{W}$ | $\left(1 - \dfrac{w_i}{W}\right)^2$ | $\dfrac{1}{f_i}\left(1 - \dfrac{W_i}{W}\right)^2$ |
|---|---|---|---|---|
| 2 | 0.18182 | 0.81818 | 0.66942 | 0.33471 |
| 5 | 0.40909 | 0.59091 | 0.34917 | 0.06983 |
| 2 | 0.40909 | 0.59091 | 0.34917 | 0.17459 |
| $\nu_2 = 9$ | 1.00000 | | | $\Lambda = 0.57913$ |

It has been shown by mathematical statisticians that when groups are of equal size, the disturbances caused by unequal group variances are less important than when groups are of varying sizes. In other words, the ordinary $F$ test becomes more insensitive to unequal variances when the group size is the same. The special method outlined in this chapter is to be used only when the group variances are widely different. This again emphasizes the importance of having groups of equal size in experimentation.

## 7. Test for equality of variance

The final question is how widely different the group variances have to be before we say they are different. The sample variances $s_i^2$ have a much larger variance than the sample mean. The seemingly different values of $s_i^2$ may not be significantly different if we take their large sampling variances into consideration. Bartlett (1937)† developed a chi-square test for the homogeneity of the group variances. His method involves the calculation of the natural logarithm of $s_i^2$. The *approximate* value of chi square with $k - 1$ degrees of freedom, designated by $X^2$, is

$$X^2 = f \log_e s^2 - \Sigma f_i \log_e s_i^2$$

where $f_i = n_i - 1$ and $f = \Sigma f_i = N - k$ for $k$ groups. As usual, $s^2$ is the pooled estimate of variance from all groups. Since our primary purpose is to illustrate the procedure of calculating this approximate chi square, we shall use the same simple data of Table 5.1 as reproduced in the preceding section. The arithmetic is shown in Table 32.5, in which the natural logarithm has been used directly. If the common logarithm is used, the procedure remains exactly the same except that the final answer should be multiplied by 2.3026 to give the appropriate chi square.

In our example the pooled estimate of variance is $s^2 = {}^{66}\!/_9 = 7.33$ as obtained in Table 8.1. The approximate chi square with 2 $df$ is

$$X^2 = 17.932 - 17.564 = 0.368 \ (0.36774 \text{ in Table } 32.5)$$

which is insignificant. ($\chi^2 = 5.99$ for 2 $df$ at the 5 per cent point.) Hence, for this set of data, the ordinary analysis of variance and $F$ test are appropriate. The value of $X^2$ as calculated from the formula above is slightly larger than the true chi square. Since $X^2$ falls below the significance value, the calculation stops here for practical purpose. If the $X^2$ is larger than the significance value, it is necessary to apply a correction factor to bring its value down before testing for significance.

† M. S. Bartlett, Properties of sufficiency and statistical tests, *Proc. Roy. Soc. (London), Series A,* **160**: 268–282 (1937).

*Table 32.5* Calculation of chi square for testing equality of variances (data from Table 5.1, also used in Table 32.4)

| Group | $f_i$ | $ssq_i$ | $s_i^2$ | $\log_e s_i^2$ | $f_i \log_e s_i^2$ | $\dfrac{1}{f_i}$ |
|-------|-------|---------|---------|----------------|--------------------|------------------|
| I     | 2     | 18      | 9.0     | 2.19722        | 4.39444            | 0.500            |
| II    | 5     | 40      | 8.0     | 2.07944        | 10.39720           | 0.200            |
| III   | 2     | 8       | 4.0     | 1.38629        | 2.77258            | 0.500            |
|       |       |         |         |                | 17.56422           | 1.200            |
| Pooled | 9    | 66      | 7.33    | 1.99244        | 17.93196           | 0.111            |
|       | $f$   | $ssq_W$ | $s^2$   |                | $X^2 = $ 0.36774   | 1.089            |

The correction factor is, from the last column of Table 32.5,

$$c = \frac{1}{3(k-1)} \left( \sum \frac{1}{f_i} - \frac{1}{f} \right)$$

in our example

$$c = \tfrac{1}{6}(1.089) = 0.1815$$

The corrected chi square is the $X^2$ calculated above divided by $1 + c$. Thus

$$\chi^2 = \frac{X^2}{1+c} = \frac{0.36774}{1.1815} = 0.311$$

Only a significantly large chi square indicates heterogeneity of the group variances and calls for special methods of analysis.

## EXERCISES

**32.1** The ten ($n_1 = n_2 = n = 10$) observations of each sample are as follows:

```
I:    17  24  21  32  31  28  17  19  25  16     Y₁ = 230
II:   20  15  13  16  13  21  22  16  18  16     Y₂ = 170
```

Perform **a** an ordinary $t$ test, **b** the $t$-like test, and the $F$-like test.
*Partial Ans.:* $F^* = 7.98 = (2.825)^2$. Compare these data with those in Table 32.2. In what respects are they different and in what respects are they similar?

**32.2** Let the weight $w_i = n_i/s_i^2$. The weighted general mean is then

$$\bar{y}^* = \frac{w_1 \bar{y}_1 + w_2 \bar{y}_2}{w_1 + w_2} \quad \text{for two groups}$$

The $F$-like statistic is defined as

$$F^* = w_1(\bar{y}_1 - \bar{y}^*)^2 + w_2(\bar{y}_2 - \bar{y}^*)^2$$

Show that it is identical with

$$t^{*2} = \frac{(\bar{y}_1 - \bar{y}_2)^2}{s_1^2/n_1 + s_2^2/n_2}$$

HINT: Substitute $\bar{y}^*$ expression and simplify.

**32.3**  In Table 32.4, where there are more than two groups, we have obtained $\Sigma w_i(\bar{y}_i - \bar{y}^*)^2 = 9.4432$ by using the value $\bar{y}^* = 6.4091$ directly. Frequently the arithmetic may be simplified somewhat by using an arbitrary but convenient number in place of the actual $\bar{y}^*$. In the accompanying table it is shown how this could be done by choosing $a = 6$, say.

| $w_i$ | $\bar{y}_i$ | $\bar{y}_i - a$ | $w_i(\bar{y}_i - a)$ | $w_i(\bar{y}_i - a)^2$ |
|---|---|---|---|---|
| 0.3333 | 6 | 0 | 0 | 0 |
| 0.7500 | 9 | 3 | 2.25 | 6.75 |
| 0.7500 | 4 | −2 | −1.50 | 3.00 |
| 1.8333 | | | 0.75 | 9.75 |

$$\Sigma w_i(\bar{y}_i - \bar{y}^*)^2 = \Sigma w_i(\bar{y}_i - a)^2 - \frac{[\Sigma w_i(\bar{y}_i - a)]^2}{W}$$

$$= 9.75 - \frac{(0.75)^2}{1.8333} = 9.4432$$

This is based on the same identity proved in Chap. 3 except for changing $w_i$ into $n_i$, $W = \Sigma w_i$ into $N = \Sigma n_i$, and $\bar{y}^*$ into $\bar{y}$. Note that the correction term

$$\frac{[\Sigma w_i(\bar{y}_i - a)]^2}{W} = \frac{[W\bar{y}^* - Wa]^2}{W} = W(\bar{y}^* - a)^2$$

Recalculate the value of $\Sigma w_i(\bar{y}_i - \bar{y}^*)^2$, using $a = 5$. The answer should be 9.4432.

**32.4**  The numerical example in the accompanying table is taken from Welch's 1951 article. Test for the equality of the sample variances by Bartlett's chi-square method. If the chi square is large enough to indicate heterogeneity of variances, proceed to perform an $F$-like test for significance of treatment effects.

| Treatment group $i$ | Size $n_i$ | Mean $\bar{y}_i$ | Variance $s_i^2$ | Weight $w_i$ |
|---|---|---|---|---|
| 1 | 20 | 27.8 | 60.1 | 0.333 |
| 2 | 10 | 24.1 | 6.3 | 1.587 |
| 3 | 10 | 22.2 | 15.4 | 0.649 |
| | | | Total | $W = 2.569$ |

*Partial Ans.:* $\Sigma w_i(\bar{y}_i - \bar{y}^*)^2 = 6.90$, $\Lambda = 0.1180$; $F^* = 3.35$; $\nu_2^* = 22.6$, nonsignificant.

# 33

# Change of scale

In making the variance-ratio test for significance (or for constructing confidence intervals) we always assume that the experimental error, the $e$ term in the linear model, is normally distributed with mean zero and variance $\sigma^2$. In addition to this preliminary requirement, two further basic assumptions are (1) all treatment groups have the same error variance (homoscedastic) and (2) treatment and environmental effects are additive, as described by the linear model. When group variances are unequal, an approximate method of testing significance or constructing confidence intervals is used. But this method, briefly presented in Chap. 32, remedies only one aspect of the failure of the observations to conform with the mathematical model. In the first place, the errors may not be normally distributed at all. Second, the various effects may not be additive. If the data do not satisfy the usual requirements, there is no single operation that can make them do so. However, experience in agricultural and biological experiments shows that if deviations from model are not too great, the analysis of variance is still applicable because the $F$ value is not sensitive to minor deviations.

Frequently, measurements in terms of one scale make the data more nearly conform with the mathematical model in certain respects than do measurements in another. Hence, by an appropriate choice of scale of measurement, we may be able to overcome some of the difficulties arising from nonnormality of error distribution, unequality of group variances, and/or nonadditivity of the various effects. The reader must at this point be warned against being too optimistic, because the phrase "an appropriate choice of scale" merely states a principle and does not tell us what specific scales are appropriate. The fact is that there is no rule of thumb by which an appropriate scale may be found for any given set of data. Indeed, as stated before, probably there is no scale that can make the data conform with *all* the requirements of the mathematical model. The problem of scale and departure from model has been well discussed in a special issue of *Biometrics* [Cochran: Some Consequences When the Assumptions for the Analysis of Variance Are Not Satisfied, **3**:22–38 (1947) and Bartlett: The Use of Transformations, **3**:39–52 (1947)], which the reader may consult for mathematical details. In this chapter we shall mention only two or three of the most commonly used changes of scale to serve as an introduction to the subject.

### 1. The nature of change of scale

To illustrate the meaning of change of scale, let us consider the relationship between $x$ and $y$ in Fig. 33.1 left-hand diagram. The variable represented by the horizontal axis is measured in $x$ units, and the variable represented by the vertical axis is measured in $y$ units. Then the relationship between $x$ and $y$ is represented by the S-shaped curve. Now, let us imagine that the $xy$ plane is a piece of very elastic rubber so that it can be freely stretched in the north-south direction. Then the rubber sheet may be so (unevenly) lengthened that the original S-shaped curve will assume the shape of a straight line, as shown in the right-hand diagram of Fig. 33.1. Let us call the new (stretched) ordinate the $u$ axis. The original marks (units) on the $y$ axis are stretched out on the $u$ axis corresponding to certain new units on the axis. This operation is called a change of scale (or transformation of variables).

The reader may ask, "What is *the* true relationship between the two variables: S-shaped or straight line?" The answer is that it is purely a matter of scale employed in actual measurement. In one scale the relationship is S-shaped; in another, it is linear. There is no such thing as *the* relationship. It may be described in various ways. And, let us always

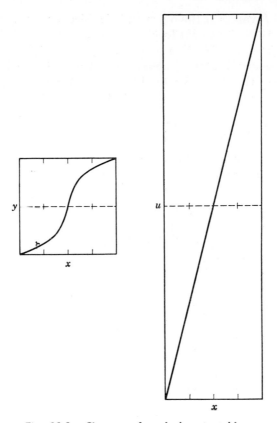

**Fig. 33.1   Change of scale by stretching.**

remember, all scales are arbitrary, although some may be more convenient for us than others.

For ease of description, the investigator, in dealing with regression problems, often adopts a scale that makes the relationship as nearly linear as possible.

A change of scale will change the mean and variance of the variable, as well as its relationship with other variables. The form of distribution of a variable also changes with scale. A nonnormally distributed variable may be rendered nearly normal by an appropriate transformation. Populations with unequal variances may be made nearly homoscedastic by an appropriate transformation. The mathematics of transformation may be quite involved, but the principle is simple enough.

The main disturbance caused by nonnormality of the experimental errors is to raise the significance level of the $F$ test. That is, when we use the tabulated *F* values at the 5 per cent significance level, we are actually using, say, the 6 or 7 per cent level.    Consequently, we shall obtain more "significant" results.    In other words, the probability of committing the mistake of pronouncing significance when there actually is none is more than 5 per cent, the risk you intend to take.    A normalizing transformation will improve the accuracy of the test.

## 2. The logarithmic transformation

Probably the most frequently used transformation is the logarithmic one. The linear model (for example, $y_{ij} = u + b_i + t_j + e_{ij}$) says that the block effect (environmental effect), treatment effect, and unaccounted error are all additive.    If the block and treatment effects are to increase or decrease the measurements by a certain percentage instead of by a certain amount, we say that the effects are multiplicative instead of additive.    In such cases, a logarithmic transformation will change the multiplicative relationship to additive, and then our linear model will be applicable to the new data.

The most important thing to remember in transformation is that the change must be made to observed single values, as if the original measurements were taken in terms of the new units, and not on any derived value such as the mean or sum of squares.    To emphasize this point, we exhibit the following numerical example, in which the transformed values are the common logarithms without the decimal point.    For instance,

$$\log_{10} 6 = 0.778$$

and it is entered as 78 for simplicity.    Subsequent calculations and analysis of variance are then based on the new set of data.    For numerical

| | | Original (Chap. 10) | | | | | Transformed (logarithms) | | | | |
| --- | --- | --- | --- | --- | --- | --- | --- | --- | --- | --- | --- |
| | *a* | *b* | *c* | *d* | | *a* | *b* | *c* | *d* | |
| I | 6 | 2 | 9 | 3 | 20 | 78 | 30 | 95 | 48 | 251 |
| II | 8 | 9 | 11 | 12 | 40 | 90 | 95 | 104 | 108 | 397 |
| III | 4 | 4 | 10 | 6 | 24 | 60 | 60 | 100 | 78 | 298 |
| | 18 | 15 | 30 | 21 | 84 | 228 | 185 | 299 | 234 | 946 |

calculation and the *F* test, it makes no difference whether the common logarithms or the natural logarithms are used, because they differ by a constant factor.

When some effects are multiplicative and some additive, the logarithmic transformation will not be as successful as when all effects are multiplicative.   Instead of a simple logarithm, it is sometimes advisable to use log $(y + 1)$ or log $(y + c)$, where $c$ is a constant.

It should be reiterated that a change of scale leads to a change of relationships between variables and thus to a change of "interpretation." To illustrate this point more concretely, let us consider the following nine observations on a $3 \times 3$ factorial experiment.   In the original set of data,

|  | Original | | | |  | Logarithm | | |
|---|---|---|---|---|---|---|---|---|
|  | $b_0$ | $b_1$ | $b_2$ | |  | $b_0$ | $b_1$ | $b_2$ |
| $a_0$ | 20 | 30 | 40 | | $a_0$ | 30 | 48 | 60 |
| $a_1$ | 30 | 45 | 60 | | $a_1$ | 48 | 66 | 78 |
| $a_2$ | 40 | 60 | 80 | | $a_2$ | 60 | 78 | 90 |

Factor effects linear; interactions present

Factor effects non-linear; interactions absent

the effect of factor $b$ is perfectly linear in all rows, and the effect of factor $a$ is also linear in all columns.   However, there is obvious interaction between the two factors.   In the first row the increment for each level of $b$ is 10; in the second row the increment is 15; and in the third, 20.   The same is true with increment of factor $a$ in the three columns.

The transformed values show an entirely different relationship.   The logarithmic values have been decoded.   For instance, $\log_{10} 20 = 1.30$, and it is entered simply as 30; and $\log_{10} 45 = 1.653$, but it is entered as 66 to compensate for the rounding errors of the other log values.   Now the increment from $b_0$ to $b_1$ is 18 and that from $b_1$ to $b_2$ is 12, showing that the effect of factor $b$ is not quite linear.   However, the increments remain the same in all rows, implying no interaction between the two factors. Thus we see that the relationships between the two factors are very different in the two scales.   For certain types of analysis the investigator prefers the scale that eliminates interactions, and for another he may prefer the scale that renders the effects linear.   The important thing to remember is that the relationship between variables is strongly influenced by the scales with which these variables are measured.   Interpretations of data are valid only with respect to the particular scale adopted in a given instance.

## 3.  The square-root transformation

When the data consist of integers arising from counts of objects, such as the number of rust spots on a leaf or the number of bacteria in a plate, the observed numbers tend to have a Poisson distribution rather than a normal one. Theoretical considerations lead to the square-root transformation of the observed numbers. This transformation usually makes the group variances more nearly equal. It is also applicable to skew distributions, for it shortens the long tail. If $y$ is the observed number, we shall use $\sqrt{y}$ for analysis and significance test. When the observed numbers are small (say, from 2 to 10), the transformation $\sqrt{y + \frac{1}{2}}$ is preferred, especially when some observed numbers are zero. Again, the transformation should be made on the originally observed single numbers, as in the example given here. Once transformed, the original data have

| | Original count, $y$ | | | | | | Transformed, $\sqrt{y + \frac{1}{2}}$ | | | | |
|---|---|---|---|---|---|---|---|---|---|---|---|
| | $a$ | $b$ | $c$ | $d$ | | | $a$ | $b$ | $c$ | $d$ | |
| I | 6 | 2 | 9 | 3 | 20 | | 2.55 | 1.58 | 3.08 | 1.87 | 9.08 |
| II | 8 | 9 | 11 | 12 | 40 | | 2.92 | 3.08 | 3.39 | 3.54 | 12.93 |
| III | 4 | 4 | 10 | 6 | 24 | | 2.12 | 2.12 | 3.24 | 2.55 | 10.03 |
| | 18 | 15 | 30 | 21 | 84 | | 7.59 | 6.78 | 9.71 | 7.96 | 32.04 |
| | General mean | | | 7.00 | | | General mean | | | 2.67 | |

no further use, and all subsequent calculations are based on the new set of data. Note that the new general mean 2.67 is *not* equal to $\sqrt{7.50} = 2.74$.

## 4. The angular transformation

Another very common data situation is that the record gives the number of individuals with a certain characteristic in a group of $n$ individuals. The simplest case is that all groups are of equal size. The number of counts tends to be binomially distributed when the treatments are such that the probability of death is not small. Let $p$ be the proportion or fraction of the $n$ individuals that possess the characteristic and $q = 1 - p$, then

$$p + q = 1$$

but

$$\sin^2 \theta + \cos^2 \theta = 1$$

where $\theta$ is an angle. The transformation

$$p = \sin^2 \theta \qquad \text{or} \qquad \theta = \sin^{-1} \sqrt{p}$$

changes a proportion or fraction into an angle; hence the term "angular transformation." The value of $\theta$ in degrees corresponding to various values of $p$ has been tabulated (Bliss, 1937), so that the angular transformation in practice is just as easy as the logarithmic and the square-root transformations are. A few of the values are shown in Table 33.1.

Note that the percentages and the derived angles have the following correspondence:

$$\text{Percentage:} \quad 0 \quad 25 \quad 50 \quad 75 \quad 100$$
$$\text{Angle, degrees:} \quad 0 \quad 30 \quad 45 \quad 60 \quad 90$$

More generally, when two percentages add up to 100, their corresponding angles will add up to 90°. Thus, the angles corresponding to 12.5 and 87.5 per cent are respectively 20.7° and 69.3°, as shown in Table 33.1.

As a numerical example, let the numbers of Table 10.1 be the counts of individuals with certain characteristics among groups of $n = 20$ individuals each. Then the transformation proceeds as shown here. All

| | Numbers in $n = 20$ | | | | Percentages | | | | Angle in degrees | | | | |
|---|---|---|---|---|---|---|---|---|---|---|---|---|---|
| | $a$ | $b$ | $c$ | $d$ | $a$ | $b$ | $c$ | $d$ | $a$ | $b$ | $c$ | $d$ | |
| I | 6 | 2 | 9 | 3 | 30 | 10 | 45 | 15 | 33.2 | 18.4 | 42.1 | 22.8 | 116.5 |
| II | 8 | 9 | 11 | 12 | 40 | 45 | 55 | 60 | 39.2 | 42.1 | 47.9 | 50.8 | 180.0 |
| III | 4 | 4 | 10 | 6 | 20 | 20 | 50 | 30 | 26.6 | 26.6 | 45.0 | 33.2 | 131.4 |
| | | | | | | | | | 99.0 | 87.1 | 135.0 | 106.8 | 427.9 |

subsequent calculations and analysis are based on the new data in degrees.

The angular transformation bears the same relation to a binomial variate as the square-root transformation does to a Poisson variate.

When groups are of unequal size, the exact analysis is very difficult and usually some approximate method is resorted to. One of these approximate methods is given in the following sections. For an experimental setup it is highly desirable to have equal or very nearly equal groups. When we start out with $n = 30$ animals in each group, it is almost unavoidable that one or two animals will be lost (through early death or

*Table* **33.1** Transformation of percentage into angle in degrees

| Percentage | Degrees | Percentage | Degrees | Percentage | Degrees | Percentage | Degrees |
|---|---|---|---|---|---|---|---|
| 2.5 | 9.1 | 27.5 | 31.6 | 52.5 | 46.4 | 77.5 | 61.7 |
| 5.0 | 12.9 | 30.0 | 33.2 | 55.0 | 47.9 | 80.0 | 63.4 |
| 7.5 | 15.9 | 32.5 | 34.8 | 57.5 | 49.3 | 82.5 | 65.3 |
| 10.0 | 18.4 | 35.0 | 36.3 | 60.0 | 50.8 | 85.0 | 67.2 |
| 12.5 | 20.7 | 37.5 | 37.8 | 62.5 | 52.2 | 87.5 | 69.3 |
| 15.0 | 22.8 | 40.0 | 39.2 | 65.0 | 53.7 | 90.0 | 71.6 |
| 17.5 | 24.7 | 42.5 | 40.7 | 67.5 | 55.2 | 92.5 | 74.1 |
| 20.0 | 26.6 | 45.0 | 42.1 | 70.0 | 56.8 | 95.0 | 77.1 |
| 22.5 | 28.3 | 47.5 | 43.6 | 72.5 | 58.4 | 97.5 | 80.9 |
| 25.0 | 30.0 | 50.0 | 45.0 | 75.0 | 60.0 | 100.0 | 90.0 |

otherwise) in some groups, so that the final count is based on 29 or 28 animals in a few groups. In such cases, the straightforward angular transformation may still be used without introducing appreciable error. Finally, when the size of the groups is large, direct analysis of the observed numbers may be permissible without the angular transformation. To summarize, both binomial and Poisson variates may be treated by the standard method of analysis of variance after an appropriate transformation of the observed numbers. For further mathematical details, see Cochran [The Analysis of Variance When Experimental Errors Follow the Poisson or Binomial Laws, *Ann. Math. Statist.*, **11**:335 (1940)].

## 5. The probit transformation

The probit transformation also deals with percentages, especially with those of mortality data. The ordinary table for normal distribution ("normal table") gives the area under the normal curve below (or above) a certain point on the abscissa. That is, it gives us the relative frequency below (or above) a specified value of the normal deviate. The so-called probit transformation is nothing but the reverse of the ordinary normal table; it gives us the value of the normal deviate for specified values of the area or relative frequency below (or above) that point. It is, of course, based on the same normal distribution, and it may easily be obtained by interpolation from a detailed normal table.

Since the (standardized) normal deviate has a mean zero, the arbitrary number 5 is to be added to the normal deviate to avoid working with negative numbers. Hence, for each specified area under the normal curve, the corresponding value is taken as

$$\text{Probit} = 5 + \text{normal deviate}$$

Figure 33.2 gives an illustration of the transformation. When the percentage is 20 per cent under the normal curve, the corresponding normal

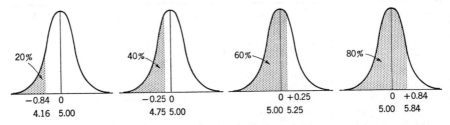

Fig. 33.2  Probit transformation. For a given percentage of the area under the normal curve the corresponding normal deviate is given and then 5.00 is added to the normal deviate. Hence the probit is a normal variate with mean 5.00 and unit variance.

deviate is $-0.84$ and the probit is taken as $5.00 - 0.84 = 4.16$. Because of the symmetry of the distribution, the normal deviates corresponding to 20 and 80 per cent have the same numerical value with opposite signs. Therefore their corresponding probit values add up to 10; for example, $4.16 + 5.84 = 10.00$. Any two percentages that add up to 100 per cent will have corresponding probit values adding up to 10. Table 33.2 gives a few of the probit values for purpose of illustration.

The rationale of using the probit transformation for data involving counts, especially those of mortality data, is very simple. In toxicological experiments, for instance, we count the number and calculate the proportion of deaths among a group of animals. The assumption is that the innate resistance of animals to the toxic agent is normally distributed. When the observed proportion of deaths is 40 per cent, say, it means those animals with resistance below normal deviate $-0.25$ (probit $= 4.75$) have died and those with higher resistance have survived. If a higher dose of the toxic agent kills 60 per cent of a similar group of animals (that is, a group with the same tolerance distribution), it kills those with resistance below probit value 5.25. Instead of using the proportions directly, we use the probit value to indicate the toxicity of the agent.

If $y$ denotes the probit value and $x$ the dosage (in some scale), the transformation will be particularly useful when it makes $y$ linearly related to $x$, thus greatly facilitating the analysis.

The usual difficulty of comparing percentages is well known. For

*Table* **33.2** Percentage, probit, and weighting coefficient

| Percentage | Probit | Percentage | Probit | Weighting coefficient |
|---|---|---|---|---|
| 10.0 | 3.72 | 90.0 | 6.28 | 0.342 |
| 12.5 | 3.85 | 87.5 | 6.15 | 0.388 |
| 15.0 | 3.96 | 85.0 | 6.04 | 0.426 |
| 17.5 | 4.07 | 82.5 | 5.93 | 0.460 |
| 20.0 | 4.16 | 80.0 | 5.84 | 0.490 |
| 22.5 | 4.24 | 77.5 | 5.76 | 0.516 |
| 25.0 | 4.33 | 75.0 | 5.67 | 0.539 |
| 27.5 | 4.40 | 72.5 | 5.60 | 0.558 |
| 30.0 | 4.48 | 70.0 | 5.52 | 0.576 |
| 32.5 | 4.55 | 67.5 | 5.45 | 0.590 |
| 35.0 | 4.61 | 65.0 | 5.39 | 0.603 |
| 37.5 | 4.68 | 62.5 | 5.32 | 0.614 |
| 40.0 | 4.75 | 60.0 | 5.25 | 0.622 |
| 42.5 | 4.81 | 57.5 | 5.19 | 0.628 |
| 45.0 | 4.87 | 55.0 | 5.13 | 0.633 |
| 47.5 | 4.94 | 52.5 | 5.06 | 0.635 |
| 50.0 | 5.00 | 50.0 | 5.00 | 0.637 |

example, decrease of mortality from 40 to 20 per cent certainly does not mean the same thing as a decrease from 20 to 1 per cent, not mentioning to 0 per cent. The probit transform will overcome the difficulty of dealing with percentages.

Although the basic reason for probit transformation is simple, the analysis is unfortunately more involved because of one disturbing factor: not all probit values carry the same weight. This is due to the fact that intermediate proportions are more accurately determined than very small or very large proportions. In the last column of Table 33.2 are listed the weighting coefficients for the various percentages. They are calculated as follows:

Let $p = 0.20$ be the observed proportion, $q = 1 - p = 0.80$, and $y = 4.16$ be the probit value corresponding to normal deviate $-0.84$. From an ordinary normal table, we find that the height (ordinate) of the normal curve at that point is approximately $h = 0.280$. The weighting coefficient (on per individual basis) is then

$$\text{Weighting coefficient} = \frac{h^2}{p \times q} = \frac{(0.280)^2}{0.20 \times 0.80} = 0.490$$

Again, owing to the symmetry of the normal curve, the weight for $p = 0.20$ is the same as that for $p = 0.80$.

The analysis of probit is practically a branch of statistical methodology by itself and it is chiefly used in bioassay problems [D. J. Finney, "Probit Analysis," Cambridge University Press, New York, (1952)]. The reader must consult special books for special subjects. However, we shall give one comparatively simple application in the following section to a type of data that arises very frequently in biological experiments.

## 6. An approximate analysis of probit

A set of hypothetical mortality data is shown in the top portion of Table 33.3. The rows and columns represent two factors of treatments. For example, the $a$'s may represent three lengths of exposure, and the $b$'s may represent four concentrations or four similar agents. The size $n$ of group varies from 20 to 60. An exact analysis being laborious, we shall describe an approximate method which is quite brief (Quenouille, 1953, chap. 14). The entry in the second row and first column is $20.0(\frac{9}{45})$, meaning that the group consists of $n = 45$ animals, of which 9 died; the percentage of death is $\frac{9}{45} = 20.0$ per cent.

The first step is to convert the percentages into probit values, using Table 33.2, and calculate the corresponding weights. Thus, 20 per cent is replaced by $y = 4.16$. The weighting coefficient is 0.490 per individual, so that the weight is $45 \times 0.490 = 22$. Proceeding the same way for all

*Table* 33.3  Percentage, probit, and weight; preliminary calculations.  Original observations: percentage (number deaths/total number)

|  | $b_0$ | $b_1$ | $b_2$ | $b_3$ |
|---|---|---|---|---|
| $a_1$ | 15.0 (9/60) | 17.5 (7/40) | 42.5 (17/40) | 55.0 (11/20) |
| $a_2$ | 20.0 (9/45) | 50.0 (13/26) | 60.0 (18/30) | 72.5 (29/40) |
| $a_3$ | 47.5 (19/40) | 50.0 (19/38) | 60.0 (15/25) | 75.0 (18/24) |

Step I  Probits ($y$) and their weights ($n \times$ wt. coeff.)

| | | | | |
|---|---|---|---|---|
| 3.96 (26) | 4.07 (18) | 4.81 (25) | 5.13 (13) | (82) |
| 4.16 (22) | 5.00 (17) | 5.25 (19) | 5.60 (22) | (80) |
| 4.94 (25) | 5.00 (24) | 5.25 (16) | 5.67 (13) | (78) |

| (73) | (59) | (60) | (48) | (240) |

Step II  Approximate proportional weight $w$

| | | | | |
|---|---|---|---|---|
| 24 | 20 | 20 | 16 | 80 |
| 24 | 20 | 20 | 16 | 80 |
| 24 | 20 | 20 | 16 | 80 |

| 72 | 60 | 60 | 48 | 240 |

Step III  Weighted probit value $wy$

| | | | | |
|---|---|---|---|---|
| 95.0 | 81.4 | 96.2 | 82.1 | 354.7 |
| 99.8 | 100.0 | 105.0 | 89.6 | 394.4 |
| 118.6 | 100.0 | 105.0 | 90.7 | 414.3 |

| 313.4 | 281.4 | 306.2 | 262.4 | 1163.4 |

other cells, we obtain the "Step I" portion of Table 33.3. This step involves no approximation except in rounding the weights into integers.

Since the 12 probits have varying weights, the effects of factors $a$ and $b$ are not orthogonal. However, we note that the total weights for the three rows are very nearly equal to 80. The total weights for the four columns are 73, 59, 60, 48, which may be regarded as 72, 60, 60, 48. Then the grand total weight, 240, is to be distributed into the 12 cells in proportion to the marginal total weights. Thus, the *approximate* weight for the first cell is to be taken as $(72 \times 80)/240 = 24$, and so on. These weights, designated by $w$, are given in the "Step II" portion of Table 33.3, and these are the weights we shall use in subsequent calculations. This step involves some real approximations. It amounts to forcing a two-way table with irregular cell frequencies into one with proportionate cell frequencies (Fig. 10.1), so that the sum of squares due to rows and columns may be calculated separately.

*Table* 33.4    Approximate analysis of
chi square (data from Table 33.3)

| Source | $df$ | $ssq = \chi^2$ |
|---|---|---|
| Factor $a$ | 2 | $R - C = 23.0$ |
| Factor $b$ | 3 | $K - C = 41.4$ |
| Interaction | 6 | (subt)    7.1 |
| Total | 11 | $A - C = 71.5$ |

The third step is to calculate the weighted probit.    For example, for
the first cell, we have $w \times y = 24 \times 3.96 = 95.0$, etc.    The row and
column totals are then obtained, and $\Sigma wy = 1{,}163.4$ is the grand total of
the weighted probits.

Subsequent analysis proceeds the same way as in ordinary analysis of
variance except that each probit value is to be weighted.    The four basic
quantities (areas) needed for the analysis of the two-way table are

$$A = \Sigma wy^2 = 24(3.96^2 + 4.16^2 + 4.94^2)$$
$$+ 20(4.07^2 + 5.00^2 + 5.00^2 + 4.81^2 + 5.25^2 + 5.25^2)$$
$$+ 16(5.13^2 + 5.60^2 + 5.67^2) = 5{,}711.1$$

$$R = \frac{(354.7)^2 + (394.4)^2 + (414.3)^2}{80} = 5{,}662.6$$

$$K = \frac{(313.4)^2}{72} + \frac{(281.4)^2}{60} + \frac{(306.2)^2}{60} + \frac{(262.4)^2}{48} = 5{,}681.0$$

$$C = \frac{(\Sigma wy)^2}{\Sigma w} = \frac{(1{,}163.4)^2}{240} = 5{,}639.6$$

The sums of squares are given in Table 33.4.    Since the probit is theoreti-
cally a normal variate with unit variance, these sums of squares should
approximately follow the chi-square distribution (Chap. 8).    Hence,
each of the sums of squares may be tested for significance by using the
$\chi^2$ table.    In our example, the main effects of the factors $a$ and $b$ are highly
significant.

## 7. Concluding remark

Here ends the introduction to experimental statistics.    We have covered
the basic principles and methods in design and analysis, and this will give
the biological and medical experimenter a working knowledge.    It must
be realized that much has been left unsaid, but those who wish to pursue
the subject further are now in a position to consult more advanced textbooks and technical papers for details.    References grouped by type are
included at the end of the book.

# References

**Classical**

Fisher, R. A. (1935 and 1960): "The Design of Experiments," Hafner Publishing Company, Inc., New York.

Yates, F. (1937): The Design and Analysis of Factorial Experiments, *Imperial Bur. Soil Sci., Tech. Bull.* 35, Harpenden, England.

**General statistics**

Anderson, R. L., and T. A. Bancroft (1952): "Statistical Theory in Research," McGraw-Hill Book Company, New York.

Brownlee, K. A. (1960): "Statistical Theory and Methodology in Science and Engineering," John Wiley & Sons, Inc., New York.

Dixon, W. J., and F. J. Massey (1957): "Introduction to Statistical Analysis," McGraw-Hill Book Company, New York.

Li, Jerome C. R. (1957): "Introduction to Statistical Inference," Edwards Brothers, Inc., Ann Arbor, Mich., and Stechert-Hafner, Inc., New York.

Ostle, B. (1963): "Statistics in Research," The Iowa State University Press, Ames, Iowa.

Snedecor, G. W. (1962): "Statistical Methods," The Iowa State University Press, Ames, Iowa.

Steel, R. G. D., and J. H. Torrie (1960): "Principles and Procedures of Statistics," McGraw-Hill Book Company, New York.

Wolf, F. L. (1962): "Elements of Probability and Statistics," McGraw-Hill Book Company, New York.

Yule, G. U., and M. G. Kendall (1950): "An Introduction to the Theory of Statistics," Hafner Publishing Company, Inc., New York.

### Experimental statistics

Cochran, W. G., and G. M. Cox (1957): "Experimental Designs," John Wiley & Sons, Inc., New York.

Cox, D. R. (1958): "Planning of Experiments," John Wiley & Sons, Inc., New York.

Federer, W. T. (1955): "Experimental Design: Theory and Application," The Macmillan Company, New York.

Kempthorne, O. (1952): "The Design and Analysis of Experiments," John Wiley & Sons, Inc., New York.

Lindquist, E. F. (1953): "Design and Analysis of Experiments in Psychology and Education," Houghton Mifflin Company, Boston.

Quenouille, M. H. (1953): "The Design and Analysis of Experiment," Hafner Publishing Company, Inc., New York.

Winer, B. J. (1962): "Statistical Principles in Experimental Design," McGraw-Hill Book Company, New York.

### Special for the medical profession

Armitage, P. (1960): "Sequential Medical Trials," Blackwell Scientific Publications, Ltd., Oxford, and Charles C Thomas, Springfield, Ill.

Bancroft, H. (1957): "Introduction to Biostatistics," Paul B. Hoeber, Inc., New York.

Council for International Organizations of Medical Sciences (1960): "Controlled Clinical Trials," Blackwell Scientific Publications, Inc., Oxford, and Charles C Thomas, Springfield, Ill.

Hill, A. B. (1961): "Principles of Medical Statistics," 7th ed., Lancet, London, and Oxford University Press, New York and London.

Mainland, D. (1963): "Elementary Medical Statistics," 2d ed., W. B. Saunders, Philadelphia.

Witts, J. L. (ed.) (1959): "Medical Surveys and Clinical Trials: Some Methods and Applications of Group Research in Medicine," Oxford University Press, New York and London.

### Paperbacks and small books

Croxton, F. E. (1953): "Elementary Statistics with Applications in Medicine and the Biological Sciences," Dover Publications, Inc., New York.

Finney, D. J. (1955): "Experimental Design and Its Statistical Basis," University of Chicago Press, Chicago.

Li, C. C. (1959): "Numbers from Experiments: A Basic Analysis of Variation," Boxwood Press, Pittsburgh.

Maxwell, A. E. (1958): "Experimental Design in Psychology and the Medical Sciences," Methuen & Co., Ltd., London, and John Wiley & Sons, Inc., New York.

Moroney, M. J. (1956): "Facts from Figures," Penguin Books, Inc., Baltimore.

Smart, J. V. (1963): "Elements of Medical Statistics," Charles C Thomas, Springfield, Ill.

Steel, R. G. D. (1962): "Elementary Mathematics for Statistical Analysis," The Technical Press, Raleigh, N.C.

Wilks, S. S. (1951): "Elementary Statistical Analysis," Princeton University Press, Princeton, N.J.

### Tables

Fisher, R. A., and F. Yates (1938–1963): "Statistical Tables for Biological, Agricultural and Medical Research," Oliver & Boyd Ltd., Edinburgh and London.

Owen, D. B. (1962): "Handbook of Statistical Tables," Addison-Wesley Publishing Company, Inc., Reading, Mass.

# Index